Air Quality: Measurement, Analysis and Monitoring Techniques

Edited by Anthony Schindler

SYRAWOOD
PUBLISHING HOUSE

New York

Published by Syrawood Publishing House,
750 Third Avenue, 9th Floor,
New York, NY 10017, USA
www.syrawoodpublishinghouse.com

Air Quality: Measurement, Analysis and Monitoring Techniques
Edited by Anthony Schindler

Cataloging-in-Publication Data

Air quality : measurement, analysis and monitoring techniques / edited by Anthony Schindler.
    p. cm.
Includes bibliographical references and index.
ISBN 978-1-68286-731-0
1. Air quality. 2. Air quality--Measurement. 3. Air quality management. 4. Air--Pollution.
5. Environmental monitoring. 6. Environmental impact analysis. I. Schindler, Anthony.
TD883 .A37 2019
363.739 2--dc23

TABLE OF CONTENTS

PREFACE

The purpose of the book is to provide a glimpse into the dynamics and to present opinions and studies of some of the scientists engaged in the development of new ideas in the field from very different standpoints. This book will prove useful to students and researchers owing to its high content quality.

Air pollution is caused by the presence of harmful substances in the air, like gases, particulates, biological molecules, etc. This can lead to various allergies, diseases and also death in humans. It also causes harm to animals and food crops. It is responsible for damage to the environment. Air quality measurement, analysis and monitoring are therefore essential for controlling air quality degradation and measuring the degree of damage. Air quality is monitored by measuring the air pollutant concentrations using specialized equipment and methods. The data obtained is then interpreted to understand the health effects associated with exposures to such concentrations. Air quality index is a measure of how polluted the air is. An increase in air quality index signifies that a larger population is vulnerable to adverse health effects. This book aims to shed light on some of the unexplored aspects of air quality measurement, analysis and monitoring. Different approaches, evaluations, methodologies and advanced studies on air quality have also been included. Students, researchers, experts and all associated with this field will benefit alike from this book.

At the end, I would like to appreciate all the efforts made by the authors in completing their chapters professionally. I express my deepest gratitude to all of them for contributing to this book by sharing their valuable works. A special thanks to my family and friends for their constant support in this journey.

Editor

Atmospheric BC is emitted as a primary species, while OC can consist of primary and secondary aerosols (Putaud et al. 2004; Pöschl 2005). Major sources of BC and OC include incomplete combustion of fossil fuels, biomass burning and traffic emissions (Bond and Sun, 2005; IPCC 2013). OC is also emitted from biogenic sources and can be formed through the oxidation of volatile organic compounds (VOCs) (Pöschl 2005). BC absorbs terrestrial long-wave radiation that has a warming effect on the atmosphere, while OC, depending on their chemical properties, could absorb or reflect incoming solar radiation (IPCC 2013).

In general it is accepted that OC has a net cooling effect (IPCC 2013). After CO_2, BC is considered to be the second most important contributor to global warming (Bond and Sun 2005; IPCC 2013). The impacts of BC are especially significant on local and regional scales, since BC has a relatively short atmospheric lifetime (days to weeks) (IPCC 2013). Greenhouse gases (GHG) spend much longer periods in the atmosphere, i.e. between 10 to 100 years (de Richter and Caillol 2011).

Although Africa is regarded as the largest source region of anthropogenic atmospheric OC and BC (Liousse et al. 1996; Kanakidou et al. 2005), it is one of the least studied continents. Within Africa, southern Africa is an important source region. Biomass burning fires (anthropogenic and natural) are endemic across this region especially during the dry season when almost no precipitation occurs (Formenti et al. 2003; Tummon et al. 2010; Laakso et al. 2012).

Biomass burning fire plumes from southern Africa are known to impact Australia and South America (Swap et al. 2004). In addition, South Africa is the economic and industrial hub of southern Africa with large anthropogenic point sources (e.g. Lourens et al. 2011). However, the relative importance of OC and BC contributions from these anthropogenic sources in Africa are still largely unknown, although some papers have been published that considered sources in west African capitals (Doumbia et al. 2012; Val et al. 2013). Venter et al. (2012) used BC data that were collected at Marikana in the North West province (South Africa) to verify that the origin of CO and PM_{10} was related to BC, while Collett et al. (2010) only presented a single diurnal plot for BC measured at Elandsfontein in the Mpumalanga Highveld. Hyvärinen et al. (2013) used BC data collected at Welgegund in the North West province to illustrate the use of a newly developed method to correct BC values measured with a multi-angle absorption photometer (MAAP), but did not go into further detail of the BC data.

Within the framework of the Deposition of Biogeochemical Important Trace Species (DEBITS)-International Global Atmospheric Chemistry (IGAC) DEBITS in Africa (IDAF) project (Galy-Lacaux et al. 2003; Martins et al. 2007), atmospheric gaseous and aerosol measurements have been performed continuously since 1994 at 7 sites in central and western Africa, as well as 3 sites in South Africa. Regarding carbonaceous aerosol, Martins (2009) determined BC and OC concentrations at two of the South African IDAF sites. However, these measurements were restricted to three two-week winter campaigns and one two-week summer campaign. This data have also not yet been published in the peer reviewed scientific domain.

In order to address the current knowledge gap, i.e. very limited OC and BC data for South Africa the main objectives of this paper are to present spatial and temporal assessments of OC and BC concentrations at the South African IDAF sites, determine the mass fractions of OC and BC of the overall aerosol mass, as well as to determine possible sources.

Experimental

Sampling sites

Aerosol samples were collected at five sampling sites in South Africa operated within the IDAF network, i.e. Louis Trichardt (LT), Skukuza (SK), Vaal Triangle (VT), Amersfoort (AF) and Botsalano (BS). The locations of these sites within a regional context are presented in Fig. 1. The South African IDAF sites are located in the north eastern part of the interior of South Africa. Mphepya et al. (2006) and Martins et al. (2007) have previously introduced LT and SK, but not the other sites. In order to contextualise all the sites a short description of each site is given in Table 1. LT, SK and BT are considered to be background sites.

In contrast, AF lies southeast of the internationally well-known NO_2 hotspot that is clearly visible from satellite observations over the Mpumalanga Highveld of South Africa (Lourens et al. 2012), while VT lies within a highly industrialised and populated area that has been proclaimed a national air pollution hotspot in terms of the South African National Environmental Management Act: Air Quality (Government Gazette Republic of South Africa 2005). Although not specified in Table 1, all the South African IDAF sites are likely to be impacted by local, as well as regional biomass burning fire emissions.

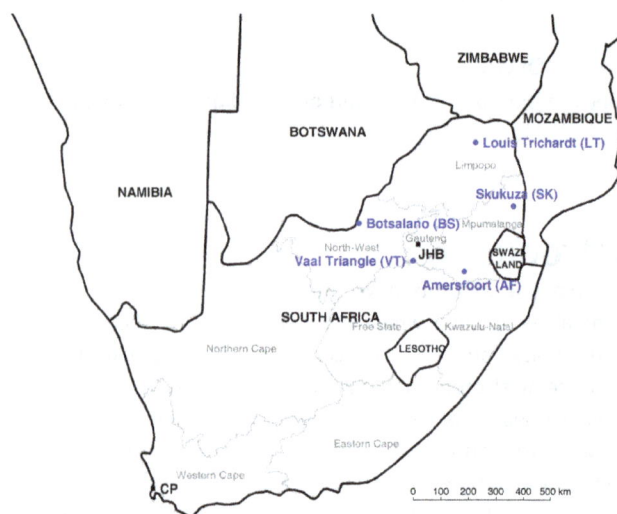

Figure 1: The location (blue dots) of the South African IDAF sampling sites where OC and BC measurements were conducted are indicated on a southern African map. Provincial borders with the provincial names within South Africa, as well as Johannesburg (JHB) and Cape Town (CP) are also included for reference

Table 1: *Geographic coordinates and short descriptions of South African IDAF sampling sites where OC and BC measurements were conducted*

Site	Location	Description
Amersfoort (AF)	27°04'13"S 29°52'02"E, 1628 m amsl	Semi-arid, within grassland biome, impacted by anthropogenic activities on the Mpumalanga Highveld
Louis Trichardt (LT)	22°59'10"S 30°01'21"E, 1300 m amsl	Semi-arid, within savannah biome, rural site predominantly used for agricultural purposes
Skukuza (SK)	24°59'35"S 31°35'02"E, 267 m amsl	Semi-arid, within savannah biome, regional background site in a protected area (Kruger National Park)
Vaal Triangle (VT)	26°43'29"S 27°53'05"E, 1320 m amsl	Semi-arid, within grassland biome, situated in the highly industrialized Vaal Triangle area, impacted by emissions from various industries, traffic and household combustion
Botsalano (BS)	25 32'28"S, 25 45'16"E	Semi-arid, within savannah biome, regional background site in a protected area (Botsalano Game Reserve)

Regional meteorology

Recently Laakso et al. (2012) and references therein gave an overview of the meteorology over the South African Highveld, as well as the interaction between meteorological patterns and pollutant levels. Therefore only a synopsis is presented here. Atmospheric circulation over the South African Highveld is dominated by an anti-cyclonic recirculation pattern throughout the year (Tyson and Preston-Whyte 2000), due to the dominance of a continental high pressure cell over the interior. This recirculation contributes significantly to the build-up of pollutants. This is especially significant during the cold dry winter (June – August) and early spring months (September – middle October) when strong inversion layers trap pollutants at several different heights inhibiting vertical mixing. This frequently causes an increase in atmospheric pollutant concentrations near the surface. In addition, the interior of South Africa is also characterised by a distinct wet and dry season. Almost all the precipitation occurs during the wet season from middle October to April, while nearly no precipitation takes place during the dry season from May to middle October. The lack of precipitation during the dry season leads to a decrease in wet deposition of pollutants and indirectly to the increase in pollutant levels due to the more frequent occurrence of large scale biomass burning fires. During the cooler autumn and cold winter months (May to August) household combustion for space heating is also a common occurrence in especially semi-formal and informal settlements (Venter et al. 2012).

Sampling

One 24-hour PM$_{2.5}$ and one PM$_{10}$ aerosol sample was collected on quartz filters (with a deposit area of 12.56 cm^2) once a month from March 2009 to April 2011 at each site. A total of 258 samples were collected, i.e. 52 samples for each site, except for BS for which only 50 samples were collected. Since both size fractions were sampled each month at each site, one half of the samples were PM$_{2.5}$ and the other half PM$_{10}$. The quartz filters were prebaked at 900°C for 4 hours and cooled down in a desiccator, prior to sample collection. MiniVol samplers developed by the United States Environmental Protection Agency (US-EPA) and the Lane Regional Air Pollution Authority were used during sampling (Baldauf et al. 2001). These samplers have a pump that is controlled by a programmable timer, which allows for the collection of samples at a constant flow rate over a pre-determined time period. In this study, samples were collected at a flow rate of 5 L/min, which was verified by using a handheld flow meter that was supplied with the MiniVol samplers. Filters were handled with tweezers while wearing surgical gloves, as a precautionary measure to prevent possible contamination of the filters. All thermally pre-treated filters were also visually inspected to ensure that there were no weak spots or flaws. After inspection, acceptable filters were weighed and packed in airtight Petri dish holders until they were used for sampling. After sampling, the filters were again placed in Petri dish holders, sealed off, bagged and stored in a portable refrigerator for transport to the laboratory. At the laboratory the sealed filters were stored in a conventional refrigerator. 24 hours prior to analysis, samples were removed from the refrigerator and weighed just prior to analysis.

OC and BC Analysis

Several methods can be used to analyse OC and BC collected on filters (Chow et al. 2001). It was decided to apply the IMPROVE thermal/optical (TOR) protocol (Chow et al. 1993; Chow et al. 2004; Environmental analysis facility 2008; Guillaume et al. 2008) by using a Desert Research Institute (DRI) thermal optical carbon analyser. In this method, filters are submitted to volatilization at temperatures of 120, 250, 450 and 550°C in a pure Helium (He) atmosphere and thereafter to combustion at temperatures of 550, 700 and 800°C in a mixture of He (98%) and oxygen (O$_2$) (2%) atmosphere. The carbon compounds that are released are then converted to CO$_2$ in an oxidation furnace with a manganese dioxide (MnO$_2$) catalyst at 932°C. Then, the flow passes into a digester where the CO$_2$ is reduced to methane (CH$_4$) on a nickel-catalysed reaction surface. The amount of CH$_4$ formed is detected by a flame ionization detector (FID), which is converted to carbon mass using a calibration coefficient. The carbon mass peaks detected correspond to the different temperatures at which the seven separate carbon fractions, which include four OC and three BC fractions, were released. These fractions were depicted as different peaks on the thermogram, of which the surface areas were proportional to the amount of CH$_4$ detected. The reflectance from the deposited sample is monitored throughout the afore-mentioned analysis. This reflectance usually decreases during the volatilization process due to the pyrolysis of OC. When oxygen is added, the reflectance is increased as the BC is burnt and removed. OC is defined as the fraction which evolves prior to re-attainment of the original reflectance (the non-absorbing light particles) and BC is defined as the fraction which evolved after the original reflectance from Martigues and Marseilles had an OC/BC ratio of 2.88. The OC/BC ratios for the two background sites, Plan d'Aups and Dupail, were 4.5 and 4.9, respectively. Chazette and Liousse (2001) determined an OC/BC ratio of 2.87 for Thessaloniki, which is the second biggest industrial city in Greece. Considering OC/BC ratio source characteristics proposed by Junker and Liousse (2008), it indicates that hard coal seems to be the dominant source type at VT, while typical OC/BC ratios associated with wood fuel, charcoal and motor gasoline correlated with OC/BC ratios determined at the other four South African IDAF sites. During SAFARI 2000, OC and BC samples were collected over the Atlantic Ocean just off the shore of Namibia and Angola from biomass burning fire plumes. The smoke of these biomass burning fires was widely distributed, and between one and two days old. In these samples the OC/BC ratios varied between 5.9 and 10.0 (Formenti et al. 2003). The higher OC/BC ratios reported by Formenti et al. (2003) can most likely be attributed to aging of the plumes and the absence of significant anthropogenic fossil fuel source contributions, which is also the case at the background South African IDAF sites, i.e. LT, SK and BS. Apart from the biomass burning fire emissions mentioned previously, these background sites are also influenced by re-circulating anthropogenic emissions from the South African Highveld.

As far as the authors could assess, the mass fractions of OC and BC as a function of the total aerosol mass, have not yet been investigated for South Africa. Therefore, in Fig. 4 box and whisker plots of the OC and BC mass fraction percentage of the total aerosol measured, in the $PM_{2.5}$ (4a) and PM_{10} (4b) size fractions, at each of the sites for the entire sampling period, are presented.

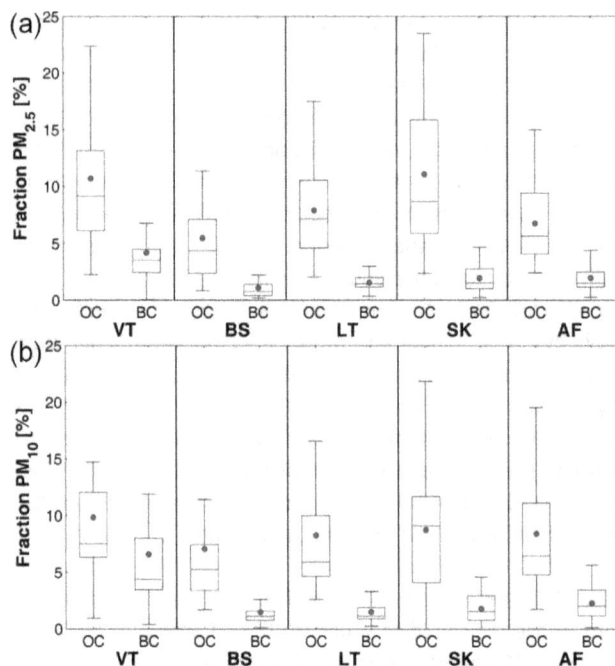

Figure 4: OC and BC as a mass fraction percentage of the total aerosol mass in $PM_{2.5}$ (a) and PM_{10} (b). The line in each box indicates the median, the dot the mean, the top and bottom edges of the box the 25th and 75th percentiles and the whiskers ± 2.7σ (99.3% coverage if the data has a normal distribution)

As expected, the OC mass fraction is higher than that of BC at all the sites for $PM_{2.5}$ and PM_{10}. The OC mass fraction percentage was up to 24%, while the BC mass fraction percentage was up to 12 % for all the sites in both size fractions. Putaud et al. (2004) reported $PM_{2.5}$ OC mass fractions to be 20-30% (rural sites) and 22-38% (near-city and kerbside sites), whereas PM_{10} OC mass fractions were 12-30% (rural sites) and 20-25% (near-city and kerbside sites) at 24 western European sites in winter. The BC contributions for $PM_{2.5}$ were reported to be 5-11% (rural sites) and 5-23% (near-city and kerbside sites), while the BC contributions for PM_{10} were 2-8% (rural sites) and 4-15% (near-city and kerbside sites). Yin et al. (2005) reported the OC mass fraction in Ireland for $PM_{2.5}$ as 30-40% (non-urban sites) and 10-30% (urban sites), while the PM_{10} OC mass fractions were 15-45% (non-urban sites) and 4-20% (urban sites). The $PM_{2.5}$ BC mass fractions were 25-30% (non-urban sites) and 8-10% (urban sites), while PM_{10} BC mass fractions were 12-22% (non-urban sites) and 3-5% (urban sites). From the afore-mentioned references it seems that the OC and BC measured at the South African IDAF sites were in the same order of magnitude, or lower than the mass fractions measured in western Europe. However, the study of Putaud et al. (2004) was biased towards winter, while all seasons are represented in the South African IDAF results. Additionally, fractional contribution to the overall aerosol load of sulphate has substantially decreased in first world countries where deSOx technology has been applied (Zhang et al. 2007). However, in South Africa sulphate is still the dominant aerosol species (Martins et al. 2007; Tiitta et al. 2014) since substantial deSOx technology have not yet been applied.

The OC mass fraction at SK and the BC mass fraction at VT were the highest in the $PM_{2.5}$ and PM_{10} size fractions. The high OC mass fraction at SK can be attributed to natural and anthropogenic sources. Mphepya et al. (2006) previously indicated the substantial impact of biomass burning fire emissions at SK. SK, LT and BS lies within the savannah biome (Mucina and Rutherford 2006), which emits more natural biogenic volatile organic compounds (BVOCs) than the Dry Highveld Grassland Bioregion wherein VT and AF lie (Mucina & Rutherford 2006). The atmospheric lifetime of BVOCs are mostly in the order of minutes to hours (Atkinson and Arey 2003), indicating that BVOCs can be transformed to less volatile species that could be collected as aerosols at sampling sites in close proximity of sources of BVOCs. In contrast, anthropogenic VOCs, which in general occur at higher concentrations than BVOCs in South Africa, have much longer atmospheric lifetimes (Jaars et al. 2014 and references therein), implying that the less volatile species are more likely to be formed further away from the source(s). SK also lies on the dominant path of air mass movement from the anthropogenic industrial hub of South Africa, which implies that primary emitted anthropogenic VOC species have enough time to oxidise to form less volatile secondary species that are measured in the OC fraction at SK. The BC high mass fraction at VT is due to this site being within a well-known anthropogenic source region as previously indicated and the nature of the sources occurring in this area.

Temporal assessment

As previously discussed, most of the OC and BC occurred in the $PM_{2.5}$ fraction, since there was not a significant difference between OC and BC concentrations in the $PM_{2.5}$ and PM_{10} fractions (Fig. 3). Therefore only the $PM_{2.5}$ measurements are presented and discussed further in this paper. A statistically meaningful seasonal temporal assessment for OC and BC concentrations could not be performed for each individual sampling site, since only one 24-hour $PM_{2.5}$ sample was obtained for each month at each site – one day cannot be construed as being representative of an entire month. However, since all 5 the South African IDAF sites are within a region with similar meteorological conditions and seasonal patterns, the results obtained at all the sites for the 2 years and 1 month measurement period were combined and statistically evaluated. Such a monthly temporal presentation of data gives a regional, rather than a site specific temporal impression. Box and whisker plot was not deemed appropriate to present the temporal statistical distribution, since there were too few data points per month, i.e. between 10 and 15. In Fig. 5 scatter plots indicating averages and standard deviations of the monthly OC (5a) and BC (5b) concentrations measured in the $PM_{2.5}$ fractions for all the sites over the entire sampling period, are presented (indicated in blue). Additionally, the VT data was also excluded from the dataset used to present the temporal pattern, which is indicated in another colour (red) in Fig. 5. This was done, since it was previously pointed out that the OC/BC ratio of the VT was substantially different compared to the other sites (Fig. 3), due to the close proximity and the nature of large point sources near the VT site.

From the data (Fig. 5), a relatively distinct seasonal pattern is observed, with higher OC and BC concentrations generally occurring during the period from May to September with the only exception being the lower OC levels observed for July. This seasonal trend was true, irrespective if VT was included or excluded from the dataset (indicated with different colours in Fig. 5). The period with generally higher OC and BC levels (May to September in Fig. 5) coincide with the onset of the dry season that is typically from May to early October in the interior of South Africa. During the wet season (middle October to April) aerosols are more frequently removed from the atmosphere through wet deposition. Furthermore, the peak frequency of biomass combustion fires occurs during the dry period in southern Africa. Fire burning frequency is especially high during late winter and early spring, i.e. August and September in southern Africa. The period with generally higher OC and BC levels also coincide with the time of the year that persistent low-level inversion layers trap pollutants close to the surface (Laakso et al. 2012), which leads to an increase in atmospheric concentrations of pollutant species. The colder months (May to August) are additionally characterised by increased household combustion for space heating (Venter et al. 2012), which could lead to higher atmospheric OC and BC concentrations. The possible contributions of sources of OC and BC associated with the meteorological conditions, as well as the occurrence of biomass burning fire events will be explored later in the paper.

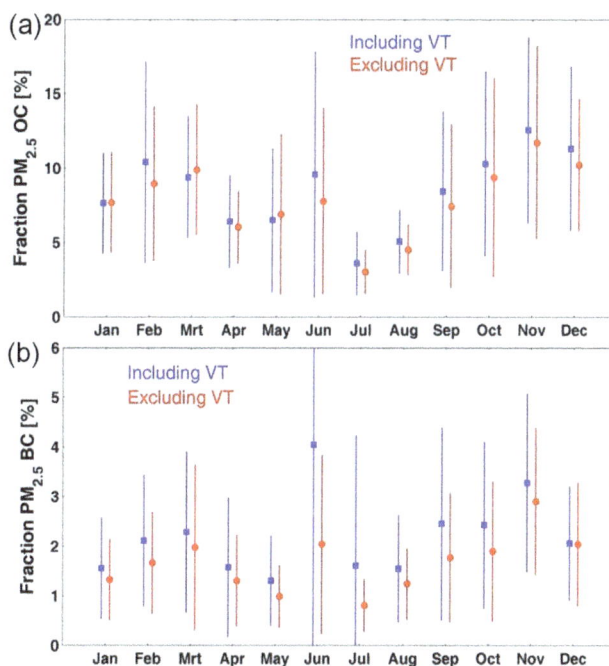

Figure 6: Temporal variations of the average and standard deviation OC (a) and BC (b) mass fraction percentage in the $PM_{2.5}$ fraction at all the sampling sites (blue) and at all the sampling sites excluding VT (red) for the entire sampling period, i.e. March 2009 to April 2011

In Fig. 6 temporal variations of the average and standard deviation OC (6a) and BC (6b) mass fractions of $PM_{2.5}$ for all the sites (including and excluding VT) over the entire sampling period are presented. With the exception of the OC and BC mass fractions of June, it is evident that the OC and BC seasonal mass fraction distribution (Fig. 6) is an inverse of the seasonal

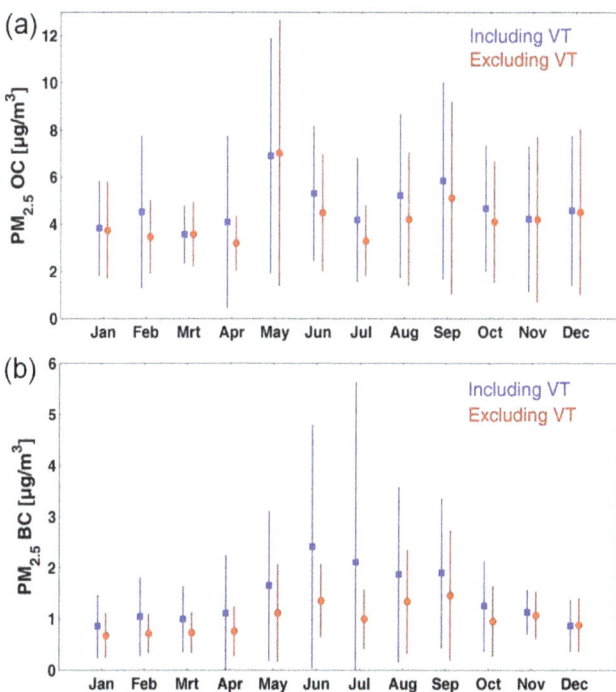

Figure 5: Temporal variations of the average and standard deviation of OC (a) and BC (b) concentrations in the $PM_{2.5}$ fraction at all the sampling sites (blue) and at all the sampling sites excluding VT (red) for the entire sampling period, i.e. March 2009 to April 2011

OC and BC concentrations presented in Fig. 5. This indicates that although OC and BC concentrations are in general higher in the dry and cold period between May to August (Fig. 5), the atmospheric aerosol load during this period is also substantially higher, resulting in a lower fractional representation of OC and BC (Fig. 6). Unfortunately complete mass closure with all possible atmospheric aerosol species included could not be conducted during this study. Therefore it cannot be stated with certainty what other aerosol species are present in higher concentrations during May to August. However, it is likely that wind-blown dust concentrations will be higher during this time of the year, since less rainfall will result in higher wind-blown dust levels (Mphepya et al. 2004). Possible higher fractional sulphate content of aerosols in the dry season (Tiitta et al. 2014) could also contribute to the lower OC and BC mass fractions. Sulphate seems to be the dominant species in South African aerosols (Martins et al. 2007; Tiitta et al. 2014).

Possible sources of OC and BC

Although it seems obvious that biomass combustion fires contribute significantly to OC and BC concentrations, this was explored further. In Fig. 7 the frequency of biomass combustion fires (obtained from MODIS burned area measurements) from March 2009 to April 2011 (correlating to the sampling period in this study) are presented for southern Africa in the region between 15 – 35 $^\circ$S and 10 – 41 $^\circ$E. The numbers of fires within a 1000 km radius from the centre of the 5 IDAF sites are also indicated. From this data (Fig. 7) it is evident that the fires mainly occurred between June and October, which correlates with the seasonal pattern observed for OC and BC (Fig. 5). Furthermore, the highest frequency of fires occurred in August and September. The highest number of fire events recorded for southern Africa (15 – 35 $^\circ$S and 10 – 41 $^\circ$E) in a single month was approximately 1.2 million in September 2010. Roughly 550 000 of these biomass burning fire events took place within the 1000 km radius from the centre of the 5 IDAF sites.

As previously described, the distances calculated between back trajectories that arrived in the middle of each sampling period and biomass burning fire events recorded for the previous 24 hours were compared to OC and BC concentrations measured.

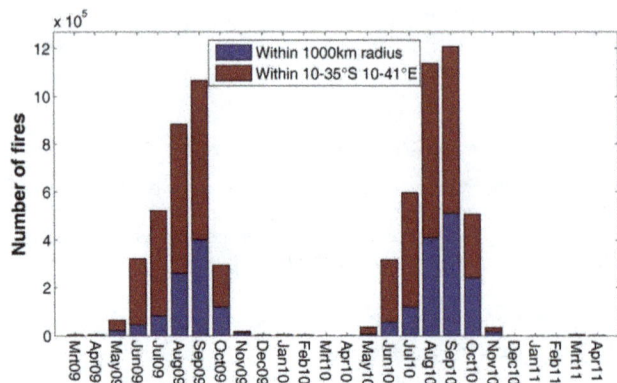

Figure 7: The frequency of biomass burning fires (obtained from MODIS burned area measurements) from March 2009 to April 2011 in southern Africa (15 – 35 $^\circ$S and 10 – 41 $^\circ$E), as well as the number of fires within a 1000 km radius from the centre of the 5 IDAF sites

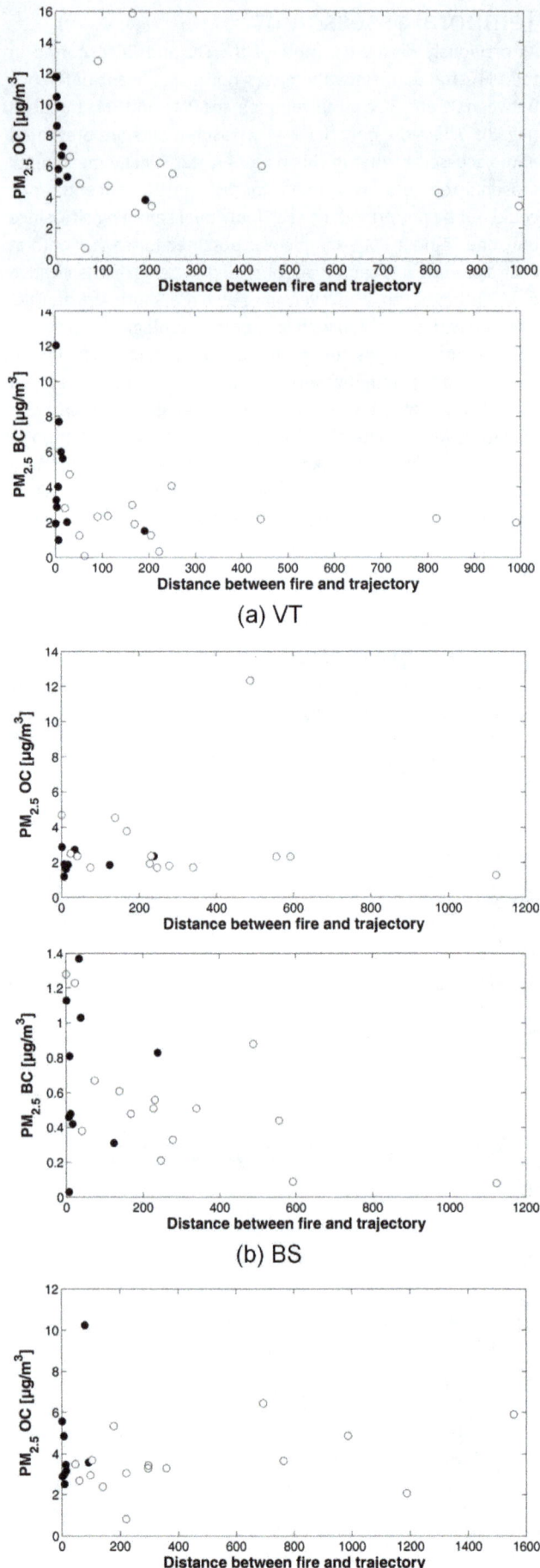

(a) VT

(b) BS

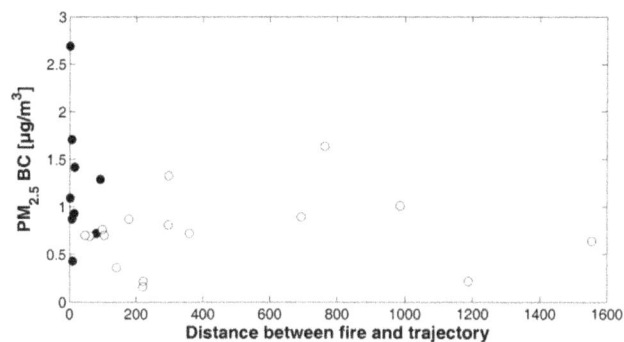

(c) LT

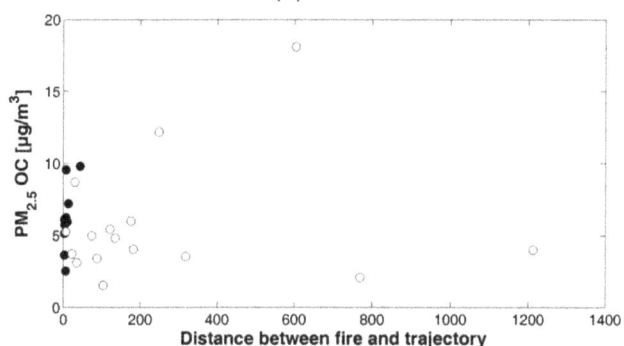

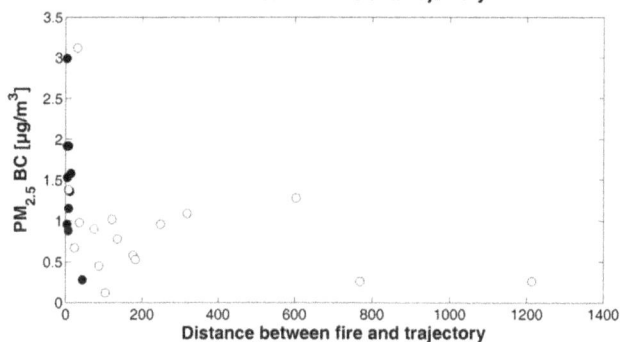

(d) SK

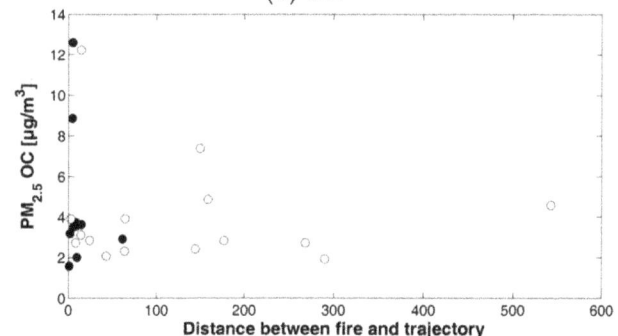

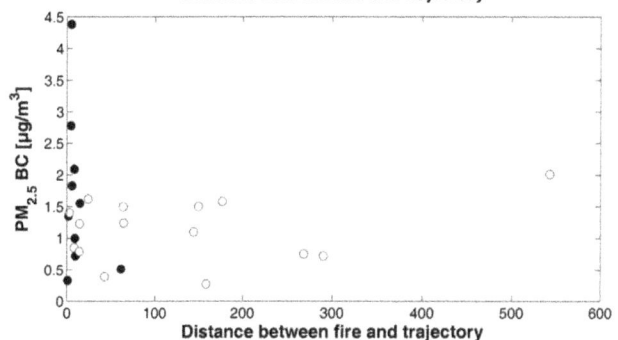

(e) AF

Figure 8: *PM$_{2.5}$ OC and BC concentrations plotted against the shortest distances between the back trajectories and the fire events at VT (a), BS (b), LT (c), SK (d) and AF (e). The black dots represent the months during which most fires occurred (June to October), while the open circles represent months with fewer fire events (November to May)*

In Fig. 8 OC and BC concentrations measured for PM$_{2.5}$ are plotted against the shortest distances between the 24-hour back trajectories and fire events for each sample at VT (8a), BS (8b), LT (8c), SK (8d) and AF (8e). The black dots indicated OC and BC concentrations measured during months that most fires occurred, i.e. June to October, while the open circles represent months with no or very few fire events (according to Fig. 7).

From Fig. 8 it seems that higher OC and BC concentrations were measured when trajectories passed closer to fires during the months when most fires occurred (June to October). In contrast, during the months with fewer fires (November to May), OC and BC concentrations did not seem to be influenced by the distances between the trajectories and fires. This indicates that fires are a major source of OC and BC during the months with high fire frequencies, while OC and BC measured during the other months were mainly emitted from other sources in this region. It has to be mentioned that the observed relationship, i.e. higher OC and BC contribution from biomass combustion fires during months with more fires, are somewhat obscured due to all the averaging periods in the data, i.e. 24-hour sampling, calculation of a single trajectory in the middle of the sampling period and 24-hour clustering of fire events by MODIS burned area measurements. However, a good indication of the influence of biomass burning fires during the months with high fire frequencies is obtained.

In an effort to further explore the regional contribution of biomass combustion fires to OC and BC, the data presented in Fig. 8 for individual sites were combined Fig. 9 for OC (Fig. 9a) and BC (Fig. 9b). These two figures therefore give a regional, rather than a site specific perspective of OC and BC concentrations as a function of distance to the closest fire for each calculated back trajectory. Since the VT sampling site was within an industrialised and residential region that could lead to bias in the data presentation, OC and BC measured at this sampling site were excluded from Fig. 9. Additionally the data points where the shortest distance to the fire for a specific back trajectory was >45 km were not included in this analysis (Fig. 9). Although not well defined, there seems to be a general trend between the OC and BC concentrations and the distance that a back trajectory had passed over a biomass burning fire, as indicated in Fig. 9. As previously mentioned, the reason for the scattered nature of the data in Fig. 9 can be attributed to multiple averaging periods that had to be applied, i.e. 24-hour sampling, calculation of a single trajectory in the middle of the sampling period and 24-hour grouping of fire events by MODIS burned area measurements. Notwithstanding these data limitations, it is evident that OC and BC are higher if a biomass combustion fire event(s) had impacted on a back trajectory air mass during the months with higher frequency in fire events. Therefore, fires seem to contribute significantly to both OC and

BC concentrations during the period when biomass burning fires frequently occur, i.e. June to October.

Similar to the data presented in Fig. 8, $PM_{2.5}$ OC and BC concentrations were plotted against the shortest distance that a back trajectory had passed over large anthropogenic point sources (e.g. pyrometallurgical smelters and coal-fired power stations) in the north-eastern region of South Africa, for each sampling site during every 24-hour sampling period. Thereafter, the individual datasets of SK, LT and BS were combined to give a regional, rather than a site specific perspective. The AF data was excluded since it had a large point source in close proximity to it, while VT was also excluded since it was within a source region with many large sources nearby. Data points where the distance between back trajectories and large point sources were >45 km were also not included in the analysis. The combined SK, LT and BS dataset is presented as a second data series in Fig. 9. From this data it is evident that $PM_{2.5}$ OC and BC concentrations were not influenced by the distances that back trajectories had passed over large point sources. This indicates that the contribution of these large point sources is not as significant compared to the biomass combustion fires on OC and BC concentrations. The observation that large point sources do not contribute significantly to regional BC concentrations is feasible, since emissions regulations in South Africa have historically regulated particulate emissions (which included BC) much more stringently than gaseous emissions.

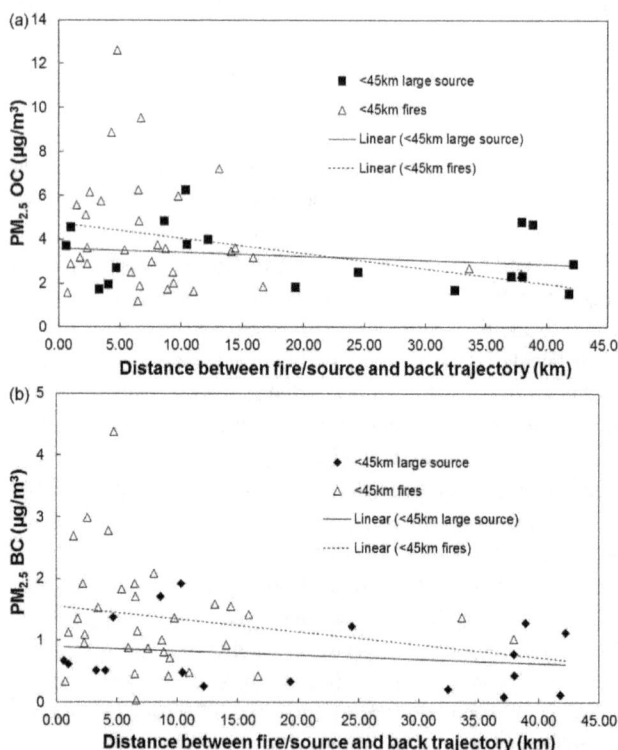

Figure 9: $PM_{2.5}$ OC (a) and BC (b) concentrations of June to October (months with higher fire frequency, according to Fig. 7) at all the sampling sites as a function of shortest distance between back trajectories and fires, excluding VT data and data where the shortest distances are >45 km. Additionally, $PM_{2.5}$ OC (a) and BC (b) concentrations of SK, LT and BS as a function of shortest distance between back trajectories and large point source, excluding data where the shortest distances are >45 km, are presented for the entire sampling period

Another possible source of OC and BC that has already been mentioned, but not yet evaluated is household combustion. Venter et al. (2012) previously suggested that household combustion in informal and semi-formal settlements in South Africa could be a significant source of BC, at least on a local scale. Household combustion generally serves two basic needs, i.e. cooking or space heating. In an effort to determine the influence of household combustion for space heating on OC and BC concentrations, the $PM_{2.5}$ OC and BC concentrations of VT were plotted against the minimum temperature measured during the actual sampling periods (Fig. 10). The VT site was specifically chosen to perform this analysis, since many informal and semi-formal settlements are present in the immediate area surrounding the measurement site.

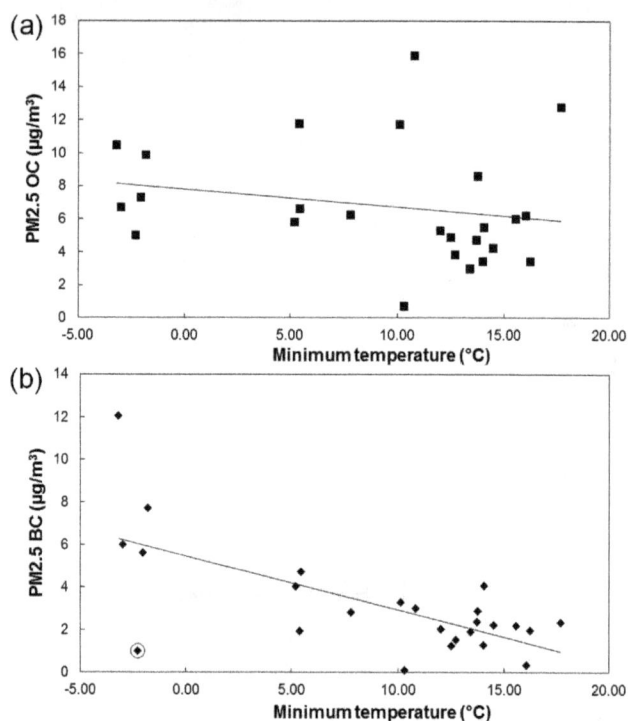

Figure 10: $PM_{2.5}$ OC (a) and BC (b) concentrations of VT plotted as a function of actual minimum temperature during the periods when the samples were collected

As is evident from Fig. 10, both $PM_{2.5}$ OC and BC concentrations seems to be inversely related to the average monthly minimum temperature. This implies that lower temperatures resulted in an increase in household combustion for space heating, which seem to result in higher OC and BC emissions. However, a weaker correlation was observed for $PM_{2.5}$ OC and minimum temperature compared to the correlation found for $PM_{2.5}$ BC and minimum temperature. This can be comprehended, since measured OC consist of primary emitted and secondary formed particles, while BC is exclusively emitted as primary particles. Therefore a better correlation between minimum temperatures and the occurrence of household combustion for space heating for BC is expected. The coefficient of determination (R^2) value for the correlation between $PM_{2.5}$ BC and minimum temperature was found to be significant, i.e. 0.45. This correlation is even more significant, i.e. 0.63, if the single circled value in Fig. 10b is excluded from the dataset.

Conclusions

The OC and BC dataset presented in this paper is one of the largest presented to date in the peer reviewed public domain for South Africa. Therefore this publication makes a contribution to a research field that is very important, e.g. BC second most important climate forcing species, but neglected. However, the dataset and methods presented here have certain limitations. Probably the most significant limitations are due to the size of the dataset and the sampling frequency. For most of the sites only one 24-hour sample per month was collected for two years and one month. This relatively small dataset makes it difficult to statistically evaluate temporal and spatial aspects, as well source contributions. Therefore the results presented should be considered as indicative and not absolute. Additionally, it also indicates the need for more comprehensive studies of this nature to be conducted. Notwithstanding the afore-mentioned limitation the following indicative deductions could be made.

OC were higher than BC concentrations at all the South African IDAF sites in the PM_{10} and $PM_{2.5}$ fractions. Most OC and BC occurred in the smaller size fraction, i.e. $PM_{2.5}$. OC and BC concentrations, as well as OC/BC ratios reflected the location of the different IDAF sites, as well as the type of sources impacting the different sites. VT, which is situated in an industrial and urban location, had the highest OC and BC concentrations with the lowest OC/BC ratio. Of the sites investigated, it was also most-likely more significantly impacted by fossil fuel combustion. AF, which is influenced by an industrial source region, had the second lowest OC/BC ratio, while the background sites (BS, LT and SK) had the highest OC/BC ratios. The background sites were also likely to be impacted by non-fossil fuel sources such as biomass burning fires. The OC and BC mass fraction percentages varied up to 24% and 12 %, respectively, for all the sites in both size fractions.

A relatively well defined seasonal pattern was observed, with higher OC and BC concentrations measured from May to October, which coincided with the dry season and highest frequency of biomass burning events in the interior of South Africa. Seasonal OC and BC mass fractions for all the sites over the entire sampling period were found to be the inverse of the seasonal OC and BC concentrations. This indicates that although OC and BC concentrations are in general higher in the dry and cold period, the atmospheric aerosol load during this period is also substantially higher, which can most-likely be attributed to wind-blown dust.

Positive correlations between OC and BC concentrations with the distance that back trajectories passed over biomass combustion fires were observed, while no such correlations were observed for the distance that back trajectories had passed over large point sources. This seems to substantiate that biomass combustion fires contribute significantly to both OC and BC concentrations on a regional scale, while this is not the case for large point sources. Correlation of OC and BC concentrations with the minimum temperatures during the sampling periods at the VT site proved that household combustion for space heating

in semi- and informal settlements contributed to OC and BC levels, at least on a local scale.

Acknowledgements

The authors acknowledge Sasol, Eskom and the National Research Foundation for financial support to the South African DEBITS-IDAF project, as well as the GDRI-ARSAIO project for financial support to scientists of France and South Africa to travel abroad. The authors would also like to thank Carin Van Der Merwe who conducted the monthly site visits and maintenance at the IDAF sites in South Africa, as well as each of the site operators, i.e. Memory Deacon (AF), Chris James (LT), Navashni Govender, Walter Kubheka, Eva Gardiner and Joel Tleane (SK), and Mike Odendaal (VT).

References

Air Resources Laboratory. (2014a). Gridded Meteorological Data Archives. http://www.ready.noaa.gov/archives.php. Accessed 24 February 2014.

Air Resources Laboratory. (2014b). Gridded Meteorological Data Archives. http://www.arl.noaa.gov/HYSPLIT_info.php. Accessed 13 February 2014.

Air Resources Laboratory. (2014c). Gridded Meteorological Data Archives. http://www.arl.noaa.gov/faq_hg17.php. Accessed 14 February 2014.

Atkinson, R., & Arey, J. (2003). Gas-phase tropospheric chemistry of biogenic volatile organic compounds: a review. *Atmospheric Environment*, 37(2), S197-S219. doi: 10.1016/S1352-2310(03)00391-1.

Baldauf, R.W., Lane, D.D., Marotz, G.A., & Wiener, R.W. (2001). Performance evaluation of the portable MiniVOL particulate matter sampler. *Atmospheric Environment*, 35, 6087-6091. PII: S1352-2310(01)00403-4.

Beukes, J.P., Vakkari, V., Van Zyl, P.G., Venter, A.D., Josipovic, M., Jaars, K., Tiita, P., Siebert, S., Pienaar, J.J., Kulmala, M., & Laakso, L. (2013). Welgegund - long-term land atmosphere measurement platform in South Africa. iLeaps Newsletter. *Integrated Land Ecosystem Atmosphere Process Study*, 12, 24-25.

Bond, T.C., & Sun, H. (2005). Can Reducing black carbon emission counteract global warming? *Environmental Science and Technology*, 39, 5921-5926.

Cachier, H., Auglagnier, F., Sarda, R., Gautier, F., Masclet, P., Besombes, J.-L., Marchand, N., Despiau, S., Croci, D., Mallet, M.,Laj, P., Marinoni, A., Deveau, P.-A., Roger, J.-C., Putaud, J.-P., Van Dingenen, R., Dell'Acqua, A., Viidanoja, J., Martins-Dos Santos, S., Liousse, C., Cousin, F., Rosset, R. Gardrat, E., & Galy-Lacaux, C. (2005). Aerosol studies during the ESCOMPTE experiment: An overview. *Atmospheric Research*, 74:547-563. doi: 10.1016/j.atmosres.2004.06.013.

Chazette, P., & Liousse, C. (2001). A case study of optical and chemical ground apportionment for urban aerosols in Thessaloniki. *Atmospheric Environment*, 35, 2497-2506. PII: S1352-2310(00)00425-8.

Chow, J.C., Watson, J.G., Crow, D., Lowenthal, D.H., & Merrifield, T. (2001). Comparison of IMPROVE and NIOSH carbon measurements. *Aerosol Science and Technology*, 34, 23-24.

Chow, J.C., Watson, J.G., Kuhns, H., Etyemezian, V., Lowenthal, D.H., Crow, D., Kohl, S.D., Engelbrecht, J.P., & Green, M.C. (2004). Source profiles for industrial, mobile, and area sources in the Big Bend Regional Aerosol Visibility and Observational study. *Chemosphere*, 54, 185-208.

Chow, J.C., Watson, J.G., Pritchett, L.C., Pierson, W.R., Frazier, C.A., & Purcell, R.G. (1993). The dri thermal/optical reflectance carbon analysis system: description, evaluation and applications in U.S. Air quality studies. *Atmospheric Environment, Part A General Topics*, 27(8), 1185-1201.

Collett, K.S., Piketh, S.J., & Ross, K. E. (2010). An assessment of the atmospheric nitrogen budget on the South African Highveld. *South African Journal of Science*, 106(5/6), 1-9. doi: 10.4102/sajs.v106i5/6.220.

de Richter, R., & Caillol, S. (2011). Fighting global warming: The potential of photocatalysis against CO_2, CH_4 N_2O, CFCs, tropospheric O_3, BC and other major contributors to climate change. *Photochemistry and Photobiology*, 12, 1-19. doi: 10.1016/j.jphotochemrev.2011.05.002.

Doumbia, E.H.T., Liousse, C., Galy-Lacaux, C., Ndiaye, S.A., Diop, B., Ouafo, M., Assamoi, E.M., Gardrat, E., Castera, P., Rosset, R., Akpo, A., & Sigha, L. (2012). Real time black carbon measurements in West and Central Africa urban sites. *Atmospheric Environment*, 54, 529-537. doi: 10.1016/j.atmosenv.2012.02.005.

Draxler, R. R., & Hess, G. D. (2004). *Description of the HYSPLIT 4 Modelling System, NOAA Technical Memorandum ERL* ARL-224.

Environmental analysis facility. (2008). DRI Standard operating procedure. 86p. Laboratoire d'Aérologie – UMR 5560. http://www.aero.obs-mip.fr/spip.php?article489. Accessed 18 July 2011.

Formenti, P., Elbert, W., Maenhaut, W., Haywood, J., Osborne, S., & Andreae, M.O. (2003). Inorganic and carbonaceous aerosols during the Southern African Regional Science Initiative (SAFARI 2000) experiment: Chemical characteristics, physical properties, and emission data for smoke from African biomass burning. *Journal of geophysical Research*, 108, D13:8488. doi: 10.1029/2002JD002408.

Galy-Lacaux, C., Al Ourabi, H., Lacaux, J.-P., Pont, V., Galloway, J., Mphepya, J., Pienaar, K., Sigha, L., & Yoboué, V. (2003). Dry and wet atmospheric nitrogen deposition in Africa. *IGAC Newsletter*,

January 2003, Issue nr. 27, 6-11.

Gauderman, W.J., Avol, E., Gilliland, F., Vora, H., Thomas, D., Berhane, K., Mcconnell, R., Kuenzli, N., Lurmann, F., Rappaport, E., Margolis, H., Bates, D., & Peters, J. (2004). The effect of air pollution on lung development from 10 to 18 years of age. *The New England Journal of Medicine*, 351(11), 1057-1067.

Government Gazette Republic of South Africa, 14 October 2005, No. 28132.

Guillame, B., Liousse, C., Galy-Lacaux, C., Rosset, R., Gardrat, E., Cachier, H., Bessagnet, B., & Poisson, N. (2008). Modeling exceptional high concentrations of carbonaceous aerosols observed at Pic du Midi in spring-summer 2003: Comparison with Sonnblick and Puy de Dôme. *Atmospheric Environment*, 42, 5140-5149. doi: 10.1016/j.atmosenv.2008.02.024.

HYSPLIT User's Guide-Version 4. Last revised: April 2013. http://www.arl.noaa.gov/documents/reports/hysplit_user_guide.pdf. Accessed 13 February 2014.

Hyvärinen, A.-P., Vakkari, V., Laakso, L., Hooda, R.K., Sharma, V.P., Panwar, T.S., Beukes, J.P., Van Zyl, P.G., Josipovic, M., Garland, R.M., Andreae, M.O., Pöschl, U., & Petzold, A. (2013). Correction for a measurement artifact of the Multi-Angle Absorption Photometer (MAAP) at high black carbon mass concentration levels. *Atmospheric Measurement Techniques*, 6, 81-90. doi: 10.5194/amt-6-81-2013.

IPCC (Intergovernmental Panel For Climate Change). (2013). *IPCC fifth assessment report: Climate change 2013*.

Jaars, K., Beukes, J.P., Van Zyl, P.G., Venter, A.D., Josipovic, M., Pienaar, J.J., Vakkari, V., Aaltonen, H., Laakso, H., Kulmala, M., Tiitta, P., Guenther, A., Hellén, H., Laakso. L., & Hakola, H. (2014). Ambient aromatic hydrocarbon measurements at Welgegund, South Africa. *Atmopheric Chemistry and Physics*, 14, 4189-4227. doi: 10.5194/acpd-14-4189-2014.

Junker, C., & Liousse, C. (2008). A global emission inventory of carbonaceous aerosol from historic records of fossil fuel and biofuel consumption for the period 1860-1997. *Atmospheric Chemistry and Physics*, 8, 1195-1207. www.atmos-chem-phys.net/8/1195/2008/. Accessed 10 March 2014.

Kanakidou, M., Seinfeld, J.H., Pandis, S.N., Barnes, I., Dentener, F.J., Facchini, M.C., Van Dingenen, R., Ervens, B., Nenes, A., Nielson, C.J., Swietlicki, E., Putaud, J.P., Balkanski, Y., Fuzzi, S., Horth, J., Moortgat, G.K., Winterhalter, R., Myhre, C.E.L., Tsigaridis, K., Vignati, E., Stephanou, E.G., & Wilson, J. (2005). Organic aerosol and global climate modelling: a review. *Atmopheric chemistry and Physics*, 5, 1053-1123. SRef-ID: 1680-7324/acp/2005-5-1053.

Laakso, L., Vakkari, V., Virkkula, A., Laakso, H., Backman, J., Kulmala, M., Beukes, J.P., van Zyl, P.G., Tiitta, P., Josipovic,

M., Pienaar, J.J., Chiloane, K., Gilardoni, S., Vignati, E., Wiedensohler, A., Tuch, T., Birmili, W., Piketh, S., Collett, K., Fourie, G.D., Komppula, M., Lihavainen, H., de Leeuw, G., & Kerminen, V.-M. (2012). South African EUCAARI measurements: seasonal variation of trace gases and aerosol optical properties. *Atmospheric Chemistry and Physics*, 12, 1847–1864. doi: 10.5194/acp-12-1847-2012.

Lazaridis, M., Semb, A., Larssen, S., Hjellbrekke, A.-G., Hov, Ø., Hanssen, J.E., Schaug, J., & Tørseth, K. (2002). Measurements of particulate matter within the framework of the European Monitoring and Evaluation Programme (EMEP) I. First results. *The Science of the Total Environment*, 285, 209-235. PII: S0048-9697(01)00932-9.

Liousse, C., Penner, J.E., Chuang, C., Walton, J.J., Eddleman, H., & Cachier, H. (1996). A global three-dimensional model study of carbonaceous aerosols. *Journal of Geophysical Research*, 105, 26871-26890.

Lourens, A.S.M., Beukes, J.P., van Zyl, P.G., Fourie, G.D., Burger, J.W., Pienaar, J.J., Read, C.E., & Jordaan, J.H.L. (2011). Spatial and Temporal assessment of Gaseous Pollutants in the Mpumalanga Highveld of South Africa. South African Journal of Science, 107(1/2), 8. Art. #269. doi: 10.4102/sajs.v107i1/2.269.

Lourens, A.S.M., Butler, T.M., Beukes, J.P., Van Zyl, P.G., Beirle, S., Wagner, T.K., Heue, K.-P., Pienaar, J.J., Fourie, G.D., & Lawrence, M.G. (2012). Re-evaluating the NO_2 hotspot over the South African Highveld. *South African Journal of Science*, 108(11/12). http://dx.doi.org/10.4102/sajs.v108i11/12.1146. Accessed 10 March 2014.

Martins, J.J. (2009). *Concentrations and deposition of atmospheric species at regional sites in southern Africa*. M.Sc. NWU Potchefstroom. 224p.

Martins, J.J., Dhammapala, R.S., Lachmann, G., Galy-Lacaux, C., & Pienaar, J.J. (2007). Long-term measurements of sulphur dioxide, nitrogen dioxide, ammonia, nitric acid and ozone in southern Africa using passive samplers. *South African Journal of Science*, 103, 336-342.

MODIS: Obtaining and processing MODIS data. http://www.yale.edu/ceo/Documentation/MODIS_data.pdf. Accessed 14 February 2014.

Mphepya, J.N., Galy-Lacaux, C., Lacaux, J.P., Held, G., & Pienaar, J.J. (2006). Precipitation Chemistry and Wet Deposition in Kruger National Park, South Africa. *Journal of Atmospheric Chemistry*, 53, 169-183. doi: 10.1007/s10874-005-9005-7.

Mphepya, J.N., Pienaar, J.J., Galy-Lacaux, C., Held, G., & Turner, C.R. (2004). Precipitation Chemistry in Semi-Arid Areas of Southern Africa: A Case Study of a Rural and an Industrial Site. *Journal of Atmospheric Chemistry*, 47, 1-24.

Mucina, L., & Rutherford, M.C. eds. (2006). *The vegetation of South Africa, Lesotho and Swaziland*. Strelitzia 19. Pretoria: South African National Biodiversity Institute. 807p.

Nasa. (2015) http://gcmd.nasa.gov/records/ GCMD_gov.noaa.ncdc.C00075.html. Accessed 28 April 2015.

Oberdörster, G., Sharp, S., Atudorei, V., Elder, A., Gelein, R., Kreyling, W., & Cox, C. (2004). Translocation of inhaled ultrafine particles to the brain. *Inhalation toxicology*, 16, 437-445. doi: 10.1080/08958370490439597.

Pöschl, U. (2005). Atmospheric Aerosols: Composition, Transformation, Climate and Health Effects. Atmospheric Chemistry: Reviews. *Angew. Chem. Int*. Ed, 44, 7520-7540. doi: 10.1002/anie.200501122.

Putaud, J.-P., Raes, F., Van Dingenen, R., Brüggemann, E., Facchini, M.-C., Decesari, S., Fuzzi, S., Gehrig, R., Hüglin, C., Laj, P., Lorbeer, G., Maenhaut, W., Mihalopoulos, N., Müller, K., Querol, X., Rodriguez, S., Schneider, J., Spindler, G., Ten Brink, H., Tørseth, K., & Wiedensohler, A. (2004). A European aerosol phenomenology-2: chemical characteristics of particulate matter at kerbside, urban, rural and background sites in Europe. *Atmospheric Environment*, 38, 2579-2595. doi: 10.1016/j.atmosenv.2004.01.041.

Riddle, E.E., Voss, P.B., Stohl, A., Holcomb, D., Maczka, D., Washburn, K., & Talbot, R.W. (2006). Trajectory model validation using newly developed altitude-controlled balloons during the International Consortium for Atmospheric Research on Transport and Transformations 2004 campaign. *Journal of Geophysical Research*, 111, D23S57. doi: 10.1029/2006JD007456.

Roy, D.P., Boschetti, L., Justice, C.O., & Ju, J. (2008). The collection 5 MODIS burned area product-Global evaluation by comparison with the MODIS active fire product. *Remote Sensing of Environment*, 112, 3690-3707p. doi: 10.1016/j.rse.2008.05.013.

Seinfeld, J. H., & Pandis, S. N. (2006). A*tmospheric Chemistry and Physics: From Air Pollution to Climate Change*. John Wiley & Sons Inc.

Slanina, S., & Zhang, Y. (2004). Aerosols: Connection between regional climatic change and air quality. IUPAC. *Pure and applied chemistry*, 76(6), 1241-1253.

Stohl, A. (1998). Computation, accuracy and application of trajectories – a review and bibliography. *Atmospheric Environment*, 32, 947–966.

Swap, R.J., Aranibar, J.N., Dowty, P.R., Gilhooly (III), W.P., & Macko, S.A. (2004). Natural abundance of ^{13}C and ^{15}N in C3 and C4 vegetation of southern Africa: patterns and implications. *Global Change Biology*, 10, 350-358. doi: 10.1046/j.1529-8817.2003.00702.x.

Tiitta, P_.,Vakkari, V., Croteau, P., Beukes, J.P., Van Zyl, P.G., Josipovic, M., Venter, A.D., Jaars, K., Pienaar, J.J., Ng, N.L., Canagaratna, M.R., Jayne, J.T., Kerminen, V.-M., Kokkola, H., Kulmala, M., Laaksonen, A., Worsnop, D.R., & Laakso, L. (2014). Chemical composition, main sources and temporal variability of PM_1 aerosols in southern African grassland. *Atmospheric Chemistry and Physics*, 14, 1909–1927. doi: 10.5194/acp-14-1909-2014.

Tummon, F., Solmon, F., Liousse, C., & Tadross, M. (2010). Simulation of the direct and semidirect aerosol effects on the southern Africa regional climate during the biomass burning season. *Journal of Geophysical Research D: Atmospheres*, 115(19). Art. no. D19206.

Tyson, P.D., & Preston-Whyte, R.A. (2000). *The Weather and Climate of Southern Africa*. Oxford University Press. South Africa. 396p. VK551.50968TYS

Val, S., Liousse, C., Doumbia, E.H.T., Galy-Lacaux, C., Cachier, H., Marchand, N., Badel, A., Gardrat, E., Sylvestre, A., & Baeza-Squiban, A. (2013). Physico-chemical characterization of African urban aerosols (Bamako in Mali and Dakar in Senegal) and their toxic effects in human bronchial epithelial cells: description of a worrying situation. *Particle and Fibre Toxicology*, 10, 10.

Vakkari, V., Kerminen, V.-M., Beukes, J. P., Tiitta, P., Van Zyl, P.G., Josipovic, M., Venter, A.D., Jaars, K., Worsnop, D.R., Kulmala, M., & Laakso, L. (2014). Rapid changes in biomass burning aerosols by atmospheric oxidation. *Submitted to Geophysical Research Letters*.

Vakkari, V., Laakso, H., Kulmala, M., Laaksonen, A., Mabaso, D., Molefe, M., Kgabi, N., & Laakso, L. (2011). New particle formation events in semi-clean South African savannah. *Atmospheric Chemistry and Physics*, 11, 3333-3346.

Venter, A.D., Vakkari, V., Beukes, J.P., Van Zyl, P.G., Laakso, H., Mabaso, D., Tiitta, P., Josipovic, M., Kulmala, M., Pienaar, J.J., & Laakso, L. (2012). An air quality assessment in the industrialised western Bushveld Igneous Complex, South Africa. *South African Journal of Science*, 10(9/10). http://dx.doi.org/10.4102/sajs.v108i9/10.1059. Accessed 10 March 2014.

Yin, J., Allen, A.G., Harrison, R.M., Jennings, S.G., Wright, E., Fitzpatrick, M., Healy, T., Barry, E., Ceburnis, D., & McCusker, D. (2005). Major component of urban PM_{10} and $PM_{2.5}$ in Ireland. *Atmospheric Research*, 78, 149-165. doi: 10.1016/jatmosres.2005.03.006.

Zhang, Q., Jimenez, J.L., Canagaratna, M.R., Allan, J.D. Coe, H., Ulbrich, I., Alfarra, M.R., Takami, A., Middlebrook, A.M., Sun, Y.L., Dzepina, K., Dunlea, E., Docherty, K., DeCarlo, P.F., Salcedo, D., Onasch, T., Jayne, J.T. Miyoshi, T., Shimono, A., Hatakeyama, S., Takegawa, N., Kondo, Y., Schneider, J., Drewnick, F., Borrmann, S., Weimer, S., Demerjian, K., Williams, P., Bower, K., Bahreini, R., Cottrell, L., Griffin, R.J., Rautiainen, J., Sun, J.Y., Zhang, Y.M., &

Worsnop, D.R. (2007). Ubiquity and dominance of oxygenated species in organic aerosols in anthropogenically-influenced Northern Hemisphere midlatitudes. *Geophysical Research Letters*, 34, L13801. doi: 10.1029/2007GL029979.

Climate change impacts on mean wind speeds in South Africa

Lynette Herbst[*1,2] **and Hannes Rautenbach**[2]

[1] Department of Engineering and Technology Management, University of Pretoria,
Cnr Lynnwood Rd and Roper Str, Hatfield, Pretoria, 0001, South Africa, lynette.herbst@up.ac.za
[2]Laboratory for Atmospheric Sciences, Department of Geography, Geoinformatics and Meteorology,
University of Pretoria, hannes.rautenbach@up.ac.za

Abstract

Climate change could potentially affect a number of variables that impact the dispersal of and human exposure to air pollutants, as well as climate dependent sectors such as wind energy. This study attempted to quantify the projected changes in seasonal daily mean wind speeds for South Africa around the mid-21st century (2051-2075) under two different atmospheric heat pathways. Seasonal daily mean wind speed increases rarely reach 6% and decreases occur to a maximum of 3% and are variable between different seasons and areas within the country. In all seasons except December-January-February, wind speeds are projected to increase in the Highveld region, suggesting that air pollution dispersing conditions could increase. Wind direction at the 850hPa-level show minor changes, except over the Western and Eastern Cape provinces.

Keywords

climate change, wind speed, air pollution, mitigation, climate models

Introduction

Changes in climate could affect local and regional air pollution concentrations by impacting atmospheric chemical reactions, transport and rates of dispersion of *inter alia* acidic materials (Bernard et al., 2001). The South African government aims to cut greenhouse gas (GHG) emissions by 42% by 2025 by implementing a range of climate change mitigation strategies, including renewable energy of which wind and solar resources are highly dependent on climate. If the country achieves this goal and successfully executes its plan to contribute 30% renewable energy to the country's energy mix by 2025, it would contribute substantially to reducing air pollution. Horton et al. (2014) suggest that large parts of South Africa could expect pollutant-dispersing conditions more often as the 21st century progresses, but their work had been conducted with global climate models (GCMs) on an annual scale.

However, variations in wind climates could occur within finer spatial scales than are represented by GCMs, and finer temporal scales than are represented by assessments at annual intervals. Therefore, the work presented in this study employed an ensemble of eight GCMs that have been dynamically downscaled to regional climate model (RCM) resolution (0.44º × 0.44º).

This study had two objectives. Firstly, to evaluate the representation of past climates by state-of-the-art climate models at a regional scale by verifying them against observational data during a 25-year reference period (1981-2005). Secondly, to establish whether notable changes exist between seasonal wind speeds at surface level and wind direction at the 850hPa-level during the reference period and a future period (2051-2075) under two atmospheric heat pathways, RCP4.5 and RCP8.5.

Data and Methods

Data

In order to address the research objectives, two CO_2 RCP pathways were considered in eight dynamically downscaled GCM simulations from the Intergovernmental Panel on Climate Change (IPCC) fifth assessment report (AR5) to determine the potential influence of global warming on South African winds. The Rossby Centre, a climate modelling research unit at the Swedish Meteorological and Hydrological Institute, has produced a substantial collection of regional climate model simulations for the African region through dynamical downscaling of a subset of eight GCMs from the CMIP5 initiative. These downscaled model simulations were produced by the Rossby Centre's RCA4 RCM.

The initiative formed part of the CORDEX-Africa (COordinated Regional Downscaling EXperiment) project. The forcing GCMs were the CanESM2, CNRM-CM5, EC-EARTH, MIROC5, HadGEM2-ES, MPI-ESM-LR, NorESM1-M, and GFDL-ESM2M coupled GCMs. Winds simulated by models are susceptible to large errors, hence the use of eight models: a multi-model ensemble allows a more robust assessment (Rasmussen et al., 2012, Pašičko et al., 2012). RCA4 RCM data are available for a historical period from 1951 to 2005, and a projected period from 2006 to 2100.

The implementation of climate change mitigation measures could also affect the manner in which wind climates behave. Fant et al. (2015) emphasise that the probability of small wind speed changes increases considerably under a climate change scenario that incorporates mitigation strategies. Therefore, the model output of this project was considered in terms of two atmospheric heat pathways (RCP4.5 and RCP8.5). In the AR5, GHG emissions pathways considered were expressed in terms of these atmospheric heat based Representative Concentration Pathways (RCPs). IPCC scenarios (SRES) used in previous assessment reports (based on CO_2 concentrations) could be updated in the AR5 to heat based pathways thanks to new information on emerging technology, economies, land use, land cover change and environmental factors of almost a decade (Moss et al., 2010). The new AR5 GHG forcing for future projections used in this study consist of CO_2 RCPs related to $4.5W.m^{-2}$ and $8.5W.m^{-2}$ atmospheric heat increases by 2100 (henceforth RCP4.5 and RCP8.5, respectively), amongst other pathways. The use of the word 'representative' resembles the fact that each RCP signifies one of numerous possible scenarios leading to particular radiative forcing characteristics (Van Vuuren et al., 2011). The word 'pathway' refers to the trajectory taken over a long time to achieve a given radiative forcing point in terms of long-term GHG concentration levels. Such time-evolving concentrations of radiatively active constituent pathways could be incorporated for driving global warming climate model simulations.

In more detail, RCP4.5 (and RCP8.5) represents a radiative forcing of ~$4.5W.m^{-2}$ at stabilisation after 2100 (>$8.5W.m^{-2}$ in 2100) and a ~650ppm CO_2-equivalent concentration at stabilisation after 2100 (>1370ppm CO_2-equivalent in 2100). RCP4.5 therefore represents a pathway that stabilises without overshoot, and RCP8.5 resembles a rising pathway. RCP4.5 was developed by the Global Change Assessment Model (GCAM) of the Pacific Northwest National Laboratory in the USA, while the RCP8.5 was developed by the Model for Energy Supply Strategy Alternatives and their General Environmental impact (MESSAGE) from the International Institute for Applied Systems Analysis in Austria (Moss et al., 2010). These RCPs are two of four which were used in AR5.

In order to identify model biases, RCA4 RCM output had to be assessed through comparison with observational data such as the ECMWF (European Centre for Medium-range Weather Forecasts) ERA-Interim global reanalysis data, which is a global atmospheric reanalysis available from 1979 to present. Its grid resolution is 0.75° × 0.75°. To examine changes in, for example, mean daily wind speeds, 30-year assessment periods are preferred to comply with the World Meteorological Organisation's definition for 'climate'. Since ERA-Interim reanalysis data is available from 1979 onwards, and the historical period for the RCA4 model output ends in 2005, a 25-

year assessment period (1981-2005) was chosen for this study, which was then compared to a projection period (2051-2075). ERA-Interim reanalysis data are particularly useful in climate change assessments as it covers numerous climate variables in a coherent structure. Observational data used in the reanalysis originate from inter alia wind profilers, pilot balloons, radiosondes, land stations, aircraft, ships and drifting buoys (Dee et al., 2011).

Methods

ERA-Interim reanalysis data
Model performance was evaluated by calculating differences between the RCA4 output and ERA-Interim reanalysis data. For this purpose, daily (00:00UCT, 06:00UCT, 12:00UCT and 18:00UCT) historical near-surface (10m above ground level (agl)) u- and v-components have been obtained for the 25-year period 1981 to 2005, across the domain 18° to 42°S and 14° to 37°E, from the ERA-Interim reanalysis databank. For comparing RCA4 RCM output (0.44° × 0.44° resolution) to ERA-Interim reanalysis data (0.75° × 0.75° resolution), ERA-Interim reanalysis fields were interpolated (bilinear) to fit the RCA4 fields. The ERA-Interim simulation domain size was also modified to correspond with the RCA4 domain. The boundaries of this domain are 19.5° to 40.5°S and 15° to 35.25°E.

Wind speed (ws) was calculated from u- and v-components as follows:

$$ws = \sqrt{u^2 + v^2} \qquad (1)$$

Wind speeds at 00:00UCT, 06:00UCT, 12:00UCT and 18:00UCT were averaged to obtain daily means, which were compatible with RCA4 RCM data: RCA4 data is provided as daily averages taken eight times a day, i.e. three-hourly (Christensen et al., 2014). The first 28 days of each month were then selected for further calculation. Residual days could not be used in the analysis, due to the fact that some model fields consist of 30-day months only, while others included leap years. A uniform month-day number was important for calculating cross-model ensemble averages. Seasonal wind speeds were then obtained after categorising daily data into four groups: December-January-February (DJF), March-April-May (MAM), June-July-August (JJA) and September-October-November (SON). From this, seasonal daily mean wind speeds for each season were calculated.

Model data
Data from eight the GCMs that were dynamically downscaled using the RCA4 RCM were obtained. A domain extending from 22° to 35°S and 16.2° and 33°E was defined for the study. Daily historical near-surface wind speeds (10m agl) were extracted for each model for the 25 years extending from 1981 to 2005. For each of these eight model files, data were grouped into seasons (DJF, MAM, JJA and SON). Days 1 to 28 were then extracted, as explained previously, for each month per season and per model.

Thereafter, ensemble means of the daily data were calculated from the eight RCA4 RCM simulations across the four seasons, from where daily mean wind speeds for each of the seasons were calculated.

To project potential diversions from dominant wind directions, the u- and v-components at the 850hPa-level were extracted for each model for the 25 years extending from 1981 to 2005 as well. The data were grouped seasonally, extracted from days 1 to 28, and ensemble means were calculated according to the same procedure followed for the near-surface wind speed data. The u- and v-component data were then used as vectors in the Grid Analysis and Display System (GrADS) to calculate and display wind directions for the historical period of 1981 to 2005.

Statistical evaluation of model performance

In order to verify RCA4 RCM performance, the Root Mean Square Error (RMSE) of seasonal daily mean wind speeds were calculated using the ERA-Interim and the RCA4 RCM data. RMSE was calculated as follows (CTEC, 2015):

$$RMSE = \sqrt{\sum_{i=1}^{n} \frac{(x_{obs,i} - x_{model,i})^2}{n}} \qquad (2)$$

where $x_{obs,i}$ represents the observed ERA-Interim values; $x_{model,i}$ represents the model values at a particular point i; and n represents the number of values.

Model evaluation against observational data

RCA4 RCM output was evaluated against observational data from six South African Weather Service (SAWS) stations distributed across the country. Data from Malmesbury, Vredendal, Greytown, Upington, Nelspruit, and Mokopane were obtained from SAWS for varying periods starting in 1981 to 2005. Wind speeds were estimated visually from pressure plate anemometers until 1992 (excluding Upington), when wind speeds began to be measured with Automatic Weather Stations (AWS), and are therefore more accurate. The data were provided as it was measured at 08:00, 14:00 and 20:00. These three times daily observations were averaged to obtain single daily averages, which were then employed in calculating seasonal average wind speeds. Model values for comparison were selected from those grid boxes in model data within which the particular weather station's coordinates lie. These coordinates are shown in Table 1, as well as the period for which data were available. Note that two stations' data were considered for the Upington area, as the periods of availability of both differ.

Table 1: SAWS weather station particulars

Station name	Coordinates	Period of data availability
Malmesbury	33.4720 S 18.7180 E	1986/02-2005/12
Vredendal	31.6730 S 18.4960 E	1981/01-2005/12
Greytown	29.0830 S 30.6030 E	1993/03-2005/12
Upington (1) Upington (2)	WK 28.4000 S 21.2670 E WO 28.4110 S 21.2640 E	1981/01-1992/04 1991/07-2005/12
Nelspruit	25.5030 S 30.9110 E	1993/07-2005/12
Mokopane	24.2050 S 29.0110 E	1995/09-2005/12

Changes in wind climate

Daily means of seasonal near-surface wind speed, in the projected period were calculated from RCA4 RCM output under conditions of the RCP4.5 and RCP8.5 pathways. Data for 850hPa-level u- and v-components were obtained for these pathways as well. The data were extracted and grouped in the same manner as in the historical period, but in this case for a 25-year period extending from 2051 to 2075.

Anomalies between RCA4 RCM output in the reference period and RCA4 RCM output in the two projections were calculated and expressed as percentage differences. Wind direction changes at the 850hPa-level were also calculated for the two RCPs by subtracting historical u/v components from projected u/v components, and then by using these anomalies to plot deviations from the dominant wind direction in the reference period.

Results

Mean seasonal wind speed

For the DJF-season, the winds in the north-western quarter of the country are captured well, showing wind speeds in the region of 4m.s^{-1} to 5m.s^{-1} in both ERA-Interim (Figure 1) and RCA4 RCM ensemble (Figure 2) runs. However, wind speeds are somewhat overestimated over the eastern escarpment by the RCA4 RCM. The ERA-Interim simulation shows that near-surface winds occur at around 1.5m.s^{-1} to 3.5m.s^{-1}, whereas the RCA4 RCM simulations project winds in this area to vary from 3m.s^{-1} to 5m.s^{-1}. Furthermore winds are projected at around 3m.s^{-1} to 5m.s^{-1} in the ERA-Interim run, but the RCA4 RCM projection ranges from 4.5m.s^{-1} to 6m.s^{-1} in the south-east of the country by the model data. In summary, the RCA4 RCM projects near-surface wind speeds at around 1.5m.s^{-1} higher than observed data shows it to be, except in the north-western quarter of the country.

In the MAM-season, the lower wind speeds occurring from central South Africa to the Highveld are well captured by the RCA4 RCM, with only a 1m.s^{-1} difference between ERA-Interim (Figure 3) and RCA4 RCM (Figure 4) output, the latter projecting the higher mean wind speed. An overestimation of wind speeds by the RCA4 RCM is observed along a west-east strip stretching from the Cape Town region to Lesotho. According to the ERA-Interim simulation, wind speeds range from 2.5m.s^{-1} to 4m.s^{-1}, and the RCA4 RCM projects wind speeds to range from 3.5m.s^{-1} to 5.5m.s^{-1}. Wind speed projection ranges therefore are estimated by the RCA4 RCM by no more than 1.5m.s^{-1} higher than ERA-Interim data in this season.

Near-surface wind speeds in the JJA-season are projected at no less than 2.5m.s^{-1} in the north-eastern quarter of the country in the RCA4 RCM ensemble run (Figure 6) – 0.5m.s^{-1} higher than the ERA-Interim run (Figure 5). The west-east strip RCA4 RCM ensemble overestimation observed from Cape Town to Lesotho in the MAM-season (Figure 4) is also present in the JJA-season:

the ERA-Interim run shows that these winds range from 3m.s^{-1} to 5m.s^{-1}, while the RCA4 RCM ensemble run shows it could range from 3.5m.s^{-1} to 6m.s^{-1}.

Finally, in the SON-season lower wind speeds in the south-eastern tip of the country are once more not well captured by the RCA4 RCM, nor is it represented along the eastern escarpment stretch (Figure 8). The simulation from the ERA-Interim data (Figure 7) suggests that winds range from 2m.s^{-1} to 5m.s^{-1}, but they range from 3.5m.s^{-1} to 6m.s^{-1} in RCA4 RCM ensemble run. On the other hand, winds over the central part of the country are well captured, as they occur at about 3.5m.s^{-1} to 4.5m.s^{-1} in both ERA-Interim and model ensemble runs.

The RMSE of seasonal mean daily near-surface wind speeds (m.s^{-1}) are shown in Figures 9 to 12. The high RMSE values on the south-eastern tip of the country demonstrate overestimations identified in previous paragraphs in all seasons. In general, it is confirmed that the RCA4 RCM performs best over central South Africa. The MAM-season has the largest area with the lowest RMSE, mostly between 0.8m.s^{-1} and 1.6m.s^{-1} (Figure 10). The highest RMSE-values of 2.8m.s^{-1} occur in the JJA-season in the Cape Town region and Lesotho (Figure 11).

Model evaluation against observational data

Mean daily wind speeds as simulated from RCA4 RCM data, plotted together with mean daily wind speeds calculated from ground station data recorded at six SAWS weather stations are shown in Figure 13. Four separate values were plotted per location according to seasons. Wind speeds from RCA4 RCM ensemble data at all of the points plotted were higher than the observational data from the SAWS weather stations. It must be borne in mind that wind speeds vary markedly on scales smaller than the grid resolution of model data of 0.44° × 0.44° (0.44° latitude translates roughly to between 45 and 40km; 0.44° longitude to 49km). Minor topographical variations, land cover and temperature variations can intensify or slow winds (Jarvis and Stuart, 2001). The model performs best at the Upington region and Mokopane regions (locations 13-16 and 21-24, respectively in Figure 13). The overestimation of winds demonstrated by the high RMSE-values in previous figures in the Western Cape region (Figures 9 to 12) are supported by the large differences in Malmesbury and Vredendal observations versus model output in Figure 13 (locations 1-4 and 5-8, respectively).

RCA4 RCM projected mean seasonal wind speed and direction changes

For the DJF-season, it is shown in Figure 14 that wind speeds are expected to increase by up to 4% along the southern parts of the Western and Eastern Cape Provinces under RCP4.5 (Figure 14). For the RCP8.5 pathway, increases in wind speeds could reach 6% in this area (Figure 15). Decreases in wind speeds of up to 1.5% might be expected in the Highveld under the RCP4.5 pathway (Figure 14) and 2.5% under the RCP8.5 pathway (Figure 15).

In the MAM-season under the RCP4.5 pathway, wind speeds are projected to increase by up to 3.5% in the north-eastern quarter of the country, and could also increase by up to 3% in the Cape Town region (Figure 16). Projected increases in wind speeds are more apparent in the RCP8.5 pathway in these regions, where it could increase by up 5% in the Limpopo province and 4.5% in the Cape Town region (Figure 17). The central part of the country could expect a slight increase in wind speeds of about 1% to 3% under the RCP8.5 pathway as well (Figure 17). Decreased wind speeds are projected under the RCP4.5 pathway in the Northern Cape and the Eastern Cape Provinces, but of only 1% (Figure 16). In the RCP8.5 pathway it is projected that the western half of the Northern Cape Province could expect decreased wind speeds of up to 2% and the Eastern Cape Province could expect wind speeds to increase by 2% along the coast, and decrease by 1% closer to the escarpment edge over the east of South Africa (Figure 17).

In the RCP4.5 pathway in the JJA-season, wind speeds are projected to increase for the majority of the country (Figure 18). Specifically the eastern half of the country could expect wind speeds to increase by up to 4.5%, but central South Africa and the Cape Town region could expect wind speeds to increase by up to 3%. Decreased wind speeds of up to 1% are projected in this same pathway over the coastal Eastern Cape Province and the West Coast. Under the RCP8.5 pathway, wind speeds are projected to increase by up to 6% in the far east of South Africa, and in the region of 1% to 3% in the interior (Figure 19). Wind speeds are projected to decease along the West Coast by up to 2%, and could decrease by 1.5% in the Eastern Cape Province.

In the SON-season, wind speeds are projected to increase by 5% in the Limpopo Province, and lesser increases are projected along the South African coast starting at KwaZulu-Natal all the way to Cape Town under the RCP4.5 pathway (Figure 20). The region around and included the Northern Cape Province could expect wind speeds to increase by 0% up to 1.5%. Under the RCP8.5 pathway, the Limpopo Province could expect wind speeds to increase by up to 6% (Figure 21). The rest of the country could expect milder wind speed increase ranging from 0.5% in parts of the Northern Cape Province, to 4% in the Cape Town region and central South Africa. Thus wind speeds are not projected to decrease under either of the pathways for the SON-season.

In the DJF-season, the north-easterlies over the Gauteng region (Figure 2) are projected to deviate towards the east in the RCP4.5 (Figure 14) and the RCP8.5 (Figure 15) pathway. Over the western half of the country, the southerlies (Figure 2) are projected to deviate in an east to south-eastern direction in the interior of the country under both pathways (Figures 14 and 15). The southerlies close to the coast in the Western Cape (Figure2) are projected to deviate to the west in both pathways as well (Figures 14 and 15). The south-easterlies simulated in the far eastern corner of the country are projected to deviate to the east in the RCP4.5 pathway (Figure 14) and are projected to deviate in an easterly to north-easterly direction in the RCP8.5 pathway (Figure 15).

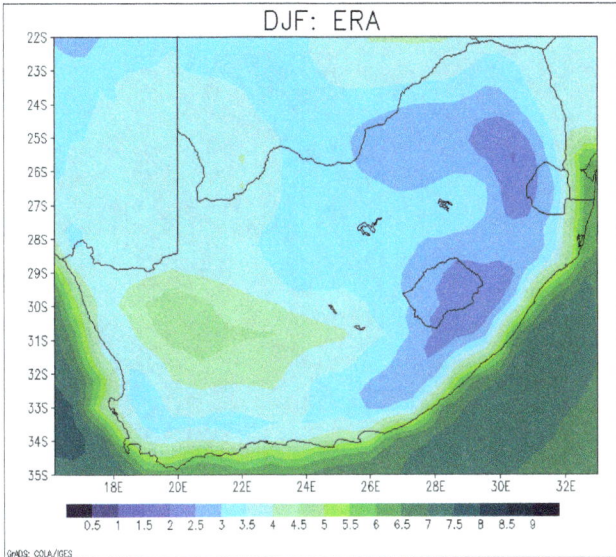

Figure 1: DJF mean seasonal wind speed (m.s⁻¹) from ERA-Interim data (1981-2005)

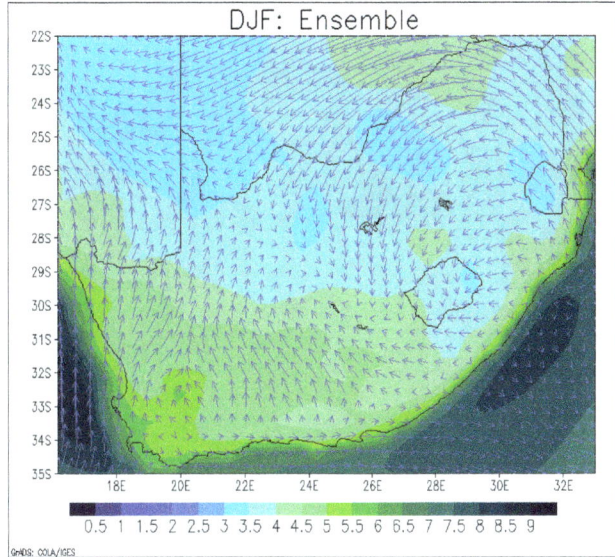

Figure 2: DJF mean seasonal wind speed (m.s⁻¹) from RCA4 RCM ensemble data (1981-2005).

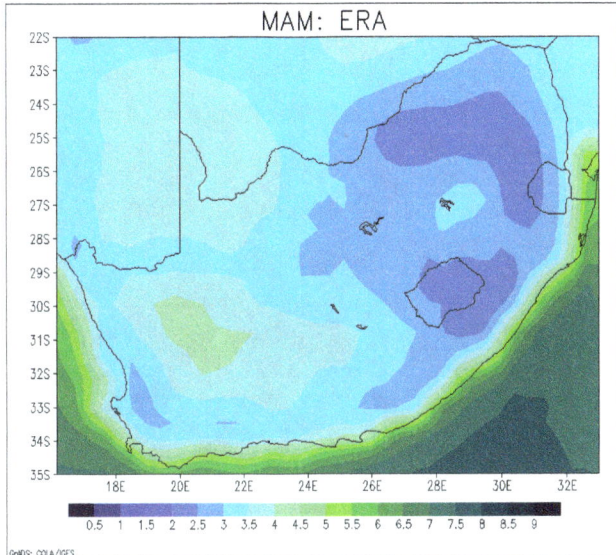

Figure 3: MAM mean seasonal wind speed (m.s⁻¹) from ERA-Interim data (1981-2005).

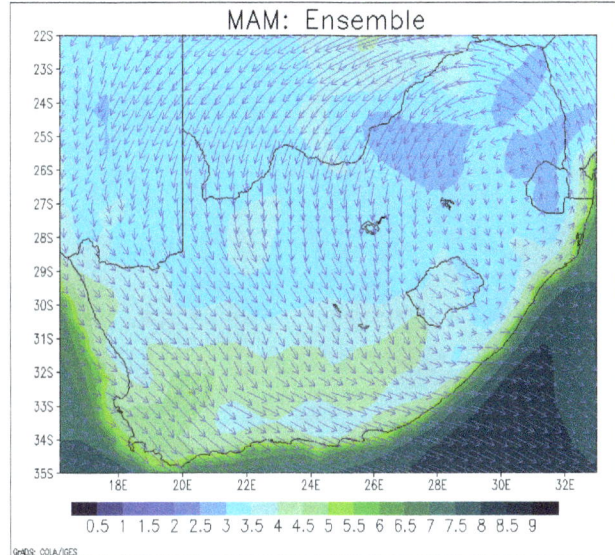

Figure 4: MAM mean seasonal wind speed (m.s⁻¹) from RCA4 RCM ensemble data (1981-2005).

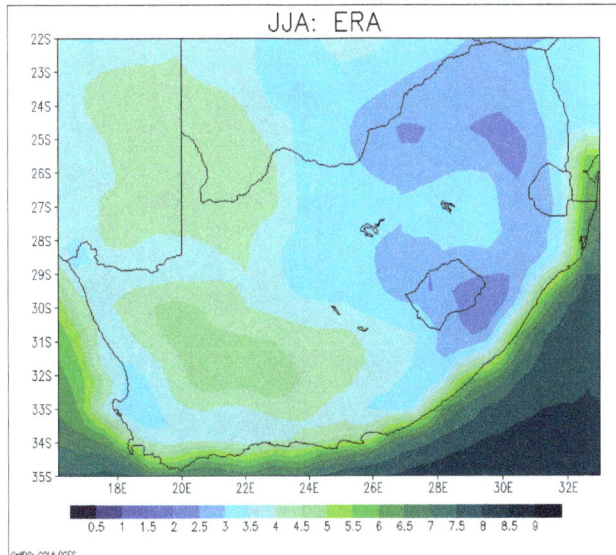

Figure 5: JJA mean seasonal wind speed (m.s⁻¹) from ERA-Interim data (1981-2005).

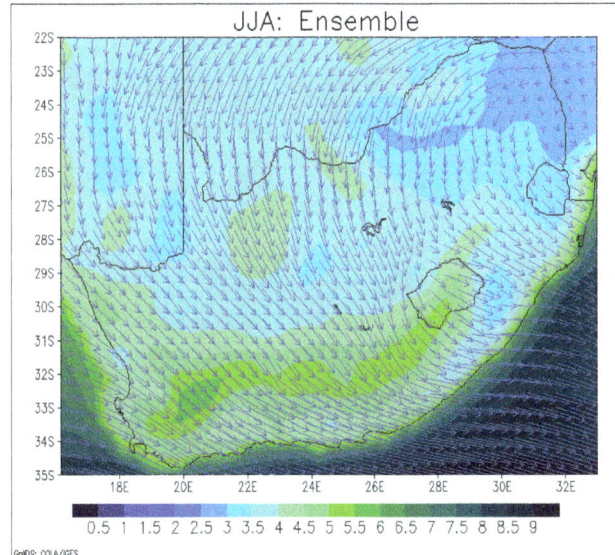

Figure 6: JJA mean seasonal wind speed (m.s⁻¹) from RCA4 RCM ensemble data (1981-2005).

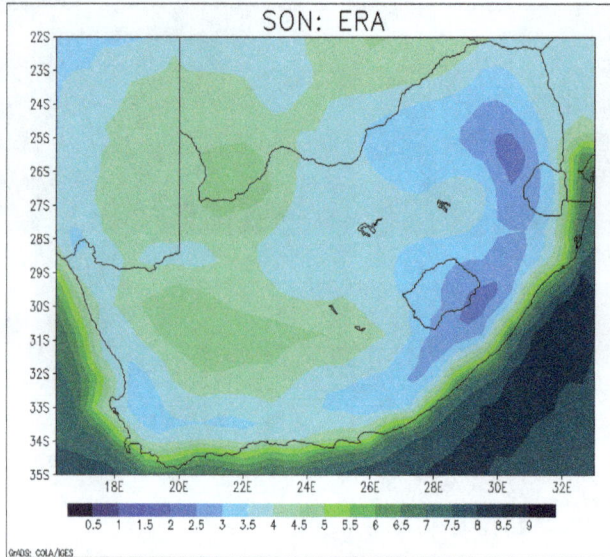

Figure 7: *SON mean seasonal wind speed (m.s⁻¹) from ERA-Interim data (1981-2005).*

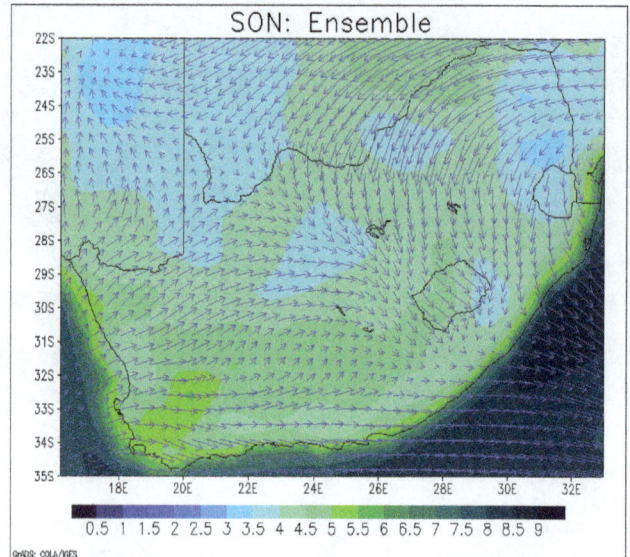

Figure 8: *SON mean seasonal wind speed (m.s⁻¹) from RCA4 RCM ensemble data (1981-2005).*

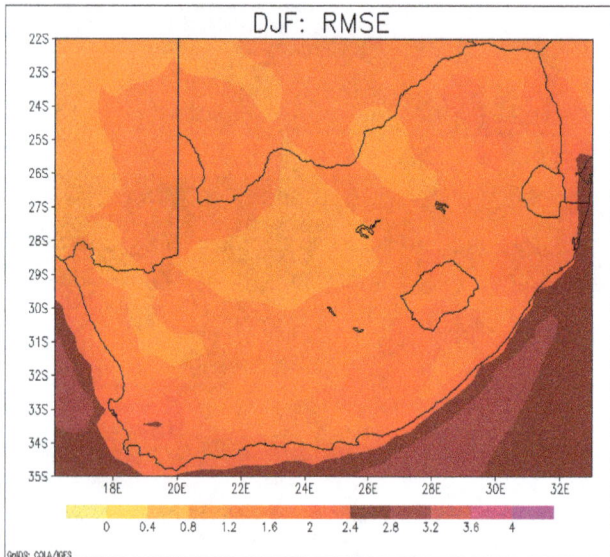

Figure 9: *Root Mean Square Error (RMSE) for the DJF wind speeds (m.s⁻¹) (1981-2005).*

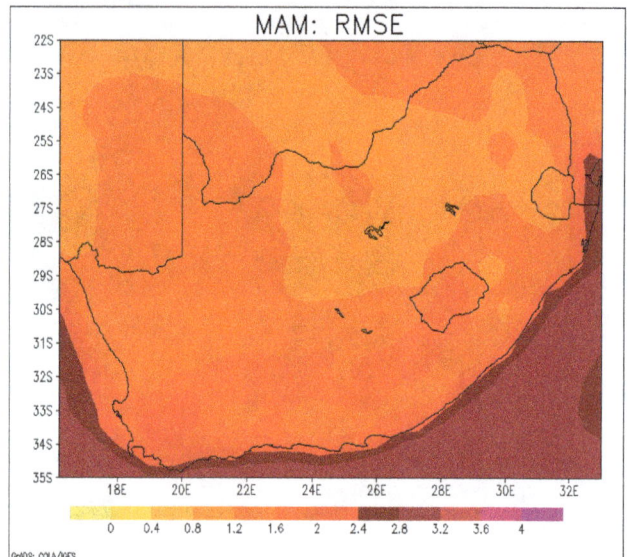

Figure 10: *Root Mean Square Error (RMSE) for the MAM wind speeds (m.s⁻¹) (1981-2005).*

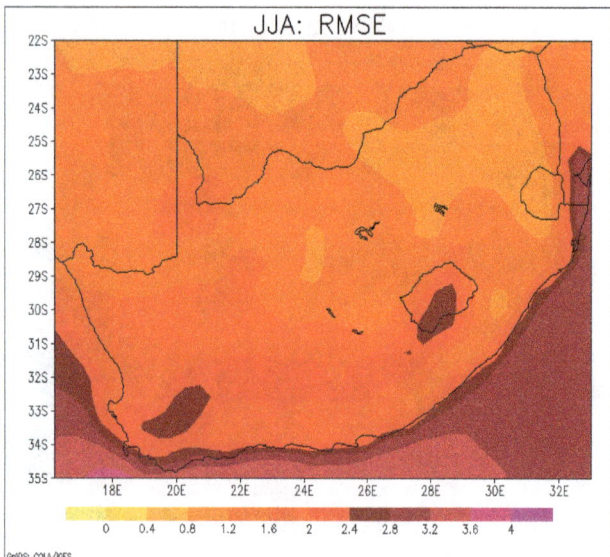

Figure 11: *Root Mean Square Error (RMSE) for the JJA wind speeds (m.s⁻¹) (1981-2005).*

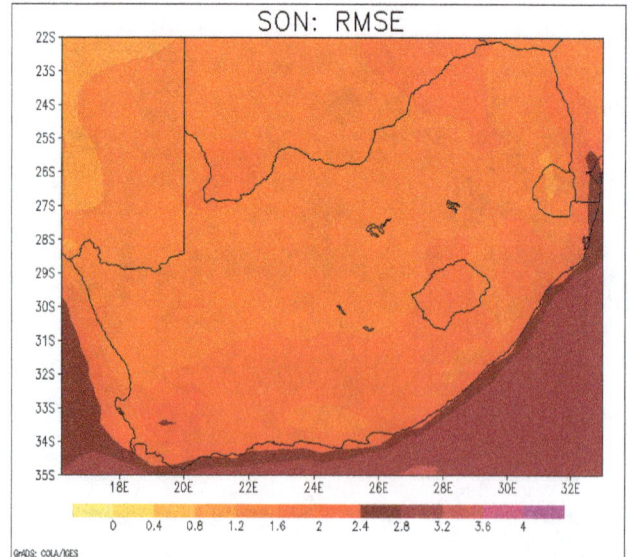

Figure 12: *Root Mean Square Error (RMSE) for the SON wind speeds (m.s⁻¹) (1981-2005).*

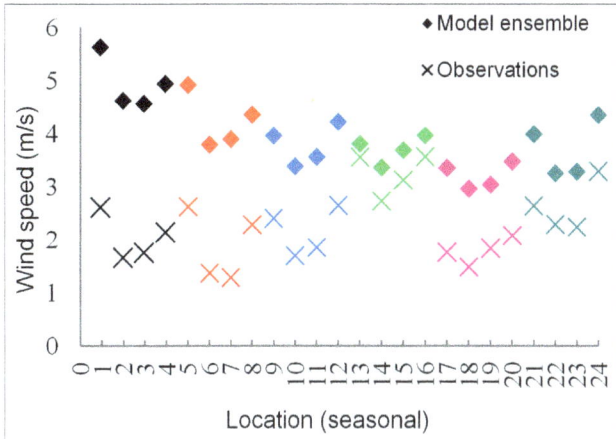

Figure 13: *Comparison of wind speeds from SAWS station data with RCA4 RCM ensemble data. Location numbers 1-4 denote Malmesbury DJF, MAM, JJA, SON; location numbers 5-8 denote Vredendal DJF, MAM, JJA, SON etc. - in the same order as in Table 1.*

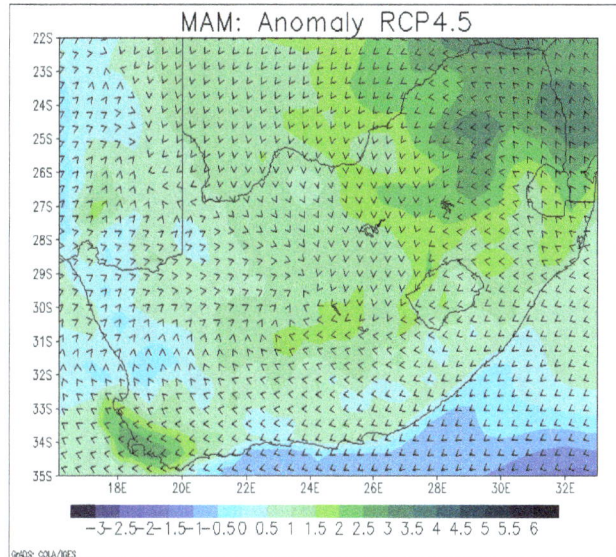

Figure 14: *Projected anomaly in mean wind speed (%) and direction for DJF (2051-2075 relative to 1981-2005) under the RCP4.5 pathway.*

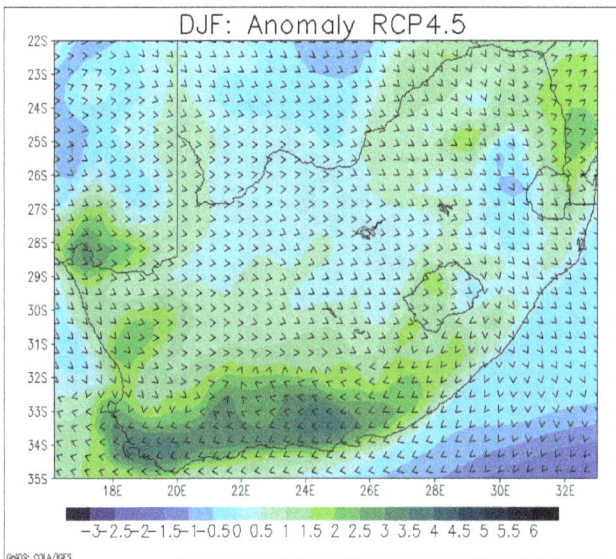

Figure 15: *Projected anomaly in mean wind speed (%) and direction for DJF (2051-2075 relative to 1981-2008) under the RCP8.5 pathway.*

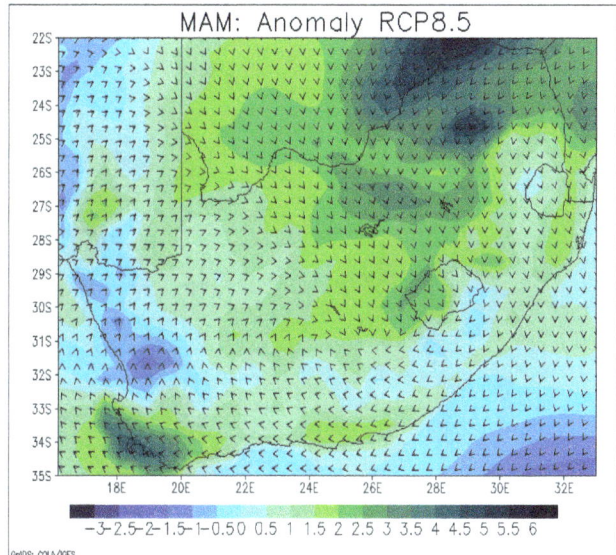

Figure 16: *Projected anomaly in mean wind speed (%) and direction for MAM (2051-2075 relative to 1981-2005) under the RCP4.5 pathway.*

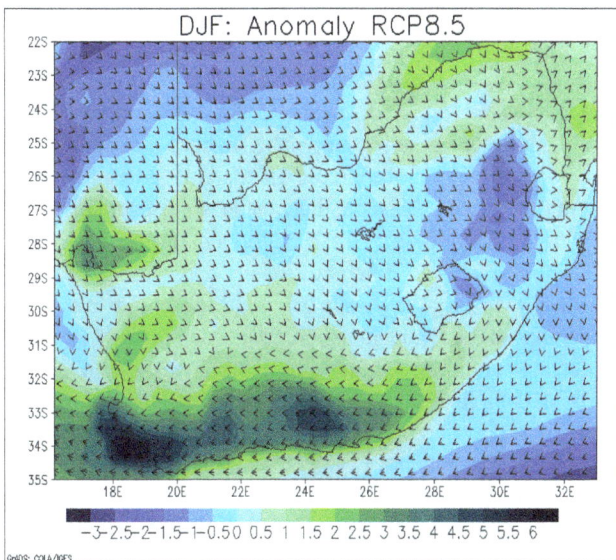

Figure 17: *Projected anomaly in mean wind speed (%) and direction for MAM (2051-2075 relative to 1981-2008) under the RCP8.5 pathway.*

In the MAM-season, the primarily southern direction in which winds blow over the Eastern Cape and Western Cape Provinces (Figure 4) are projected to deviate in a western direction in the RCP4.5 (Figure 16) and the RCP8.5 (Figure 17) pathways where current wind farm developments are underway. In the RCP4.5 pathway, winds are projected to deviate (Figure 16) minimally from their dominant directions (Figure 4) in central South Africa. However, in the western expanses of the country, the northerlies (Figure 4) are projected to deviate to an eastern direction (Figure 16). Winds in the Limpopo area are projected to remain fairly unchanged in the both the RCP4.5 (Figure 16) and RCP8.5 (Figure 17) pathways. North-westerlies along the Eastern Cape coast (Figure 4) are projected to deviate in the opposite direction i.e. to the southwest in both pathways (Figures 16 and 17).

The north-easterlies in Limpopo (Figure 6) are projected to deviate in a southern direction in both the RCP4.5 (Figure 18) and RCP8.5 (Figure 19) pathways in the JJA-season. The north-

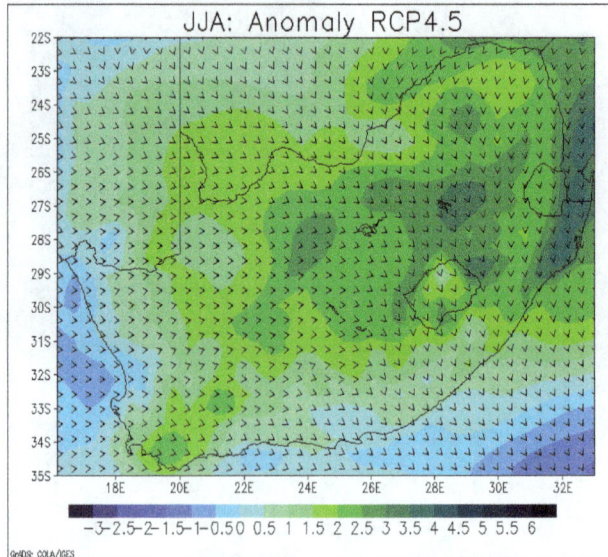

Figure 18: Projected anomaly in mean wind speed and direction (%) for JJA (2051-2075 relative to 1981-2005) under the RCP4.5 pathway.

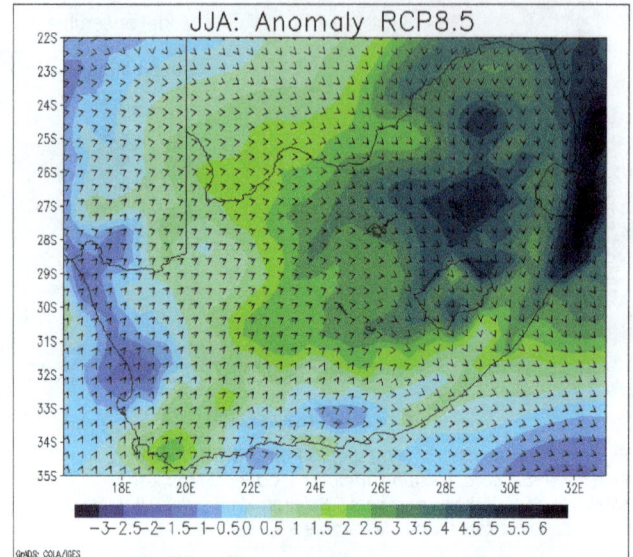

Figure 19: Projected anomaly in mean wind speed and direction (%) for JJA (2051-2075 relative to 1981-2005) under the RCP4.5 pathway.

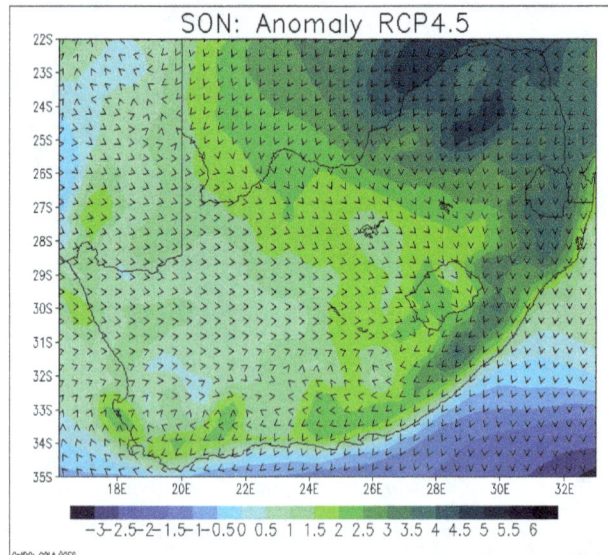

Figure 20: Projected anomaly in mean wind speed (%) and direction for SON (2051-2075 relative to 1981-2005) under the RCP4.5 pathway.

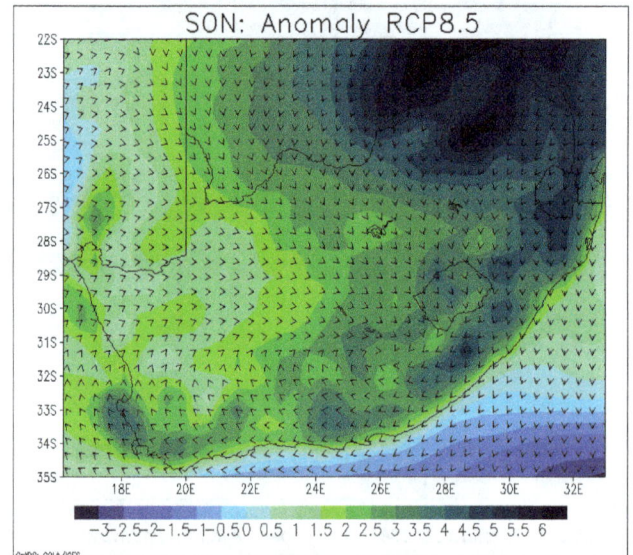

Figure 21: Projected anomaly in mean wind speed (%) and direction for SON (2051-2075 relative to 1981-2008) under the RCP8.5 pathway.

westerlies over the Western Cape and Eastern Cape Provinces (Figure 6) are projected to deviate very little in the RCP4.5 pathway (Figure 18), but are projected to deviate in a northern direction under the RCP8.5 pathway (Figure 19).

While wind directions in the eastern half of the country (Figure 8) are projected to remain relatively unchanged in the SON-season under both the RCP4.5 (Figure 20) and RCP8.5 (Figure 21) pathways, they are projected to deviate to the west along the coasts of the Western Cape and Eastern Cape Provinces under both pathways (Figures 20 and 21). South-westerly winds in the Northern Cape Province (Figure 8) are projected to deviate in an eastern direction in both pathways (Figures 20 and 21).

Conclusions
Numerous studies are beginning to clarify the impacts of climate

change on various physical climatic variables (Fant et al., 2015). Two such variables, wind speed and direction, were assessed in this study to gauge its potential changes in South Africa within the mid-21st century (2051-2075), relative to a historical period (1981-2005). Seasonal daily mean wind speed increases rarely reach 6% and decreases occur to a maximum of 3% and are variable between different seasons and areas within the country. In all seasons except DJF, wind speeds are projected to increase in the Highveld region, suggesting that air pollution dispersing conditions could increase. Dominant wind direction at the 850hPa level is projected to remain unchanged, except over the Western and Eastern Cape provinces in most seasons.

RCPs impact the severity of changes in wind speeds, showing that climate change mitigation measures could curb drastic changes to wind climates in South Africa (Fant et al., 2015). Climate change impact studies such as this one could also contribute to understanding its potential effects on various other sectors such as wind energy (Herbst & Lalk, 2014). The findings of the study are consistent with those of Jury (2013). In his 20th century analysis, he found no trend in surface zonal winds, but future projections from circulation models suggested intensified easterly flow along the south coast of South Africa.

References

Bernard S.M., Samet J.M., Grambsch A., Ebi K.L. & Romieu I. 2001, 'The potential impacts of climate variability and change on air pollution-related health effects in the United States,' *Environmental Health Perspectives* 109:109-209.

Christensen O.B., Gutowksi W.J., Nikulin G. & Legutke S. 2014, CORDEX Archive Design. CORDEX Experiment Guidelines. Available: http://cordex.dmi.dk/joomla/images/CORDEX/cordex_archive_specifications.pdf [Date Accessed 15 July 2014].

CTEC 2015, Model evaluation methods. Available http://www.ctec.ufal.br/professor/crfj/Graduacao/MSH/Model%20evaluation%20methods.doc [Date Accessed 2 February 2015].

Dee D.P., Uppala S.M., Simmons A.J., Berrisford P., Poli P., Kobayashi S., Andrae U., Balmaseda M.A., Balsamo G., Bauer P., Bechtold P., Beljaars A.C.M., Van de Berg L., Bidlot J., Bormann N., Delsol C., Dragani R., Fuentes M., Geer A.J., Haimberger L., Healy S.B., Hersbach H., Hólm E.V., Isaksen L., Kållberg P., Köhler M., Matricardi M., McNally A.P., Monge-Sanz B.M., Morcrette J.-J., Park B.-K., Peubey C., De Rosnay P., Tavolato C., Thépaut J.-N. & Vitart F. 2011, 'The ERA-Interim reanalysis: configuration and performance of the data assimilation system,' *Quarterly Journal of the Royal Meteorological Society* 137:553-597.

Fant C., Schlosser C.A. & Strzepek K. 2015, 'The impact of climate change on wind and solar resources in southern Africa,' *Applied Energy*, Article in Press [Available from:http://dx.doi.org/10.1016/j.apenergy.2015.03.042].

Herbst L. & Lalk J. 2014, 'A case study of climate variability effects on wind resources in South Africa,' *Journal of Energy in Southern Africa* 25:2-10.

Horton D.E., Skinner C.B., Singh D. & Diffenbaugh N.S. 2014, 'Occurrence and persistence of future atmospheric stagnation events,' *Nature Climate Change* 4:698-703.

Jarvis C.H. & Stuart N. 2001, 'A Comparison among Strategies for Interpolating Maximum and Minimum Daily Air Temperatures. Part I: The Selection of "Guiding" Topographic and Land Cover Variables,' *Journal of Applied Meteorology* 40:1060-1074.

Jury M. 2013, 'Climate trends in southern Africa,' *South African Journal of Science* 109(1/2):11 pages.

Moss R.H., Edmonds J.A., Hibbard K.A., Manning M.R., Rose S.K., Van Vuuren D.P., Carter T.R., Emori S., Kainuma M., Kram T.,

Meehl G.A., Mitchell J.F.B., Nakicenovic N., Riahi K., Smith S.J., Stouffer R.J., Thomson A.M., Weyant J.P. & Wilbanks T.J. 2010. 'The next generation of scenarios for climate change research and assessment,' *Nature*, 463:747-756.

Pašičko R., Branković Č. & Šimić Z. 2012, 'Assessment of climate change impacts on energy generation from renewable sources in Croatia,' *Renewable Energy* 46:224-231.

Rasmussen D.J., Holloway T. & Nemet G.F. 2011, 'Opportunities and challenges in assessing climate change impacts on wind energy - a critical comparison of wind speed projections in California,' *Environmental Research Letters* 6 [Available from: DOI 10.1088/1748-9326/6/2/024008].

Van Vuuren D.P., Edmonds J., Kainuma M., Riahi K., Thomson A., Hibbard, K., Hurtt G.C., Kram T., Krey V., Lamarque J., Masui T., Meinshausen M., Nakicenovic N., Smith S.J. & Rose S.K. 2011. 'The representative concentration pathways: an overview,' *Climatic Change* 109:5-31.

3

Human health risk assessment of airborne metals to a potentially exposed community: a screening exercise

M.A. Oosthuizen[1*], C.Y. Wright[2,3], M. Matooane[1] and N. Phala[1]

[1]Council for Scientific and Industrial Research, Natural Resources and the Environment, Climate Studies, Modelling and Environmental Health Research Group, P.O. Box 395, Pretoria, 0001, South Africa.
[2]Environment and Health Research Unit, South African Medical Research Council, Pretoria, South Africa.
[3]Department of Geography, Geoinformatics and Meteorology, University of Pretoria, Pretoria, South Africa.
*Corresponding author. Email: roosthui@csir.co.za

Abstract

Exposure to high concentrations of inhalable particulate matter (PM) is a known human health risk, depending on the chemical composition of the PM inhaled. Mogale City (Gauteng) is known for having several sources of airborne PM, however, less is known about the metals in the airborne PM. The aim of this study was to determine the metals in measured PM at Kagiso, Mogale City. An independent PM_{10} monitor was installed at the municipality's existing monitoring site. This monitor continuously monitored PM_{10} between 23 August and 9 October 2013 and simultaneously sampled particles below 20 μm in diameter onto a glass fibre filter. This filter was replaced once towards the middle of the monitoring period. These two filters were chemically analysed to determine their metal content (30 metals) by the South African Bureau of Standards accredited laboratory at the Council for Scientific and Industrial Research by means of Inductively Coupled Plasma Spectroscopy (ICPS) based on the US EPA Method IO-3.1. To provide an estimate of possible health risk, the metal concentrations were used in a screening US-EPA human health risk assessment (HHRA). Since the analysed metals were reportedly below the detection limit, three hypothetical exposure scenarios (S) based on US-EPA recommendations were created for the HHRA. In S1, concentrations were considered to be the same as the detection limit for each metal; S2 assumed concentrations to be 50% of the detection limit; and S3 put concentrations at 10% of the detection limit. Potential risks (should pollution worsen) of developing respiratory and neurological effects were identified depending on the hypothetical scenarios. Continuous long-term monitoring and chemical characterisation are necessary to confirm these preliminary findings.

Keywords

Human Health Risk Assessment, South Africa, mining, metals, PM_{10}, air pollution

Introduction

Metals are natural components of the Earth's crust. Many of these metals are needed in the human body in small amounts, such as iron (Fe) which is contained in haemoglobin, copper (Cu) and manganese (Mn) which are in enzymes, and chromium (Cr), which is a co-factor in the regulation of sugar levels (CDC, 2011). However, depending on concentrations, these trace elements may have detrimental health effects. Heavy metals such as lead (Pb), cadmium (Cd) and mercury (Hg) are detrimental to human health as they may bio-accumulate in the body, while others are carcinogenic, such as arsenic, beryllium, cadmium and nickel (CDC, 2011). Metals emitted from mining activities accumulate in soil in surrounding areas and contaminated soil then poses a hazard to human health (Kumar et al., 2014), as the metals may be absorbed by vegetables grown in the contaminated soil, leach into underground water sources, or become airborne through wind-blown dust. Humans may therefore, be exposed to these metals via inhalation, ingestion and/or dermal contact. Mogale City Local Municipality (MCLM) in Gauteng Province

(South Africa) has a long history of gold mining. As a result, it has a number of mine dumps, some of which have not been rehabilitated. Activities within the MCLM that may contribute to concentrations of particulate matter (PM) in air include mining of minerals, quarrying of stone, extraction of clay and sandpits, use of motor vehicles, various heavy and light industrial activities, as well as domestic fuel burning (AQMP, 2013).

Focus-group discussions with MCLM residents in 2013 revealed that residents' perceived dust (PM) emissions in the area was responsible for most of their illnesses, including respiratory illness and cancer, and that dust was soiling their properties and damaging their appliances (Phala et al., 2012; Wright et al., 2014). One of the study recommendations was that the metal content of the PM should be characterised and possible health impacts from PM inhalation quantified using Human Health Risk Assessment (HHRA).

HHRA links environmental exposure to potential human health

effects. The potential for detrimental health effects is assessed based on the United States Environmental Protection Agency (US-EPA) Human Health Risk Assessment Framework (US-EPA, 2014) as it relates to the physical and/or chemical properties of air pollutants and their concentrations. The framework comprises the following steps: (1) Hazard identification to determine whether exposure to a particular substance may result in detrimental human health effects; (2) Exposure assessment to determine environmental concentrations through source and emissions characterization, monitoring, and / or environmental fate, transport, and deposition modelling, and to estimate the magnitude, duration, and frequency of human exposure; (3) Dose-response assessment to estimate the relationship between dose, or level of intake of a substance, and the incidence and severity of an effect (Several agencies such as the US-EPA (IRIS, 2015), the World Health Organization (WHO) (WHO, 1999) and the Centre for Disease Control (CDC) (ATSDR, 2015) have developed databases for benchmark values, which are used to describe the dose-response relationships determined for various chemicals); and (4) Risk characterisation which combines all the information obtained in the previous three steps to describe whether a risk to public health is predicted.

Therefore, the aim of this article was to describe the chemical composition of PM measured at Kagiso, MCLM and to apply the HHRA in a screening exercise to estimate possible human health risks with the purpose of identifying metals for further investigation.

Methods

Study area
Kagiso, the community surrounding the monitoring station, formed the study area (Figure 1). The population density in Kagiso varies from 700 to 20 500 people per km^2 (AQMP, 2013). Nearly 30% of the population of 362 422 were considered vulnerable because they were either below 15 years or above 65 years of age (StatsSA, 2011). The majority (54.8%) of households had access to piped water inside the dwelling; also to electricity (85.9%) and weekly refuse removal (79.7%) (StatsSA, 2011).

Meteorological data from 2007 to 2012 indicated that the prevailing wind direction during spring (August and September), the period in which the current study was conducted, was from a north-westerly to northerly direction (AQMP, 2013). Meteorological data further indicated that the wind in the MCLM is calm for about 40% of the time and when the wind is blowing it is mostly at a speed of < 3 ms^{-1} (AQMP, 2013). There are mine dumps within four to five kilometres to the north and the north-west, as well as the south of the monitoring site, and two industrial areas (Factoria and Chamdor) resides to the north and north-west, respectively.

Monitoring and chemical analyses
The West Rand District Municipality operates an air quality monitoring station in MCLM and monitors air pollution including

PM_{10}, however, chemical composition of PM_{10} is not routinely analysed. This site was chosen as the measurement location in this study for security reasons and to access electricity. A TOPAS Sira MC 090158/00 PM_{10} monitor was installed from 23 August 2013 to 9 October 2013. This instrument continuously monitored PM_{10} concentration. It also simultaneously sampled PM_{20} onto a glass fibre filter, recorded the volume of air pumped through as well as the mass deposited on the filter. The filter was changed once during this period to have two filters for analyses. The filters and a blank were chemically analysed for their metal content by the South African Bureau of Standards accredited laboratory at the CSIR, based on the US EPA Method IO-3.1 and using Inductively Coupled Plasma Spectroscopy (ICPS). The instrument also heated to 60 °C to drive off moisture and volatiles. The concentrations of each metal in air were subsequently calculated and used to determine potential exposure in a HHRA.

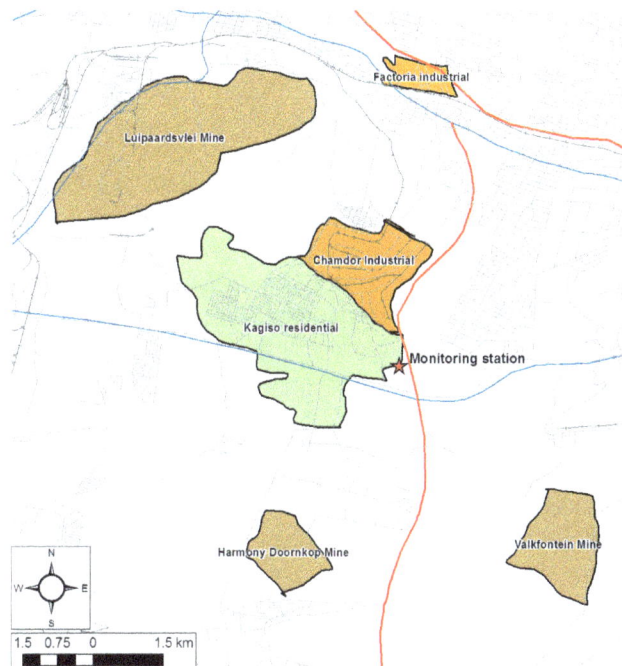

Figure 1: *Location of Kagiso, other communities, mines and industrial areas in relation to the location of the measurement instrument used in this study located at the MCLM monitoring station.*

Applying the HHRA framework
Three types of risk estimates were calculated in this quantitative HHRA: The (1) Hazard Quotient (HQ), which describes the potential for developing detrimental effects (other than cancer) from exposure to a hazardous substance; (2) Incremental (over and above the background prevalence) Cancer Risk, which is the probability of individuals developing cancer from exposure to a hazardous substance; and (3) Hazard Index (HI) was calculated to determine the total incremental cancer risk for the area (community).

For Hazard Identification, reliable databases, including those of the US-EPA, were reviewed to determine whether the elements identified during chemical analyses of the PM samples may be detrimental to human health. It was assumed in a worst-case scenario, that the Kagiso community, specifically, in the MCLM study area were continuously exposed (for example, mothers looking after small children at home) to the concentrations of PM_{10} and metals in air as determined by the monitoring and analyses of the short-term PM samples. Air was the medium and inhalation the route of exposure assessed. Where concentrations of metals were found to be below detection limits, hypothetical scenarios of exposure were created based on recommendations by the US-EPA (US-EPA, 2000), as follows: (a) Scenario 1 (S1): concentrations are the same as the detection limit for each metal; (b) Scenario 2 (S2): concentrations are 50% of the detection limit and; (c) Scenario 3 (S3): concentrations are 10% of the detection limit.

To obtain relevant benchmarks for the identified elements, the focus was on the most recently published standards or guidelines from reliable databases, including the US-EPA, WHO and CDC. A benchmark value is a value of exposure to pollutants that is believed to not be detrimental to even sensitive individuals in a population. In the case of air pollutants, the benchmark value is a "safe" concentration expressed as mass per volume.

A quantitative risk characterisation was performed, providing a numeric estimate of the potential for public health consequences from exposure to the metals concerned. This quantified potential was expressed as an HQ, which is unitless. The HQ is determined by the ratio between the expected exposure concentration of the metal and the benchmark value, which is an exposure that is assumed not to be associated with detrimental health effects. In the case of carcinogens, the incremental (over and above the background) cancer risk, which is a function of the 'Inhalation Concentration' and the 'Inhalation Unit Risk', was quantified. The Inhalation Unit Risk (risk per $\mu g/m^3$) is the unitless upper bound estimate of the probability of tumour formation per unit concentration of a chemical (Mitchell, 2004).

Results are given for the TOPAS measured PM_{10} concentrations and chemical composition followed by the results for each HHRA step and risk estimates for S1, S2 and S3.

Results and discussion

Particulate monitoring

PM_{10} concentrations measured by the TOPAS instrument exceeded the then South African 24-h ambient air quality standard of 120 $\mu g/m^3$ three times during the monitoring period of 23 August 2013 to 9 October 2013 (Figure 2). The current standard (since January 2015) of 75 $\mu g/m^3$ was exceeded 14 times during the same period. Sharp peaks were detected on 24 and 30 August 2013.

The South African Air Quality Information System (SAAQIS) website (www.saaqis.org.za) was consulted for results of monitored PM_{10} data from the municipality's monitoring station. Raw data were not available but a graph for the period

1 August to 30 September 2013 (Figure 3) showed incomplete data and although the 24-h standard of 120 $\mu g/m^3$ was not exceeded during the specified period, the current standard of 75 $\mu g/m^3$ was exceeded on at least six occasions. The highest peak of about 110 $\mu g/m^3$ was around the 30th of August, which coincided with the TOPAS peak, although lower in magnitude.

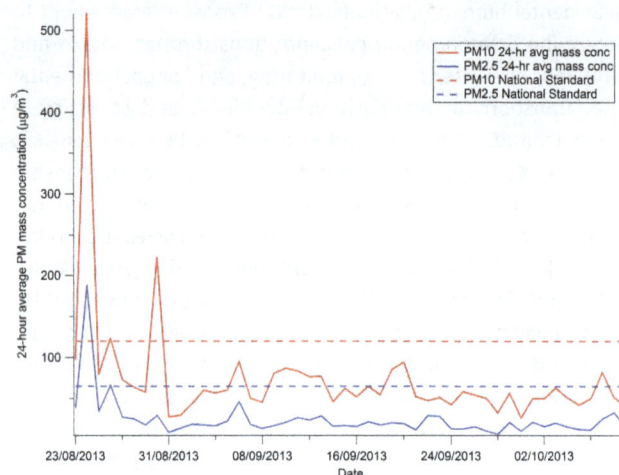

Figure 2: PM_{10} concentrations monitored during the period 23 August 2013 to 9 October 2013.

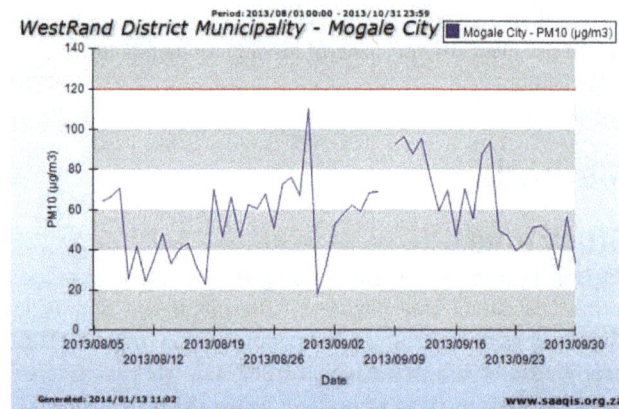

Figure 3: PM_{10} concentrations monitored by the municipality monitoring station during the period 1 August to 30 September 2013 (graph generated in SAAQIS www.saaqis.org.za).

Metal analyses

The two filters and a blank were chemically analysed for the presence of 30 metals, expressed in mg/kg. The results obtained were all below the detection limits for the individual metals (namely, calcium, magnesium, total phosphorous, potassium, nickel, silver, tin, titanium, bismuth, sodium, sulphur, aluminium, antimony, arsenic, boron, barium, beryllium, cadmium, cobalt, chromium, copper, iron, lead, lithium, manganese, selenium, silica, strontium, vanadium and zinc). The total weight of PM_{10} accumulated on the filter over each monitoring period was divided by the total volume of air sampled during that period to obtain a concentration ($\mu g/m^3$) per metal. Due to the fact that concentrations were below the detection limits, these concentrations were considered to be at the detection limit as a worst case scenario (S1). Two other scenarios were also assessed, namely concentrations at 50% of the detection limit (S2) and at 10% of the detection limit (S3).

Health risks assessed

Hazard identification
Non-cancer health effects caused by the metals analysed are mostly respiratory-related (metals included bismuth, boron, calcium, chromium, copper, iron, phosphorus, sulphur, titanium, tin, aluminium, antimony, beryllium, cadmium, cobalt, lithium, silica and vanadium, zinc and nickel), while some may cause neurological effects (lead, manganese and selenium) (CDC, 2011). Silver may cause argyria (bluish-grey skin), while arsenic, beryllium, cadmium and nickel have been identified as being carcinogenic (CDC, 2011).

Exposure assessment
The communities of concern were considered to be those around the monitoring station in the MCLM, with specific emphasis on Kagiso since it was the closest to the monitoring site. The concentrations for each scenario used in this HHRA are given in Table 1. In this worst-case scenario exercise, it was assumed that individuals were constantly (24 hours per day) exposed to the varying concentrations calculated for the different exposure scenarios. In terms of cancer risk assessment, it was assumed that the calculated concentrations were the concentrations that individuals were exposed to over a lifetime. It is not possible to completely avoid contact with these metals naturally found in the environment. The smaller the particles, the deeper they will penetrate into the lungs. The amount absorbed into the bloodstream will again depend on how well the particles dissolve. The more easily the particles dissolve, the more easily they may enter the blood stream. If the particles do not dissolve easily, they may remain in the lungs for longer periods of time. Some of the particles may leave the lungs through the normal clearing process while some may be swallowed.

Dose-response assessment
Numeric benchmark values obtained from reliable databases were used to describe the dose-response relationships. The preferred benchmark values used were those set on the basis of health effects in human beings and not those incorporating economic or social factors. Since people are normally exposed to metals in the air, predominantly in an occupational environment, and since the main route of exposure to metals is ingestion, a number of the elements did not have benchmark values for inhalation. Where no benchmark value could be found from reliable databases surveyed, the South African Occupational Exposure Limit (SA OEL) (SA Occupational Standards, 1995) was considered to indicate the level of toxicity of the element. These OELs are indicated in bold in Table 1. These were for aluminium, arsenic, antimony, barium, beryllium, bismuth, calcium, copper, iron, lithium, magnesium, silica, silver, selenium, sulphur, tin and zinc. In cases where there was no OEL for the element but only for a species of that element, that species is given in brackets in Table 1.

It must be noted that the application of occupational standards is not technically applicable in this case, because the samples were taken in ambient air and for a period longer than eight hours. In addition, occupational standards are set with healthy workers in mind, who are only exposed for eight hours a day and forty hours a week. Therefore, occupational standards are higher than ambient standards and they will not protect sensitive individuals (such as asthmatics and children) which ambient standards are supposed to do. However, it was decided to keep the pollutants for which only occupational standards were available as part of this assessment, because if the HQs of any of these pollutants with relatively high benchmark values were >1 it would indicate possible drivers of risk that can then be further investigated. Those metals for which no benchmark values could be found, namely sodium, potassium and strontium, were excluded from the HHRA.

Risk characterisation
Non-cancer risk estimates, expressed as HQs, were calculated for the metals and are presented in Table 2. HQ values > 1 indicate that the likelihood of detrimental non-cancer effects is enhanced while HQ values <1 indicate that the potential for detrimental health effects is minimal. The HQs of most (about 80%) of the metals analysed were below one, indicating that non-cancer detrimental health effects were unlikely. For six of the analysed metals (i.e. cadmium, cobalt, manganese, nickel, lead and vanadium), the HQs indicated a risk to human health in one or more of the exposure scenarios.

(a) Risk characterisation for S1
In S1, where the concentrations were assumed to be at the detection limit, the HQs indicated a risk of developing respiratory effects from cadmium, cobalt, nickel and vanadium and neurological effects from manganese and lead.

(b) Risk characterisation for S2
In S2, HQs indicated a health risk for respiratory effects due to exposure to cadmium, cobalt and nickel as well as neurological effects due to exposure to manganese and lead. The HQ for vanadium did not indicate a risk of respiratory effects in S2.

(c) Risk characterisation for S3
In S3, assuming that the exposure concentrations were 10% of the detection limit, only cadmium and cobalt posed a risk of respiratory effects as only the HQs for these two metals were above one.

(d) Cancer risk characterisation
The incremental cancer risks (Table 3) were estimated using the determined concentrations and the inhalation unit risk for each metal (from the US-EPA) known to be a confirmed human carcinogen. The HI is also presented in Table 3. The total incremental risk to develop cancer ranges from 1.73 in 1 000 in S3, to 17.7 in 1 000 in S1. Arsenic was found to be the main driver of the risk to cancer. The calculated incremental cancer risks are thus not only above 1 in a million, but also above 1 in 10 000. Evaluation of these risks against the US-EPA air office criteria

which "strives to reduce risk for as many people as possible to 1 in a million, while assuming that the maximally-exposed individual is protected against risks greater than 1 in 10 000" indicates that these acceptable risks have been exceeded here.

Uncertainties and limitations

Several limitations and uncertainties should be considered when interpreting these findings. HHRA is a predictive process, therefore it only estimates what could occur. Due to the fact that concentrations were below the detection limits, these concentrations were considered to be at the detection limit as a worst case scenario at 50% of the detection limit and at 10% of the detection limit. While these scenarios are not representative of the status quo, should current air pollution interventions and management change, and PM levels increase, such scenarios may be experienced. Being prepared by having predicted, likely worst-case scenario health impacts supports the need for continuous improvement in air quality management and air pollution control in this area.

It is acknowledged that PM_{10} was monitored only during a short period and at one site only. Personal monitoring and time-activity data are the optimal method for exposure assessment, however, they were not possible in this small study. Although the instrument was calibrated, concentrations may not be representative for all the communities, particularly those furthest away from the monitoring site. However, this screening study aimed to understand the metals present in the dust. It did not aim to collect information on pollution sources. Although the monitoring data from the municipality at the same site may give additional data on PM_{10} exposure, the simultaneous capturing of the PM onto a filter for metal analysis was not possible from the instrumentation used by the municipality, and this was crucial for this exercise.

To address model uncertainty in this study, equations from the US-EPA were used, and applied benchmark values were based on national and international standards and guidelines which were set based on human health effects. In terms of selection of the pollutants of concern, the pollutants were pre-identified as those metals which the accredited laboratory could analyse for and which was present in the PM. Finally, in terms of the exposure pathway and route used, the risk assessment was based on inhalable PM and therefore the inhalation route was selected.

Conclusions

PM_{10} concentrations measured in this study exceeded the previous South African 24-h ambient air quality standard of 120 µg/m³ three times during the monitoring period and the current standard (since January 2015) of 75 µg/m³ 14 times. Although the concentrations of metals determined in this study were below detection limits, and were not representative of all exposure periods and all seasons, or for all individuals in

the local municipality, using US-EPA recommended scenarios, screening for possible carcinogenic and non-carcinogenic risks based on concentration and toxicity from metal exposure through inhalation was performed. Results provide an indication of which metals may drive human health risks in the area. Those pollutants identified as being potential risks should be investigated further.

Acknowledgements

This project was funded by a CSIR Strategic Research Platform Grant. We thank Mogale City Local Municipality for providing us with access to their air quality monitoring site and to Mr. D. Otto from Exito Environmental Projects for operation of the TOPAS instrument.

References

ATSDR, 2015. Agency for Toxic Substances and Disease Registry. Toxic Substances Portal. Available at: http://www.atsdr.cdc.gov/substances/indexAZ.asp Accessed on 3 June 2015.

AQMP. 2013. Mogale City Local Municipality Air Quality Management Plan; Baseline report 2013.

CDC, 2011. Centre for Disease Control. http://www.atsdr.cdc.gov/substances/indexAZ.asp. Accessed 20 May 2015.

IRIS, 2015. US EPA Integrated Risk Information System A to Z list of substances. Available at http://cfpub.epa.gov/ncea/iris/index.cfm?fuseaction=iris.showSubstanceList Accessed on 3 June 2015.

Kumar B, Verma VK, Naskar AK, Sharma CS and Mukherjee DP. 2014. Bioavailability of Metals in Soil and Health Risk Assessment for Populations near an Indian Chromite Mine Area. *Human and Ecological Risk Assessment*, 20: 917-928.

Mitchell K. (2004). Quantitative Risk Assessment for Toxic Air Pollutants; An Introduction Course presented at the NACA conference Indaba Hotel, Johannesburg 4-8 October 2004.

MCLM (Mogale City Local Municipality). 2011. Darft 5-year integrated development plan 2011-16: Spatial development framework 2011. http://www.mogalecity.gov.za/your-council/documents/integrated-development-plan/processes.

Phala N, Matooane M, Oosthuizen MA and Wright CY. 2012. It's not all about acid mine drainage on the West Rand – there is dust too. *Government Digest* August 2012: 65-66.

US-EPA 2014. Framework for Human Health Risk Assessment to Inform Decision Making. EPA/100/R-14/001April2014. Available: http://www.epa.gov/raf/files/hhra-framework-final-2014.pdf.

The size of the particles collected for chemical analysis was ≤ 20 µm.

US-EPA. 2000. Assigning Values to Non-Detected/Non-Quantified Pesticide Residues http://www.epa.gov/oppfead1/trac/science/trac3b012.pdf

SA Occupational Standards 1995. Government Gazette number 16596; 25 August 1995.
South African Air Quality Information System (SAAQIS) 2013. http://saaqis.org.za

StatsSA (Statistics South Africa). 2007. Community survey of 2007. http://www.statssa.gov.za/.

StatsSA (Statistics South Africa). 2010. Labour force survey. http://www.statssa.gov.za/.

StatsSA (Statistics South Africa). 2011. Census 2011 Municipal Fact Sheet. Available:
https://www.statssa.gov.za/Census2011/Products/Census_2011_Municipal_fact_sheet.pdf.

WHO, 1999. Air Quality Guidelines, Geneva, World Health Oganization.

Wright CY, Matooane M, Oosthuizen MA and Phala N. 2014. Risk perceptions of dust and its impacts among communities living in a minig area of the Witwatersrand, South Africa. Clean Air Journal 24(1):17-22.

Table 1: *Concentrations of compounds of potential concern and their benchmark values.*

Element	Scenario 1		Scenario 2		Scenario 3	
	Concentration (µg/m³)	Benchmark Value (µg/m³)	Concentration (µg/m³)	Benchmark Value (µg/m³)	Concentration (µg/m³)	Benchmark Value (µg/m³)
Aluminium (respirable)	2.70	**5000**	1.36	**5000**	0.27	**5000**
Arsenic	2.70	**10**	1.36	**10**	0.27	**10**
Antimony (and compounds)	6.90	**500**	3.43	**500**	0.69	**500**
Barium compounds	6.90	**500**	3.43	**500**	0.69	**500**
Beryllium (and compounds)	1.4	**2**	0.68	**2**	0.14	**2**
Bismuth (undoped)	6.90	**10000**	3.43	**10000**	0.69	**10000**
Boron	10.20	300*	5.08	300*	1.02	300*
Cadmium	1.40	0.03*	0.68	0.03*	0.14	0.03*
Calcium oxide	68.40	**2000**	34.20	**2000**	6.84	**2000**
Cobalt	1.40	0.10*	0.68	0.10*	0.14	0.10*
Chromium III	1.40	5*	0.68	5*	0.14	5*
Copper	6.90	**1000**	3.43	**1000**	0.69	**1000**
Iron oxide (dust)	1.40	**5000**	0.68	**5000**	0.14	**5000**
Lead	2.70	0.5**	1.36	0.5**	0.27	0.5**
Lithium hydride	6.90	25	3.43	25	0.69	25
Magnesium	13.70	5000	6.83	5000	1.37	5000
Manganese (and compounds)	1.40	0.03*	0.68	0.03*	0.14	0.03*
Nickel	1.40	0.20*	0.68	0.20*	0.14	0.20*
Phosphorus (white)	13.70	20*	6.83	20*	1.37	20*
Silica	13.70	**100**	6.83	**100**	1.37	**100**
Silver	6.90	**100**	3.43	**100**	0.69	**100**
Selenium (and compounds)	1.40	**100**	0.68	**100**	0.14	**100**
Sulphur	13.70	**6000 000**	6.83	**6000 000**	1.37	**6000 000**
Tin	6.90	**200**	3.43	**200**	0.69	**200**
Titanium	6.90	10*	3.43	10*	0.69	10*
Vanadium	1.40	0.8*	0.68	0.8*	0.14	0.8*
Zinc	1.40	**500**	0.68	**500**	0.14	**500**

*Note. *MRL   **SA National Standard   SA-OEL figures in bold.*

Table 2: Non-cancer risk estimates of the elements analysed for the different exposure scenarios.

Element	Scenario 1		Scenario 2		Scenario 3	
	Concentration (µg/m³)	Hazard Quotient (HQ)	Concentration (µg/m³)	Hazard Quotient (HQ)	Concentration (µg/m³)	Hazard Quotient (HQ)
Aluminium	2.70	0.0005	1.36	0.0003	0.27	<0.0001
Antimony	6.90	0.01	3.43	0.007	0.69	0.001
Arsenic	2.70	0.27	1.36	0.14	0.27	0.03
Barium	6.90	0.01	3.43	0.007	0.69	0.001
Beryllium (and compounds)	1.40	0.68	0.68	0.34	0.14	0.07
Bismuth	6.90	0.0007	3.43	0.0003	0.69	<0.0001
Boron	10.20	0.03	5.08	0.02	1.02	0.003
Cadmium	1.40	45.48	0.68	22.70	0.14	4.55
Calcium	68.40	0.03	34.20	0.02	6.84	0.003
Cobalt	1.40	13.64	0.68	6.82	0.14	1.36
Copper	6.90	0.007	3.43	0.003	0.69	0.0007
Chromium III	1.40	0.27	0.68	0.14	0.14	0.03
Iron	1.40	0.0003	0.68	0.0001	0.14	<0.0001
Magnesium	13.70	0.003	6.83	0.001	1.37	0.0003
Manganese (and compounds)	1.40	4.55	0.68	2.27	0.14	0.45
Nickel	1.40	6.82	0.68	3.41	0.14	0.68
Phosphorus (white)	13.70	0.68	6.83	0.34	1.37	0.07
Selenium	1.40	0.01	0.68	0.007	0.14	0.001
Silica	13.70	0.14	6.83	0.07	1.37	0.01
Silver	6.90	0.07	3.43	0.03	0.69	0.007
Sulphur	13.70	0.002	6.83	0.001	1.37	0.0002
Tin	6.90	0.03	3.43	0.02	0.69	0.003
Titanium	6.90	0.69	3.43	0.34	0.69	0.07
Lead	2.70	5.46	1.36	2.73	0.27	0.55
Lithium	6.90	0.27	3.43	0.14	0.69	0.03
Vanadium	1.40	1.71	0.68	0.85	0.14	0.17
Zinc	1.40	0.003	0.68	0.001	0.14	0.0003

Table3: Incremental cancer risks estimates for the different exposure scenarios.

Element	Scenario 1			Scenario 2			Scenario 3		
	Concentration (µg/m³)	Inhalation Unit Risk	Incremental cancer risk	Concentration (µg/m³)	Inhalation Unit Risk	Incremental cancer risk	Concentration (µg/m³)	Inhalation Unit Risk	Incremental cancer risk
Arsenic	2.70	4.3×10^{-3}	11.6 in 1000	1.36	4.3×10^{-3}	5.8 in 1000	0.27	4.3×10^{-3}	1.1 in 1000
Beryllium	1.40	2.4×10^{-3}	3.3 in 1000	0.68	2.4×10^{-3}	1.6 in 1000	0.14	2.4×10^{-3}	0.3 in 1000
Cadmium	1.40	1.8×10^{-3}	2.5 in 1000	0.68	1.8×10^{-3}	1.2 in 1000	0.14	1.8×10^{-3}	0.3 in 1000
Nickel	1.40	2.4×10^{-4}	0.3 in 1000	0.68	2.4×10^{-4}	0.2 in 1000	0.14	2.4×10^{-4}	0.03 in 1000
Total HI			**17.7 in 1000**			**8.8 in 1000**			**1.73 in 1000**

4

Measurement of atmospheric black carbon in the Vaal Triangle and Highveld Priority Areas

Gregor T. Feig[1], Beverley Vertue[2], Seneca Naidoo[3], Nokulunga Ncgukana[4], Desmond Mabaso[5]

South African Weather Service 442 Rigel Ave South Erasmusrand Pretoria South Africa
[1]gregor.feig@weathersa.co.za, [2]beverley.barnes@weathersa.co.za, [3]seneca.naidoo@weathersa.co.za, [4]lunga.ncgukana@weathersa.co.za, [5]desmond.mabaso@weathersa.co.za

Abstract

Atmospheric black carbon is an important atmospheric pollutant; it has impacts on human health and a strong climate impact. Black carbon particles are functionally defined by their optical properties (viz. characteristics in light absorption). As a result, black carbon particles are derived from a wide range of sources, but are largely the result of incomplete combustion processes. In order to quantify the atmospheric load of black carbon particles, multi angle absorption photometer (MAAP) instruments have been installed in 8 of the ambient air quality monitoring stations in the Vaal Triangle and Highveld Priority areas. Three of the instruments have been in operation since 2012 and the other 5 were installed in August 2013. This paper presents an analysis of the initial black carbon monitoring data. The impacts of seasonality and meteorological conditions as well as the relationship of the black carbon concentration to PM_{10} and $PM_{2.5}$ concentrations are discussed.

Keywords

Vaal Triangle Priority Area, Highveld priority area, black carbon

Introduction

Black carbon (BC) is a component of the atmospheric aerosol that is highly absorbent of visible light and is resistant to chemical transformation (Petzold et al. 2013). Black carbon is formally defined through its optical properties as "ideally light absorbing substances comprised of carbon", this definition does not take into account the formation processes. Black carbon is predominantly formed through the incomplete combustion of organic materials; however pyrolysis and dehydrogenation of wood and sugars under anaerobic conditions may also result in its formation (Petzold et al. 2013). The fact that BC is largely formed through combustion processes makes it a useful indicator for combustion sources of particulate matter.

Black carbon makes up an important component of the particulate matter less than 2.5µm in aerodynamic diameter ($PM_{2.5}$) fraction and therefore is implicated in the health impacts of $PM_{2.5}$. It has been suggested that the BC concentration is a better indicator of the risk to human health from PM than the PM_{10} or $PM_{2.5}$ mass concentration (Janssen et al. 2012). Further evidence suggests that $PM_{2.5}$ mixtures with a large BC component have a more adverse health effect than other mixtures (Anenberg et al. 2011).

The absorption of solar radiation by BC significantly enhances the heating of the atmosphere. The influence of BC can cause changes in cloud cover and surface albedo, therefore affecting the Earth's radiative budget both directly and indirectly. These aerosol particles not only influence atmospheric temperature but cause considerable changes in atmospheric chemistry. Black carbon particles have a relatively short lifespan within the atmosphere and for this reason tend to have more localised effects, impacting the regions closest to the source (Bauer et al. 2010).

Since atmospheric BC has such strong human health and climatic impacts, and is an indicator of combustion sources of particulate matter, it was deemed necessary to monitor the concentrations of BC at locations in the Vaal Triangle and Highveld Priority areas. This study examines a 12 month period (September 2013 to August 2014) for the measurement of BC in the Vaal Triangle and Highveld Priority Areas with the aim of characterising the ambient BC concentrations in terms of the seasonal, diurnal and air flow patterns.

Methods

During 2012, Multi Angle Absorption Photometer (MAAP; Thermo Scientific) instruments Model 5012 for the measurement of atmospheric BC were installed in the Witbank, Secunda and Zamdela monitoring stations. During August 2013, further instruments were installed at the monitoring stations in Diepkloof, Sharpeville, Sebokeng, Three Rivers and Kliprivier in the Vaal Triangle Network. The specific instrumentation used for this study is presented in Table 1. The instruments report to the South African Air Quality Information System (SAAQIS)

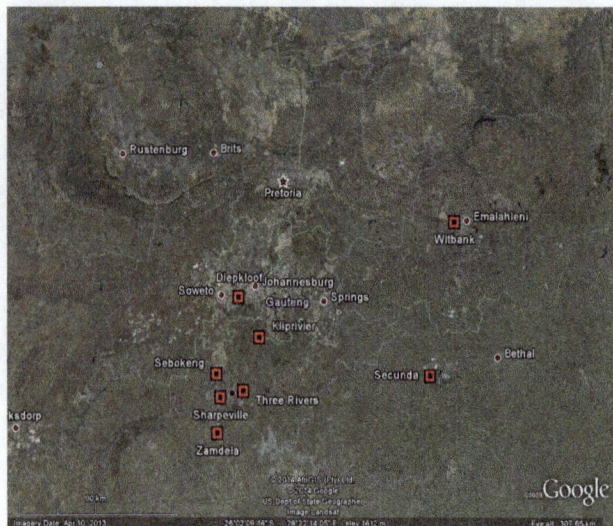

Figure 1: *Google Earth image showing locations of the monitoring stations*

(www.saaqis.org.za) at a 5 minute temporal resolution. In addition to the measurement of BC, all stations in the Vaal Triangle and Highveld Priority Area networks are instrumented for the measurement of PM_{10}, $PM_{2.5}$, SO_2, NO_x, CO, O_3, Benzene, Toluene and Xylene (BTX), and the meteorological parameters wind speed, wind direction, rainfall, temperature, pressure, humidity and solar radiation. All the data from the Vaal Triangle and Highveld networks are available from the SAAQIS, and has been validated to remove calibration periods, instrument drifts and spikes, all validation processes are detailed in the monthly network reports for the Vaal Triangle and Highveld Priority Area networks. These reports are available online on the SAAQIS website.

For this study, BC, PM_{10}, $PM_{2.5}$ mass concentration data and meteorological data for the period 1 September 2013-31 August 2014 was downloaded from the SAAQIS at an hourly temporal resolution for the Diepkloof, Sharpeville, Sebokeng, Zamdela, Three Rivers, Kliprivier Witbank and Secunda stations (Figure 1). The site and instrument specifications are presented in Table 1. The data was analysed using Excel and the "openair" package of R (Uria-Tellaetxe and Carslaw 2014).

It has been reported that there is an artefact in the MAAP instrument at high BC concentration (Hyvärinen et al. 2013). The correction suggested by Hyvärinen et al. (2013) was not applied for this study as not all of the parameters required for implementing the correction were logged, and the manufacturers did not recommend the implementation of the correction when inquiry was made. At high concentrations the instrument may under-report the BC concentration.

Results

The results for this study are divided according to the seasonal and diurnal effects; the impact of the wind direction and speed, and the relationship between the mass concentrations of BC and the other PM classes measured at the sites.

Table 1: *Site and instrument specifications*

Site	BC instrument	PM instrument	Site characteristics
Diepkloof	Thermo Model 5012 MAAP	Thermo FH62C14	Location in school in middle to low income residential with impacts from traffic light industry and domestic combustion
Sebokeng	Thermo Model 5012 MAAP	Thermo FH62C14	Location in community centre in low income residential with impact from metallurgical industry, and domestic combustion
Sharpeville	Thermo Model 5012 MAAP	Thermo FH62C14	Location in school in low income residential area with impact from metallurgical industry, and domestic combustion
Sharpeville	Thermo Model 5012 MAAP	Thermo FH62C14	Location in school in low income residential area with impact from chemical and petrochemical industry, and domestic combustion
Three Rivers	Thermo Model 5012 MAAP	Thermo FH62C14	Location in school in middle income residential area with impact from a coal power plant
Kliprivier	Thermo Model 5012 MAAP	Thermo FH62C14	Location in Police station in low income residential area with impact from traffic and domestic combustion
Witbank	Thermo Model 5012 MAAP	GRIMM EDM180	Location in school in low income residential area with impact from metallurgical industry, coal power generation and domestic combustion
Secunda	Thermo Model 5012 MAAP	GRIMM EDM180	Location in sports centre in low income residential area with impact from chemical and petrochemical industry, and domestic combustion

Average Concentrations

The average concentration of atmospheric BC is between 2.5 and 4.5 µg/m³ for all stations, however hourly values of up to 20 µg/m³ occurred at all sites except Three Rivers (Figure 2).

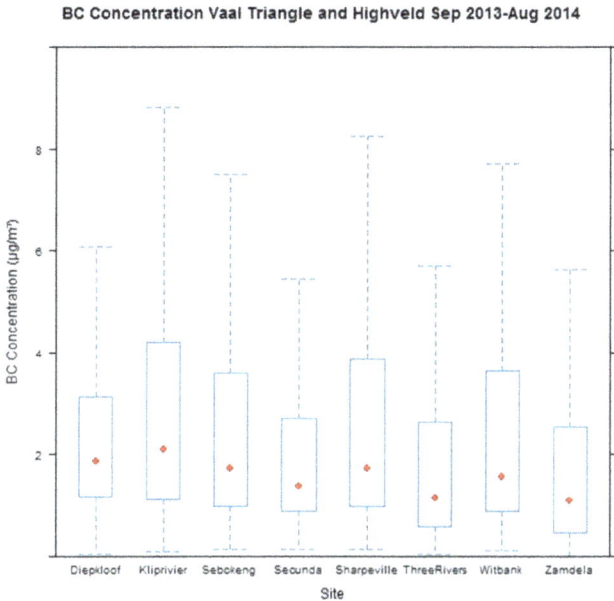

Figure 2: *Average hourly BC concentration (µg/m³) for all the sites for the period September 2013-August 2014*

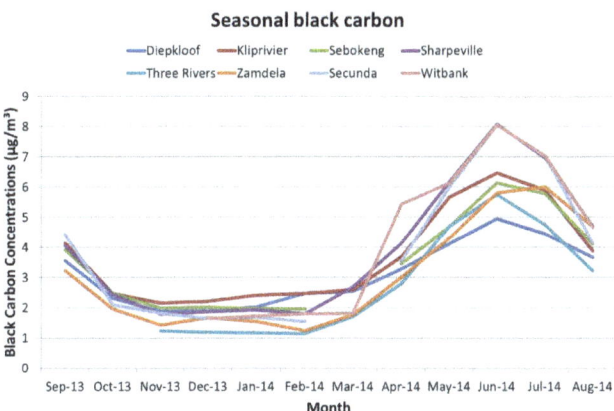

Figure 3: *Monthly average concentration of BC for the eight sites for the monitoring period*

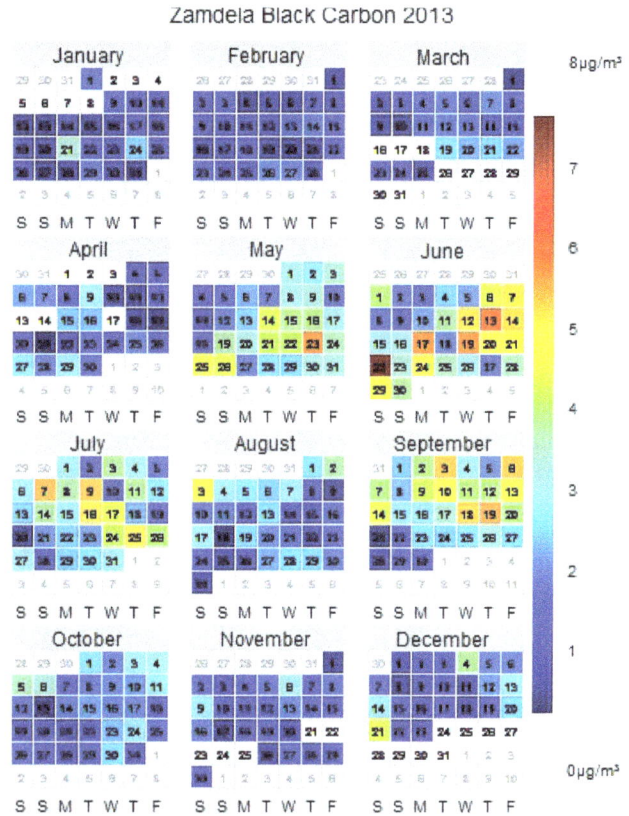

Figure 4: *Calendar plot of the daily BC concentrations recorded at Zamdela for 2013*

Seasonal and diurnal effects

A strong seasonal trend in the ambient concentration of atmospheric BC was observed (Figure 3). The concentration of BC increased significantly with the onset of winter. Increases in the ambient concentrations were observed during the months of April, May and June 2014, with reductions occurring from August. During the warmer months (October – March), the monthly average concentration remained fairly constant and similar between the stations. During the cooler months (April – September), stronger differences are observed between the sites, with higher concentrations being observed in the Witbank, Secunda and Sharpeville stations.

The "openair" calendar plot function for the Zamdela station during 2013 (Figure 4) illustrates that the BC mass concentration peaks during the cooler period (May – July as well as September) and remains fairly low during the warmer months. During the June and July period the daily average concentrations of BC vary with days of relatively low concentration following

periods of high concentration, many of these periods of low BC concentration are associated with the passing of a cold front over the interior of the country, particularly on the 2 June 2013, 8 June 2013, 3 July 2013 and 28 July 2013. During September 2013 there is an increase in the BC concentrations; this is potentially due to long range transport of biomass burning emissions.

The hourly profile of the ambient BC concentrations (for the entire time period) recorded at all the stations show a strong bimodal distribution (Figure 5) with peaks occurring in the mornings (5-8 am) and in the evenings (6-8 pm). The concentrations remain elevated during the night and then reduce during the day time. This pattern of increased concentrations is fairly typical of domestic burning emissions. This can be seen in more detail in Figure 6 which is a time variation plot of BC in Zamdela for the entire time period. The diurnal pattern of high BC concentration remains consistent across the days of the week, but the peak concentrations are reduced over the weekends and the morning peaks are spread out over a longer period on Saturday and Sunday, presumably due to people starting their activities later in the morning.

Impact of air flow

The "Polar Plot" function from "openair" plots the concentration of BC (in colour) in relation to the wind speed and wind direction (Figure 7). All the stations considered in this study show that local sources are important and high concentrations occur when there is a fairly low wind speed. Five of the stations in the

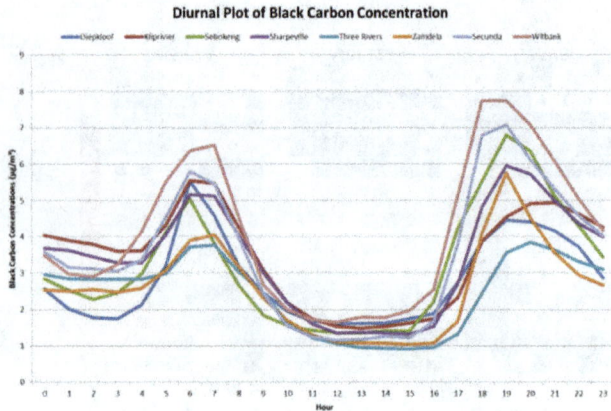

Figure 5: *Diurnal characteristics of the ambient BC concentration*

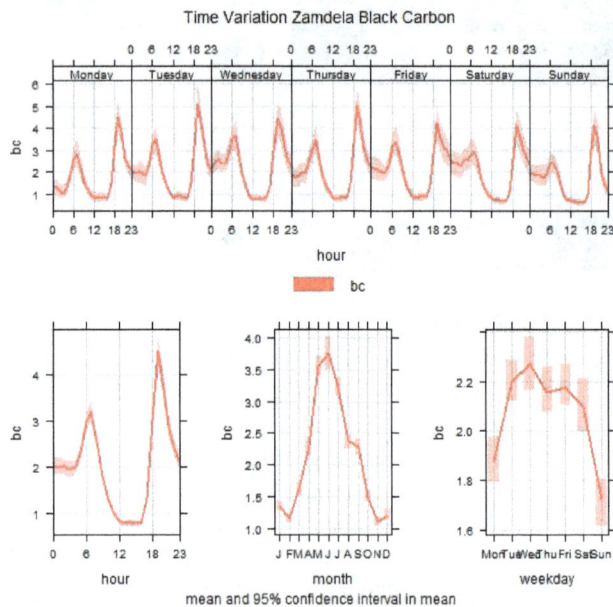

Figure 6: *Time Variation Plot Zamdela BC. In this plot the top panel shows the diurnal pattern for each day of the week, while the lower panel shows the diurnal pattern for the year, the monthly average and the day of week concentrations. The mean for each time period is indicated by the red line with the 95% confidence interval in the mean shown in the shaded area.*

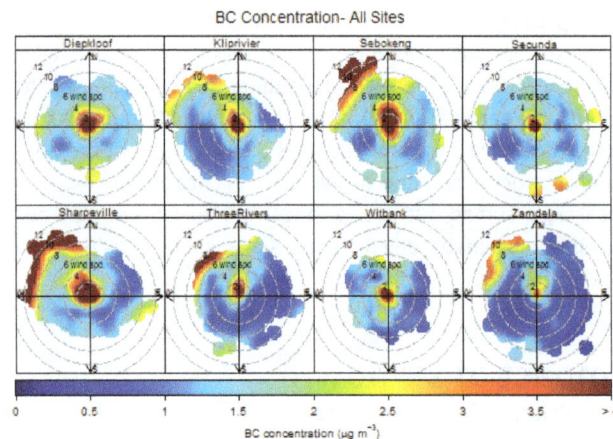

Figure 7: *Polar plot of site specific BC concentration and meteorological conditions*

Vaal Triangle (Kliprivier, Sebokeng, Three Rivers, Sharpeville and Zamdela) also show high concentrations of BC associated with strong winds from the north-westerly directions sources to the north west of these stations may include the gold fields of Randfontein/ Carletonville and the Bojanala platinum belt, further analysis is required to identify potential sources. Zamdela shows very low BC concentrations associated with winds from the south and easterly sectors where there is very little industrial activity.

Relationship between BC, PM$_{10}$ and PM$_{2.5}$

There is a strong relationship between the 1-hr mass concentrations of BC, and the concentrations of PM$_{10}$ and PM$_{2.5}$ as shown in Figure 8. The mass concentrations of PM$_{2.5}$ and PM$_{10}$ are plotted against each other while the BC concentration is represented by the colour of the point. It can be seen that as the concentrations of PM$_{10}$ and PM$_{2.5}$ increase so do the concentrations of BC, however the BC concentration tracks more closely with the PM$_{2.5}$ values.

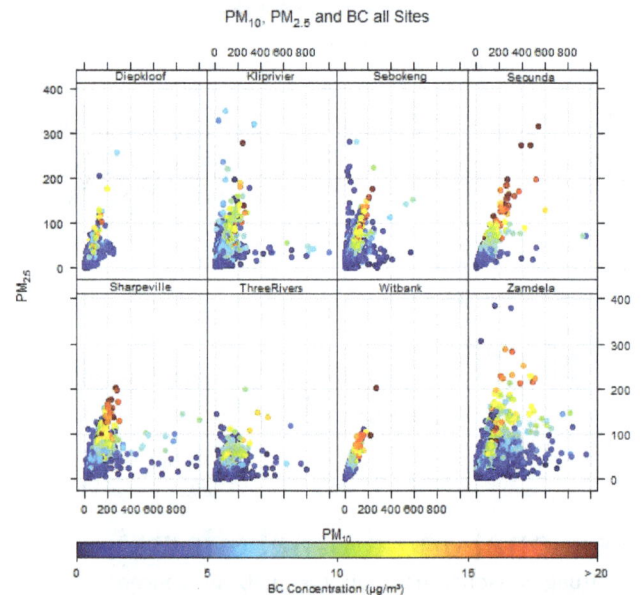

Figure 8: *Scatter Plot 1-hr averaged mass concentrations of PM$_{10}$, PM$_{2.5}$ and BC*

Using the full time period of measurements (July 2012-June 2014) at Zamdela the monthly linear relationship between BC concentration and the PM$_{2.5}$ concentration was plotted using the linear relation function from "openair" (Figure 9). The "linear relation" function looks at the relation between two pollutants over differing time periods (in this case monthly) the error bars represent the 95% confidence interval while the red trend line represents the long term trend in the relationship. The ratio between BC and PM$_{2.5}$ changes seasonally with increases in the BC component during the winter months. The value of the plotted linear relationship is the ratio of BC:PM$_{2.5}$. While in general there is fairly good correlation between the concentrations of PM$_{2.5}$ and BC, the BC makes up a small portion of the total PM$_{2.5}$ concentration, typically less than 10%.

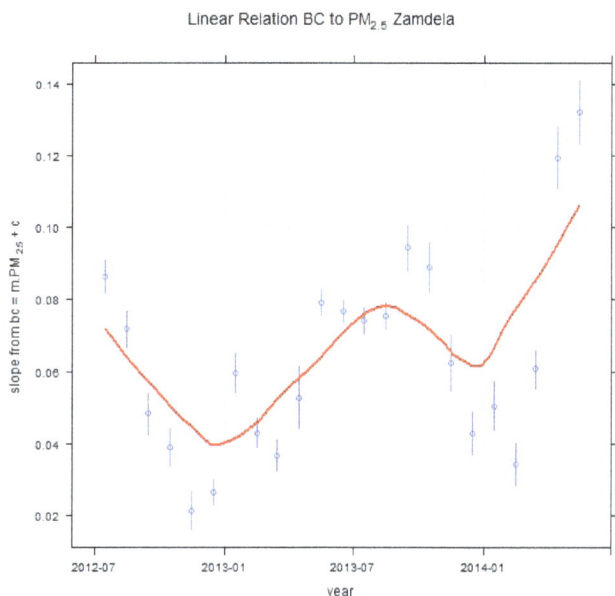

Figure 9: *Time series of the monthly linear relation between black carbon and PM$_{2.5}$ at Zamdela for the period July 2012 to May 2014 blue points and error bars represent the monthly linear relationship between the parameters while the red line represents the trend in the monthly linear relationship observed*

Discussion

The ambient concentrations of BC are monitored at 8 ambient air quality monitoring stations in the Vaal Triangle and Highveld Priority Areas. The concentrations of BC show a distinct seasonal pattern, with higher concentrations occurring in the cooler months. This is similar to what has been reported previously in the North West Province (Venter et al. 2012). This observed seasonal trend is likely linked to greater emissions of BC from domestic burning and biomass burning sources, and the presence of highly stable atmospheric conditions, which reduce mixing. The diurnal profile shows a strong bimodal distribution with peaks in the early morning and evening. Such concentration profiles are indicative of either domestic combustion and/or traffic sources. Since the maximum BC concentrations occur between 5:00 and 7:00 in the morning and between 18:00 and 20:00 in the evening the predominant source may be domestic combustion as the maximum traffic flows are expected to occur later in the morning and finish earlier in the evening. This is confirmed when looking at the weekday diurnal profiles as the morning peak in BC concentration is spread over a greater time period as people tend to start their daily activities later on Sundays.

The analysis of the BC concentration in relation to the air flow indicates that for most of the stations local sources are important, however, the stations in the Vaal Triangle show high BC concentrations associated with strong winds from the north-west. In a previous study of ozone concentrations in the Vaal Triangle, high ozone concentrations were associated with biomass burning events and the approach of a cold front, drawing in air masses from the north west (Feig et al. 2014).

Black carbon accounts for approximately 6%-12% of the mass

concentration of PM$_{2.5}$. The proportion of BC in the PM$_{2.5}$ fraction is impacted by season with a higher BC contribution occurring in the winter months, presumably due to the greater emissions from domestic and biomass burning sources.

References

Anenberg, S. C., K. Talgo, S. Arunachalam, P. Dolwick, C. Jang, and J. J. West. 2011. "Impacts of Global, Regional, and Sectoral Black Carbon Emission Reductions on Surface Air Quality and Human Mortality." *Atmospheric Chemistry and Physics* 11 (14) (July 25): 7253–7267. doi:10.5194/acp-11-7253-2011. http://www.atmos-chem-phys.net/11/7253/2011/.

Bauer, S. E., S. Menon, D. Koch, T. C. Bond, and K. Tsigaridis. 2010. "A Global Modeling Study on Carbonaceous Aerosol Microphysical Characteristics and Radiative Effects." *Atmospheric Chemistry and Physics* 10 (15) (August 10): 7439–7456. doi:10.5194/acp-10-7439-2010. http://www.atmos-chem-phys.net/10/7439/2010/.

Feig, Gregor, Xolile Ncipha, Beverley Vertue, Seneca Naidoo, Desmond Mabaso, Nokulunga Ngcukana, Cheledi Tshehla, and Njabulo Masuku. 2014. "Analysis of a Period of Elevated Ozone Concentration Reported over the Vaal Triangle on 2 June 2013." *Clean Air Journal* 24 (1): 10–16.

Hyvärinen, A.-P., V. Vakkari, L. Laakso, R. K. Hooda, V. P. Sharma, T. S. Panwar, J. P. Beukes, et al. 2013. "Correction for a Measurement Artifact of the Multi-Angle Absorption Photometer (MAAP) at High Black Carbon Mass Concentration Levels." *Atmospheric Measurement Techniques* 6 (1) (January 11): 81–90. doi:10.5194/amt-6-81-2013. http://www.atmos-meas-tech.net/6/81/2013/.

Janssen, NAH, ME Gerlofs-Nijland, T Lanki, RO Salonen, F Cassee, G Hoek, P Fischer, B Brunekreef, and M Krzyzanowski. 2012. "Health Effects of Black Carbon." *Office*. Copenhagen.

Petzold, A., J.A. Ogren, M. Feibig, P. Laj, S.M. Li, U. Baltensperger, T. Holzer-Popp, et al. 2013. "Recommendations for Reporting 'Black Carbon' Measurements." *Atmospheric Chemistry and Physics* 13 (16) (August 22): 8365–8379. doi:10.5194/acp-13-8365-2013. http://www.atmos-chem-phys.net/13/8365/2013/.

Uria-Tellaetxe, Iratxe, and David C. Carslaw. 2014. "Conditional Bivariate Probability Function for Source Identification." *Environmental Modelling & Software* 59 (September): 1–9. doi:10.1016/j.envsoft.2014.05.002. http://linkinghub.elsevier.com/retrieve/pii/S1364815214001339.

Venter, Andrew D, Ville Vakkari, Johan P Beukes, Pieter G Van Zyl, Heikki Laakso, Desmond Mabaso, Petri Tiitta, et al. 2012. "An Air Quality Assessment in the Industrialised Western Bushveld Igneous Complex, South Africa." *S Afr J Sci* 108: 1–10.

5

Indoor and outdoor particulate matter concentrations on the Mpumalanga highveld

Bianca Wernecke*[1] Brigitte Language[1], Stuart J. Piketh[1] and Roelof P. Burger[1]

*Eskom Holdings SOC Ltd, Megawatt Park, Maxwell Drive, Sunninghill, 2001
[1] Unit for Environmental Sciences and Management, North West University, Potchefstroom, 2520, South Africa, wernecb@eskom.co.za, bl23034149@gmail.com, Stuart.Piketh@nwu.ac.za, Roelof.Burger@nwu.ac.za

Abstract

The household combustion of solid fuels, for the purpose of heating and cooking, is an activity practiced by many people in South Africa. Air pollution caused by the combustion of solid fuels in households has a significant influence on public health. People most affected are those considered to be the poorest, living in low-income settlements, where burning solid fuel is the primary source of energy. Insufficient data has been collected in South Africa to quantify the concentrations of particulate emissions that people are exposed to, especially the respirable fraction, associated with the combustion of solid fuels. The aim of this paper is to gain an understanding of the particulate matter (PM) concentrations a person living in a typical household in a low income settlement in the South African Highveld is exposed to. It also seeks to demonstrate that the use of solid fuels in the household can lead to indoor air pollution concentrations reaching levels very similar to ambient PM concentrations, which could be well in excess of the National Ambient Air Quality Standards, representing a major national public health threat. A mobile monitoring station was used in KwaDela, Mpumalanga to measure both ambient particulate concentrations and meteorological conditions, while a range of dust/particulate monitors were used for indoor and personal particulate concentration measurements. Indoor and personal measurements are limited to the respirable fraction (PM_4) as this fraction contributes significantly to the negative health impacts. The sampling for this case study took place from 7-19 August 2014. Highest particulate matter concentrations were evident during the early mornings and the early evenings, when solid fuel burning activities were at their highest. Indoor and personal daily average PM_4 concentrations did not exceed the 24h National Ambient $PM_{2.5}$ Standard of 65 µg/m³ nor did they exceed the 24h National Ambient PM_{10} Standard of 75 µg/m³. The outdoor $PM_{2.5}$ concentrations were found to be below the standards for the duration of the sampling period. The outdoor PM_{10} concentrations exceeded the standards for one day during the sampling period. Results indicate that, although people in KwaDela may be exposed to ambient PM concentrations that can be non-compliant to ambient standards, the exposure to indoor air, where solid fuel is burnt, may be detrimental to their health.

Keywords

particulate matter exposure, indoor air quality, ambient air quality, personal exposure

Introduction

The household combustion of solid fuels (coal, wood, dung, and crop waste), for the purpose of heating and cooking, is an activity practiced by approximately 3 billion people around the world (Chafe et al. 2014). Air pollution caused by the combustion of these solid fuels has a significant influence on public health, attributing to more than 4 million premature deaths globally in 2012 (Bruce et al., 2015). People most affected are those living in low- income settlements in developing countries, where burning solid fuels is the primary source of domestic energy (Xie et al. 2015).

In South Africa, the low level burning of solid fuels such as coal and wood contributes significantly to the high levels of ambient air pollution in the country (Terblanche et al. 1992).Many people in rural communities and in townships utilise solid fuels for cooking and heating. Exposure to the emissions, in particular to the respirable aerosols stemming from these burning practices is known to cause a large number of health problems (Smith 2000). Various literature sources have acknowledged that ambient pollution levels are not necessarily indicative of the concentrations of air pollution that humans are exposed to on an everyday basis, as most people tend to spend most of their time indoors (Lim et al. 2012, Diapouli et al. 2011, Ferro et al. 2004, Smith 2002). The Medical Research Council Burden of Disease Research Unit ranked indoor air pollution at number 15 for South Africa, higher than urban air pollution (MRC 2008, Norman et al. 2007).

This paper aims to evaluate the level of indoor and outdoor

particulate matter exposure within a typical household in a low-income settlement in Mpumalanga, South Africa (KwaDela) in the winter period, the time of year in which low level burning practices are particularly prevalent, and to demonstrate that the use of solid fuels in the household level can lead to indoor air pollution concentrations reaching well in excess of the National Ambient Air Quality Standards, representing a major national public health threat.

The following questions were answered (i) what is the outdoor, indoor, and personal exposure of residents; (ii) what is the relationship between outdoor, indoor and personal mass concentration measurements; and (iii) what are the associated diurnal patterns of exposure of residents during the sampling period.

Experimental Method

Sampling Site

KwaDela (26°27'47.53"S; 29°39'51.73"E) is situated in the Gert Sibande Disctrict Municipality of Mpumalanga, South Africa, which lies in the Highveld Priority Area. Located approximately 200 km South-East of Johannesburg. According to census data, Kwadela has a population of about 3777 (Census 2011). In 2014, 79.6% of KwaDela's residents made use of solid fuel burning for daily activities such as heating and cooking.

Instrumentation

A mobile monitoring station was used to obtain ambient concentrations of particulate matter (PM$_{10}$ and PM$_{2.5}$ were measured using a MetOne BAM 1020, MetOne E-Bam and MetOne E-Samplers) and meteorological data (temperature, humidity, pressure, wind speed and direction, rainfall). The ambient monitoring site was located at the Secondary School close to the centre of KwaDela. Additionally two E-samplers were used to measure ambient PM concentrations (PM$_{2.5}$ and PM$_{10}$) in separate locations of the settlement.

The household considered in this study was semi-randomly chosen for indoor monitoring. Indoor particulate concentrations (PM$_4$) were measured using the TSI DustTrak (Models 8520 and 8530) photometric monitors, while the personal exposure of one of the residents in the household (PM$_4$) was monitored using the TSI SidePak AM510 photometric monitor. Temperature iButtons were placed in strategic locations within the sampling household to help better understand the indoor temperature dynamics. This included measuring temperatures in various rooms of the household (bedroom, kitchen, by the stove, and outside the house).

Sampling was conducted as part of a larger sampling campaign in KwaDela in various seasons (winter 2013 and 2014 and summer 2014 and 2015). This paper focuses on the measurements taken in one household in the winter 2014 campaign. This study was approved by the North-West University Research Ethics Committees (NWU-00066-13-S3).

Kwadela, Mpumalanga

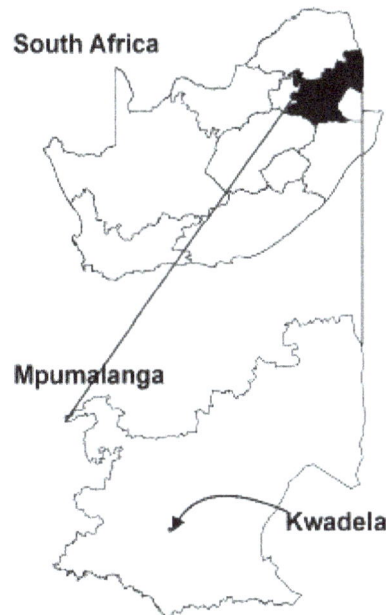

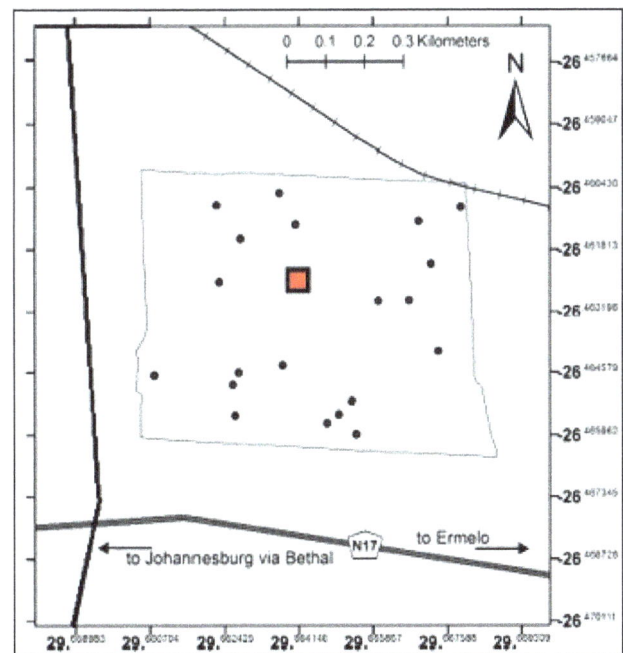

Figure 1: Map of Kwadela showing the distribution of household and mobile monitoring sites.

Data Processing and Analysis

The data collected by the various instruments was merged into one overarching data set by synchronising the sampling intervals of the various instruments into an hourly data set. The instruments within the mobile monitoring station were checked once a week during the sampling. The indoor instruments were zero calibrated and flow checked as per manufacturer's instructions. The personal monitoring instruments were flow checked once a week and zero calibrated each day before sampling.

Simple time series were plotted to identify the average diurnal PM patterns of the specific household. Furthermore, correlations between the indoor and outdoor and indoor and personal particulate concentration levels were found. Lastly, frequency distributions illustrated the 99th percentile of indoor and outdoor particulate concentrations.

The indoor and personal PM_4 measurements have been corrected according to the specific photometric calibration factors obtained for the DustTrak and SidePak instruments. It is noted that, as a possible limitation of this study, PM_4 and $PM_{2.5}$ are included as smaller fractions/ subsets of PM_{10}.

Results and Discussion

Exposure of residents to particulate matter

Mean outdoor $PM_{2.5}$ and PM_{10}, indoor PM_4 and personal PM_4 concentrations were 27±18 and 21±122 and 17±23 and 16±7 μg/m³ respectively (Table 1). The variability of the particulate matter was highest for outdoor PM concentrations. 99th percentile values for outdoor $PM_{2.5}$, outdoor PM_{10}, indoor PM_4 and personal PM_4 concentrations were 81, 303, 126 and 30 μg/m³ respectively. Frequency distributions indicate that the majority of all PM concentrations measured fall between 0 and 50 μg/m³ (Figures 2-5).

Table 1: Descriptive Statistics of Outdoor, Indoor and Personal PM Measurements in (ug/m³).

	N	M	SD	Min	Max
O PM_{10} (ug/m³)	178	48	122	1	1518
O $PM_{2.5}$ (ug/m³)	274	27	18	2	196
I PM_4 (ug/m³)	291	17	23	1	154
P PM_4 (ug/m³)	7	16	7	10	112
N - Sample size M – Mean SD – Standard deviation					
Min – Minimum value Max – Maximum value					
O – Outdoor I – Indoor - P – Personal					

Relationship between indoor, outdoor and personal PM concentrations

The value of R^2 in the regression analyses in Figures 6 and 7 indicate a weak correlation between indoor and outdoor $PM_{2.5}$ concentrations ($R^2 = 0.087$) and between indoor and outdoor PM_{10} concentrations ($R^2 = 0.11$). However, the correlation between personal and indoor PM is stronger at $R^2 = 0.93$ (Figure 8). This result limited by the fact that merely 7 measurements were available for this particular case. Regression analyses between the personal PM_4 and outdoor (PM_{10} and $PM_{2.5}$) measurements are not displayed here as there are too few data points for personal PM_4 to be representative of the true relationship.

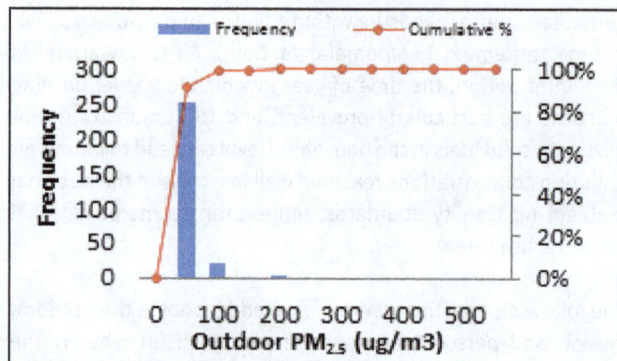

Figure 2: Distribution of mean 1-hourly outdoor $PM_{2.5}$ mass concentrations in μg/m³.

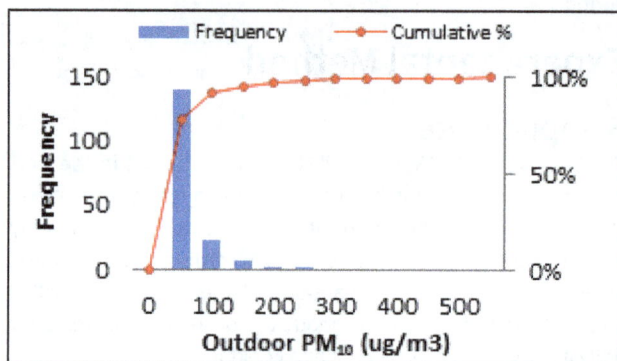

Figure 3: Distribution of mean 1-hourly outdoor PM_{10} mass concentrations in μg/m³.

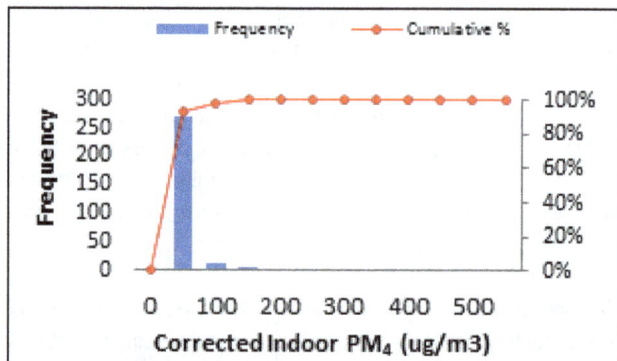

Figure 4: Distribution of mean 1-hourly indoor PM_4 mass concentrations in μg/m³.

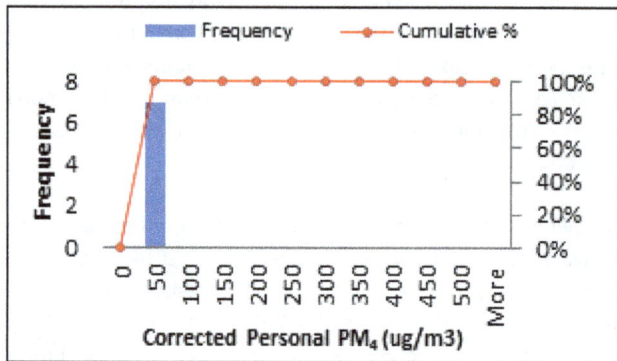

Figure 5: Distribution of mean 1-hourly personal PM_4 mass concentrations in μg/m³.

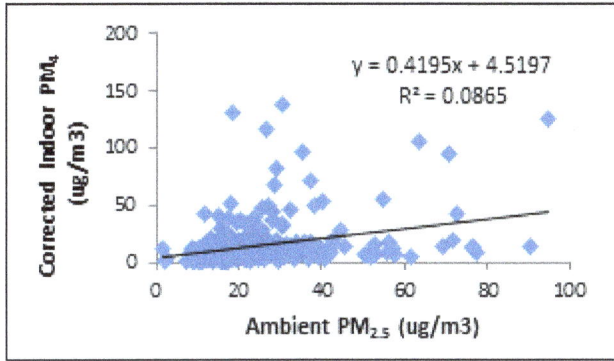

Figure 6: *Regression analysis of indoor (PM$_4$) and outdoor (PM$_{2.5}$) concentrations in μg/m³.*

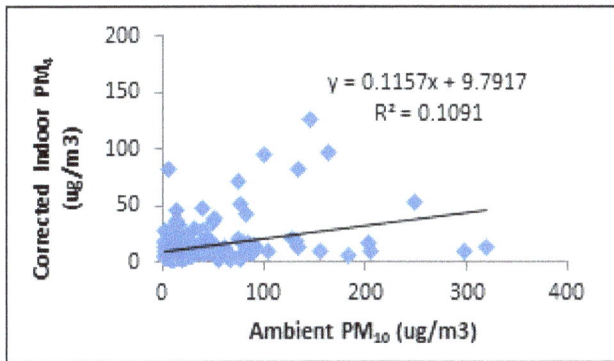

Figure 7: *Regression analysis of indoor (PM$_4$) and outdoor (PM$_{10}$) concentrations in μg/m³.*

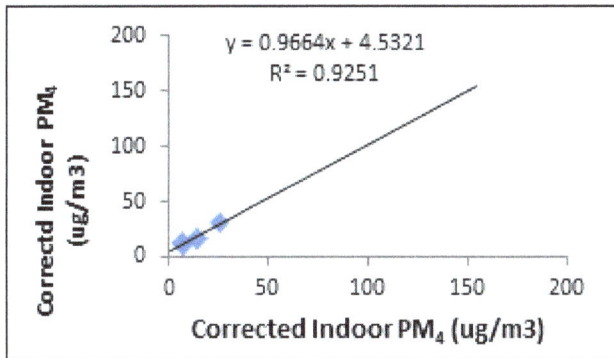

Figure 8: *Regression analysis of personal and indoor PM$_4$ concentrations in μg/m³.*

Diurnal patterns of exposure

Lower ambient temperatures and morning and evening time periods correspond with outdoor PM peaks (Figure 9). Higher stove temperatures link to cooking and heating activities in the early morning and afternoon hours as well as the evening hours with elevated indoor PM concentrations (Figure 10). Morning and evening peaks are also likely to be caused by vehicle emissions.

Indoor and outdoor PM concentrations follow a similar diurnal trend throughout the day, however, the mean PM$_{10}$ hourly average concentrations lie above the mean outdoor PM$_{2.5}$ and indoor PM$_4$ concentrations. The visible diurnal trend is a signature trend for low level burning practices, indicating that all measured PM levels are directly influenced by solid fuel burning in the household and most likely by surrounding households. The differences in outdoor PM$_{2.5}$ and outdoor PM$_{10}$ concentrations can most likely be attributed to the fact that the two monitoring instruments were located in different areas of KwaDela, being exposed to different ambient PM concentrations entirely.

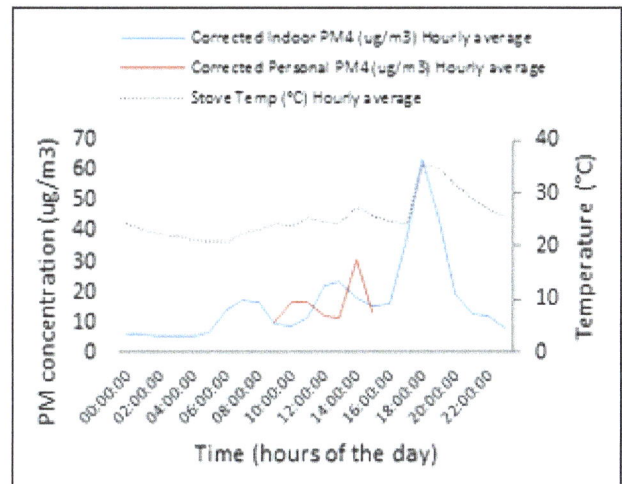

Figure 10: *Hourly concentrations of indoor PM$_4$ averaged during the winter period 7-19 August 2014.*

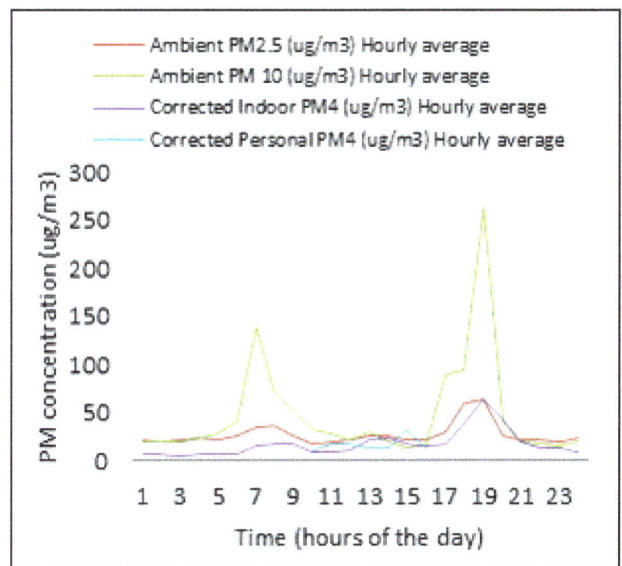

Figure 11: *Hourly concentrations of indoor and outdoor PM concentrations averaged during the winter period 7-19 August 2014.*

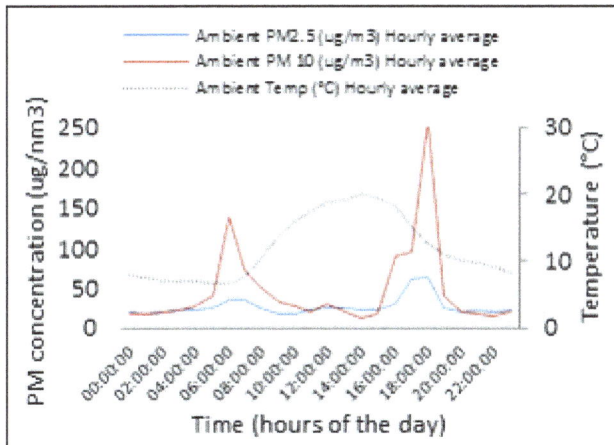

Figure 9: *Hourly concentrations of outdoor particulate matter averaged during the winter period 7-19 August 2014.*

Indoor and personal daily average PM_4 concentrations did not exceed the 24h National Ambient $PM_{2.5}$ Standard of 65 µg/m³ nor did they exceed the 24h National Ambient PM_{10} Standard of 75 µg/m³. The outdoor $PM_{2.5}$ concentrations were found to be below the standards for the duration of the sampling period. The outdoor PM_{10} concentrations exceeded the standards for one day during the sampling period (Figure 12).

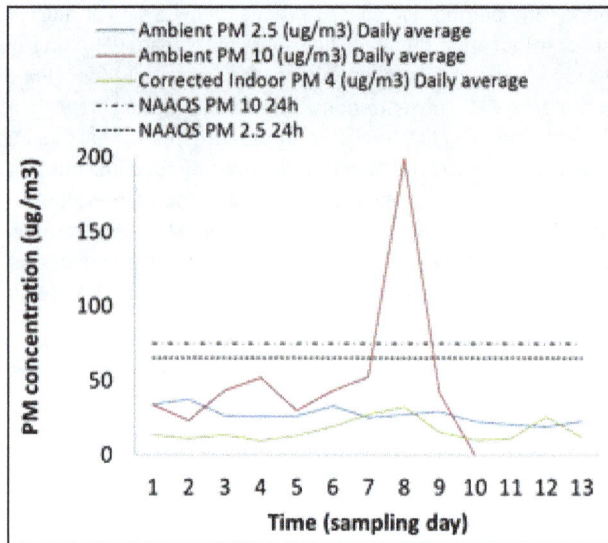

Figure 12: Daily average for PM concentrations measured against the NAAQS for PM_{10} and $PM_{2.5}$ over the sampling period.

Conclusion

This study indicates that, in this household and during this sampling campaign, PM concentrations experienced outdoors were on average higher than those experienced indoors. Outdoor mean concentrations were 48 µg/m³ and 27 µg/m³ for PM_{10} and $PM_{2.5}$ and indoor mean concentrations were 17 µg/m³ and 16 µg/m³ for indoor and personal measurements respectively. Extreme events were evident in outdoor PM_{10} measurements where a maximum hourly value of 1518 µg/m³ was measured. Maximum hourly values for outdoor $PM_{2.5}$, indoor PM_4 and personal PM_4 were found to be 196 µg/m³, 15µg/m³ and 122 µg/m³ respectively.

Even though this study outlines the PM concentrations experienced at only one household, it gives an indication of the average indoor PM_4 concentrations an average person experiences in the Highveld, Mpumalanga, where the combustion of solid fuels is a daily practice occurring in the majority of households. This study indicates that people living in KwaDela are exposed to high PM concentrations which may exceed ambient PM standards outdoors, but that they are exposed to high PM concentrations even indoors, which may be detrimental to their health.

Acknowledgments

The data collection was conducted by Richhein Du Preez, Corné Grové and Brigitte Language from North West University and The NOVA Institute under the supervision of Professor Stuart Piketh for Sasol's air quality offset pilot study.

References

Bruce, N., Pope, D., Rehfuess, E., Balakrishnan, K., Adair-Rohani, H. and Dora, C. 2015, 'WHO indoor air quality guidelines on household fuel combustion: strategy implications of new evidence on interventions and exposure – risk functions', *Atmospheric Environment* 106: 451-457.

Diapouli, E., Eleftheriadis, K., Karanasiou, A., Vratolis, S., Hermansen, O., Colbeck, I. and Lazaridis, M. 2011, 'Indoor and Outdoor Particle Number and Mass Concentrations in Athens. Sources, Sinks and Variability of Aerosol Parameters', *Aerosol and Air Quality Research* 11:632–642.

Chafe, Z.A., Brauer, M., Klimont, Z., Van Dingenen, R., Mehta, S., Rao, S., Riahi, K., Dentener, F. and Smith, K.R., 2014, 'Household Cooking with Solid Fuels Contributes to Ambient $PM_{2.5}$ Air Pollution and the Burden of Disease', *Environmental Health Perspectives* 122:1314-1320.

Ferro, A.R., Kopperud, R.J., Hildemann, l.M. 2004, 'Elevated personal exposure to particulate matter from human activities in a residence', *Journal of Exposure Analysis and Environmental Epidemiology* 14:34–40.

Lim S.S, Vos T, Flaxman AD, Danaei G, Shibuya K, Adair-Rohani H, et al. 2012, 'A comparative risk assessment of burden of disease and injury attributable to 67 risk factors and risk factor clusters in 21 regions, 1990–2010: a systematic analysis for the Global Burden of Disease Study 2010', *The Lancet* 380:2224-60.

Medical Research Council, 2008, 'South Africa Comparative Risk Assessment' *Summary Report*.

Norman R, Barnes B, Mathee A, Bradshaw D and the South African Comparative Risk Assessment Collaborating Group. 2000, 'Estimating the burden of disease attributable to indoor air pollution in South Africa in 2000', *South African Medical Journal*, 97:764-771.

Smith, K.R. 2000, 'National burden of disease in India from indoor air pollution', *PNAS* 24:13286–13293.

Smith, K.R. 2002, 'Indoor air pollution in developing countries: Recommendations for research', *Indoor Air* 12:198-207.

Terblanche, A.P.S., Nel, R., Reinach, G., and Opperman, L. 1992, 'Personal exposures to total suspended particulates from domestic coal burning in South Africa', *The Clean Air Journal* 8(6):15-17.

Xie, Y., Zhao, B., Zhang L. and Luo, R. 2015, 'Spatiotemporal variations of $PM_{2.5}$ and PM_{10} concentrations between 31 Chinese cities and their relationships with SO_2, NO_2, CO and O_3', *Particuology* 20:141–149.

6

The use of fine water sprays to suppress fume emissions when casting ferromanganese

Sarel J. Gates[*1], Gerrit Kornelius[1], Steven C. Rencken[1], Neil M. Fagan[1], Peter Cowx[2], Luther Els[3]

[1]University of Pretoria, Dept of Chemical Engineering,Environmental Engineering Group, Private Bag X20 Hatfield, Pretoria, South Africa, 0028, gerrit.kornelius@up.ac.za
[2]Eramet Norway, Sauda, Norway, peter.cowx@erametgroup.com
[3]Resonant Environmental Technologies, P.O. Box 12225, Centurion, South Africa, 0046, luther@resonant.co.za

Abstract

During the casting of ferromanganese alloys from electric arc furnaces into sand beds at temperatures of up to 1800°C a considerable amount of very brown fumes are generated when the alloy fume is oxidized in the atmosphere. The fume is difficult to capture because of the large flux of gas that is generated. Possible reasons for this flux include the high evaporation rate of Mn at elevated temperatures, the large surface area of the casting beds and the large thermal plumes over the furnace tapholes and casting beds. It has been found that the use of fine water sprays along the edge of the roof that covers the casting bed resulted in a significant reduction in visible emissions. This paper describes research into the kinetics of the fume to improve the design of the capture hoods, as well as the mechanism of suppression by the water sprays by using CFD analysis. It is shown that the oxidation reaction produces less than 20% of the energy content of the plume over the arc furnace taphole, and also that radiation heat transfer may play an important role in increasing the energy content of the taphole plume. The capture of fume particles by fine spray droplets is shown to have limited efficiency, while the heat sink that is caused by evaporation does not materially contribute to the circulation of fume through the spray. It is postulated that the increased moisture content of the air over the casting beds may be instrumental in reducing the oxygen partial pressure or in the formation of an oxide layer, both of which would reduce metal evaporation and, therefore fume formation. The exact mechanism requires further investigation.

Keywords

ferromanganese, secondary fume, water sprays, fume capture hoods, fume extraction, ferro-alloy tapping

Introduction

Eramet Sauda, in Norway, operates two ferromanganese furnaces producing high carbon ferromanganese (HCFeMn) as well as a Manganese Oxygen Refining (MOR) unit to produce medium and low-carbon ferromanganese (LCFeMn). Secondary fume emissions occur at the tapholes of the arc furnaces as well as during post-taphole operations. The current secondary fume capturing system has good capacity, but emissions do still escape from the furnace building. Due to their small particle size, manganese oxides that are present in these fumes pass through the trachea and bronchi to the lungs (de Nevers 2010). Health risks include manganism, a serious and irreversible brain disease, and various lung disorders. It is therefore important to either suppress or capture the fumes formed during casting operations (Goodfellow et al. 2001). The formation mechanism of the fumes was investigated in order to understand its possible contribution to the fume emission volume and energy content. A study was then conducted to investigate whether a fine water spray can suppress fume emissions from casting bed operations and, consequently, mitigate Mn_3O_4 pollution without enclosing the sand beds or significantly increasing the fume capture capacity. To achieve this, flat jet sprayers were placed on the edge of the shed roof that cover the casting beds. These sprayers sprayed horizontally away from the shed to form a water curtain on the outside of the shed without water accumulation in the shed.

Theoretical Background

Fume Energy Contributions

Mechanisms and Fume Formation

Little information on the fume generation mechanisms in the ferromanganese industry could be found, but studies in other similar applications suggest that the predominant mechanisms in steelmaking are bubble bursting and volatilisation (Guézennec et al. 2004; Huber et al. 2000; Gonser et al. 2011). There is little bubble formation during the metal flow, from one casting pocket to the next. Due to the high metal temperatures and the low relative boiling point of Mn, evaporation and oxidation to Mn_3O_4 is considered the most significant mechanism of the fume generation. Mn_3O_4 has the highest Gibbs Free energy of the manganese oxides, making it the most likely product to form at the elevated temperatures found during casting operations (Els et al. 2013).

The energy generated by the reaction involved in the Mn_3O_4 formation, can be assumed to contribute to the fume's energy. A discussion of the reaction kinetics involved in the Mn_3O_4 formation will help in the understanding of the unexpected high energy content of the fumes, the fumes' rise velocity and the necessary extraction required to ensure that the fumes are captured.

Lee et al. (2005) considered various reaction-limiting factors for the fume formation in the casting of the high carbon FeMn produced in the arc furnace. They found that the Mn in the melt is essentially lost through evaporation and oxidation to form MnO mist, which is further oxidised to Mn_3O_4 particulate. Turkdogan et al. (1963) studied the diffusion-limited rates of vaporization of metals and showed that at high oxygen partial pressures, metal evaporation rates approach those in a vacuum and can be predicted by the Langmuir equation:

$$E_a = p_a \left(\frac{M_a}{2\pi RT} \right)^{0.5} \tag{1}$$

where R is the gas constant, T the absolute temperature, E_a and M_a the evaporation rate and the molar mass of a respectively (Turkdogan et al. 1963; Dennis et al. 2001)[1]. This effect is known as oxidation enhanced vaporization and is caused by the oxidation of the metal vapour above the liquid surface (Turkdogan et al. 1963) to form the MnO mist, reducing the Mn concentration in the gas-liquid interface and promoting further evaporation of Mn into this sink.

In reality the O_2 concentrations may not be high enough to cause O_2 enhanced oxidation. Lee et al. (2005) suggest that no MnO mist and, therefore, no Mn_3O_4 forms when the O_2 partial pressure is below 17kPa, which may be the case where decarburisation of HCFeMn occurs during tapping.

The possible rate limiting factors are the mass transfer of Mn in the melt to the gas-liquid interface, evaporation of Mn at the interface, Mn vapour transport away from the interface and transport of O_2 to the interface. The high metal temperature ensures that the evaporation of Mn is fast. Because of the relative abundance of Mn in the melt, it is not likely to be depleted at the metal surface. Therefore the MnO fume formation rate is considered to be controlled by the counter-diffusion of Mn and O_2 in the boundary layer above the metal surface (Dushman et al. 1962), which can be expressed mathematically with:

$$J_{Mn} = J_{O_2} = \frac{h_{Mn}}{RT} \left(p_{Mn}^{sat} - p_{Mn} \right) \tag{2}$$

where J_{Mn} is the evaporation flux of the Mn, J_{O_2} the flux of O_2 to the gas-liquid interface, $h_{Mn} = D_{Mn}/l$ the average mass transfer coefficient of the Mn vapour and p_{Mn} the partial pressure of Mn at the top of the mass transfer boundary layer (l).

[1] Unless indicated otherwise, all equations are expressed in SI units

According to Lee et al. (2005), the affinity of O_2 with Fe (at high temperatures) is approximately two orders of magnitude less than that of Mn, rendering the formation of FeO negligible.

Radiation Effects

The net rate of radiation can be expressed with the following equation (Çengel et al. 2011; Welty et al. 2009):

$$Q = \sigma \varepsilon A_s T_s^4 - \sigma \alpha A_s T_{surr}^4 \tag{3}$$

where σ is the Stefan-Boltzmann constant (5.67×10^{-8} $W/(m^2 \cdot K^4)$), ε the emissivity of the surface, A_s the surface area, α the absorptivity of the surface, T_s and T_{surr} the temperatures of the surface and surroundings, respectively.

Due to the significant difference between the temperature of the melt and its surroundings, the amount of energy released from the melt in the form of thermal radiation is high. The energy transfer rate varies between 100kW/m² and 300kW/m², depending on the metal's surface temperature and the emissivity. It is improbable that the melt will absorb a significant amount of energy (Els et al. 2013).

The generated fumes primarily consists of Mn_3O_4 particulates, which results in the hazy appearance. Non-polar gasses are virtually unaffected by radiation effects, whilst polar molecules are capable of absorbing radiation (Çengel et al. 2011; Modest 2003). With the above-mentioned in mind, the thermal radiation will be dependent on the Mn_3O_4 particulate content of the fumes. Energy may be reflected by the particulates and may further be re emitted. The exact effects of the particulate presence in the plume are at this stage difficult to quantify.

Natural Convectional Effects

As Els et al. (2010) describe, the majority of fume extraction systems are designed based on flow rates calculated from thermal updrafts which are caused by convection. The convective heat transfer rate is calculated by using *Equation* (4) (Çengel et al. 2011):

$$\Phi = h A_s (T_s - T_\infty) \tag{4}$$

where h is the convective heat transfer coefficient, A_s the surface area of the melt, T_s and T_∞ the surface temperature and surrounding air temperature respectively (Çengel et al. 2011).

Furthermore, there is also natural convective heat transfer between the air and the droplets from the water spray, where a spraying system is used.

Water Spray System

Overview

To attempt to reduce the amount of fugitive secondary fumes formed during casting of ferromanganese on casting beds, Els et al. (2014) studied the possibility of implementing air curtains, enlarging the extraction volume or extending the tapping shed structure. As expected, the efficiency increases when the operational area is enclosed or the extraction volume is increased. Some of the experimental runs that use air curtains also show promising results. In an attempt to find a more efficient alternative, water sprays were installed at the casting beds at Eramet Sauda. The implementation of a water curtain was found to visibly reduce the fume concentration over the casting beds. Two mechanisms were thought to influence the secondary fume.

The first involves the suppression of fume emissions because of the water spray acting as a heat sink, which enhance convectional effects. The air density in the vicinity of the water droplets tends to increase as the air temperature decreases, resulting in buoyancy effects playing a role. The change in circulatory pattern and/or the increased humidity of the air is believed to be involved in the enhanced rate of formation of an oxide layer on the molten metal, thereby reducing the rate by which fumes form. This will be discussed further under *Heat Transfer Effects* on this page.

The second postulated effect of the water spray may comprise the mass transfer of Mn_3O_4 particles from the fumes to the water droplets. It is expected that the Mn_3O_4 particulates will be captured by the water droplets, and then settle on the floor. Thereafter appropriate processing actions may be taken to manage the Mn_3O_4 particles.

Computer Fluid Dynamics (CFD) Modelling

Due to the complex flow patterns of the fumes, a computational model is necessary to efficiently determine the air flow patterns (Witt et al. 2006). CFD modelling was therefore used to simulate the fumes rising off the casting beds and the effect of the water spray. The simulations were performed using FloEFD 14.1.0 (Mentor Graphics 2015). Figure 1 provides a CFD simulation of the temperature profile over a casting bed, where the water spray system (situated at the edge of the roof over the casting bed) is inactive. The CFD simulation delivered comparable results to the on-site measurements of the fugitive fumes' temperature. The air flow patterns of this base case were compared to the air flow patterns when the spray system was activated and the spray acted as a heat sink.

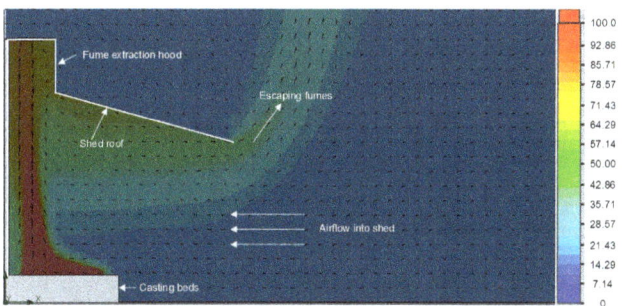

Figure 1: Base case CFD showing the temperature with the scale on the left.

Spray Pattern Development

The amount of heat and mass transfer effects of the spray depends on the spray area. Spray area determination requires modelling of the droplet trajectory. Droplets were assumed to be homogenous in size and shape after dispersion from the spray. It is reasonable to assume that heat transfer effects will have the largest effect at the top of the shed where the highest temperature occurs. The initial horizontal velocity of the droplet can be determined by *Equation* (5):

$$v_d = \frac{Q_d}{A} \tag{5}$$

where v_d is the velocity Q_d the water volumetric flowrate and A the open area of the spray nozzle. Once dispersed, the horizontal velocity of the droplets decreases exponentially due to drag force effects. Two possible relationships between C_d and the Re_d are expressed in *Equations* (6) and (7) (de Nevers 2010):

$$C_d = \frac{24}{Re_d}, Re_d < 0.3 \tag{6}$$

$$C_d = \frac{24}{Re_d}(1 + 0.14R_d^{0.7}), 0.3 \le Re_d \le 1000 \tag{7}$$

where Re_d is the dimensionless Reynolds number which can be calculated with (Çengel et al. 2011):

$$Re_d = \frac{\rho D_d v_d}{\mu} \tag{8}$$

where ρ is the fluid density, D_d the droplet diameter and μ the fluid viscosity.

Since the decrease in the droplets' velocity is much greater than its decrease in the diameter, it is assumed that the droplet's diameter remains constant when the area of water dispersion is calculated. The horizontal velocity of the droplets was modelled in time increments of 0.1s until the horizontal component of the velocity has decreased to zero. This implies that only vertical forces are still active after this point.

For reasons of simplicity, it is assumed that the time it takes until the droplet is influenced only by vertical forces is negligible in comparison with the total falling time. For the transfer of particulate to the droplets, a homogenous dispersion of droplets is assumed to fall vertically from the roof height to the floor over the entire spray area. Finally, the spray area is calculated as the product of the shed roof length and the horizontal distance that the droplets travel.

Heat Transfer Effects

Balances:

By assuming that the heat which is transferred to the water is the same as the heat that is lost through hot air escaping from underneath the shed roof in a steady state operation, and that the kinetic and potential energies are negligible, the energy balance reduces to *Equation* (9):

$$\sum M_{in} C_{p_{in}} T_{in} = \sum M_{out} C_{p_{out}} T_{out} \tag{9}$$

where C_p is the specific heat capacity at constant pressure, M the mass, and T the absolute temperature.

Noting that the particulate concentration is low, the mass balance becomes:

$$M_{a_{in}} + M_{\mathcal{H}_{in}} + M_{w_{in}} = M_{a_{out}} + M_{\mathcal{H}_{out}} + M_{w_{in}} \qquad (10)$$

where M_a, M_H and M_w represents the mass of dry air, water vapour in the air and water from the sprayers respectively.

Humidity

Absolute humidity is defined as the mass of water vapour per unit of dry air and can be expressed mathematically for an air-water system as in *Equation* (11) (Green et al. 2007):

$$\mathcal{H} = \frac{0.622 p_v}{P - p_v} \qquad (11)$$

where p_v is the vapour pressure of water at a given temperature and P the total pressure of the air-water system. The evaporation rate of the water droplets is dependent on the humidity of the air.

Drop Diameter

It is further assumed that the water and air are at thermal equilibrium as it leaves the control volume. Therefore, the temperature of the water and air will be equal. According to Holterman (2003), a droplet's temperature will decrease as it falls through the air as a result of evaporation and will continue to decrease until the water reaches the wet-bulb temperature. The difference in temperature of the droplets and the air results in energy transfer, with the droplets acting as heat sinks. This transfer of heat may affect the droplet's diameter significantly. Various models describing the change in droplet diameter have been developed by Ranz et al. (1952), Goering et al. (1972) and by Williamson et al. (1974). Williamson (*ibid.*) proposed using *Equation* (12) to determine the change in droplet diameter:

$$\frac{dD_d}{dt} = -\frac{4 MW_L D_{v,f}}{D_d \rho_d R T_f} \Delta p \left(1 + 0.276 R e_d^{\frac{1}{2}} Sc^{\frac{1}{3}} \right) \qquad (12)$$

where MW_L is the molecular weight of the evaporating liquid, ρ_L the density of the drop, $D_{v,f}$ the average diffusion coefficient for the vapour molecules in the saturated film around the drop, T_f the average absolute temperature in the film, Re the Reynolds number, Sc the Schmidt's number, Δp the difference between the vapour pressure near the drop and at the ambient atmosphere and R the gas constant.

It is assumed that the heat transfer effects will result in the droplets to evaporate uniformly. Diffusivity dependence on temperature is given by *Equation* (13) (Welty et al. 2009):

$$D_{AB} \propto T^{\frac{3}{2}} \qquad (13)$$

Schmidt's number is defined as (Çengel et al. 2011):

$$Sc = \frac{\mu_{a,f}}{\rho_{a,f} D_{v,f}} \qquad (14)$$

where the subscript f indicates that the given properties is at the film temperature.

Mass Transfer Effects

Mass transfer occurs when Mn_3O_4 particulates come into contact with the water droplets. The droplet size as well as the particle characteristics influence the capture efficiency. Furthermore, the water curtain is dependent on the nozzle type (Grant et al. 2000; Nuyttens et al. 2007). Flat jet K 1590 type nozzles are installed at Eramet Sauda. The control volume for the material balance is illustrated in Figure 2, adapted from de Nevers (2010)

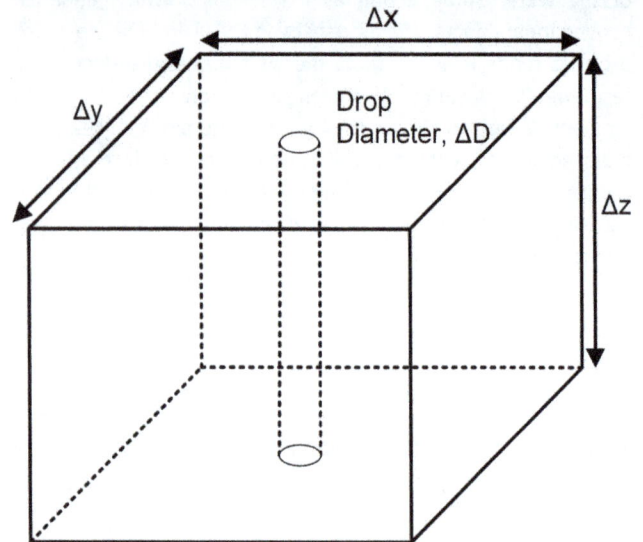

Figure 2: *Control volume (de Nevers 2010)*

Particulate Distribution

A chemical analysis of the fumes showed that Mn_3O_4 formed about 97% of the total mass. The particle size distribution (PSD), shown in Figure 3, was measured by laser diffraction. The particles have a mean diameter of about 0.6µm.

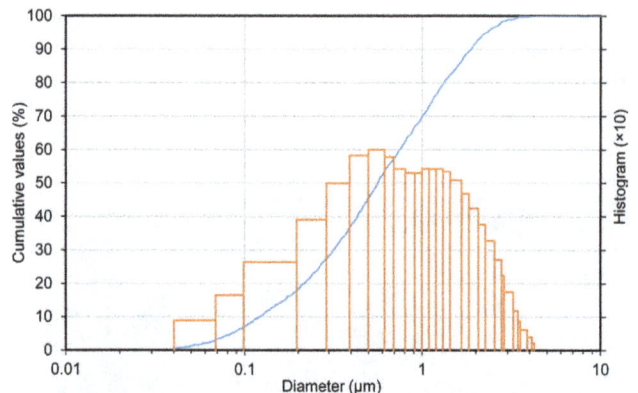

Figure 3: *Particle size distribution for the fume*

Water Droplet Characteristics

It is assumed that for each Δz increment, the droplet characteristics remain constant in terms of diameter and droplet shape. Further assuming that the droplet is spherical, the volume through which the droplet falls can be calculated by (de Nevers 2010):

$$V_{swept} = \frac{\pi}{4} D_d^2 \Delta z \qquad (15)$$

where V_{swept} is the sweeping volume, D_d the droplet diameter and Δz the incremental length.

Assuming spherical droplets and uniform distribution of the fume particulate in the air, the efficiency of particle capture, as illustrated graphically in Figure 4 (Kopita 1955).

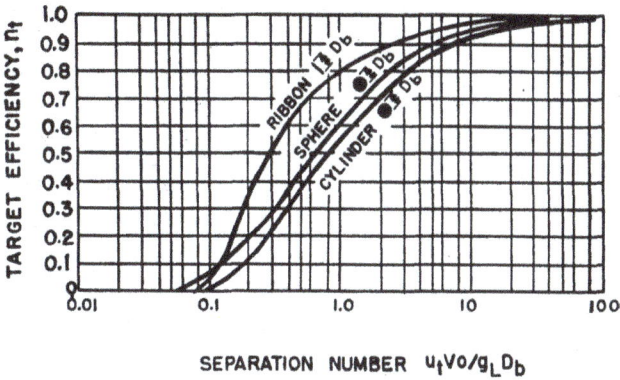

Figure 4: Target efficiency correlation (Kopita 1955)

The separation number can be calculated with *Equation* (16) adapted from de Nevers (2010):

$$N_s = \frac{\rho_p D_p^2 v_{d,t}}{18 \mu_g D_d} \qquad (16)$$

where D_p is the diameter of the particle, μ_g the viscosity of the plume and $v_{d,t}$ the terminal velocity of the droplets. Once the target efficiency is known, the mass of particles absorbed can be calculated with *Equation* (17) (de Nevers 2010):

$$M_{transfered} = \frac{\pi}{4} D_d^2 \Delta z c \eta_t \qquad (17)$$

where c represent the particle concentration which can be calculated with *Equation* (18) (de Nevers 2010):

$$\frac{dc}{dt} = -\frac{1.5 c \eta_t Q_d}{D_d A} \qquad (18)$$

where η_t is the target efficiency, Q_d the volumetric flowrate of the water droplets, D_d the droplet diameter and A the spray area.

The initial particle concentration is estimated as the concentration that is measured when the sprayer system is inactive. If the initial concentration is known, the above equation can be solved without difficulty. The remaining concentration is a function of time, which is determined by the water droplet's settling velocity. Because the droplets are small, they will rapidly reach their terminal settling velocity; the assumption is therefore made that the droplets reach terminal settling velocity at the edge of the shed roof.

Acting Forces on the Water Droplet
Considering the forces acting on the water droplets, Newton's second law of motion reduces to *Equation* (19) (de Nevers 2010).

$$ma = \rho_d \left(\frac{\pi}{6}\right) D_d^3 g - \rho_{fluid} \left(\frac{\pi}{6}\right) D_d^3 g - F_d \qquad (19)$$

where m is the mass of a water droplet, a its acceleration, ρ_d its density, g the gravitational constant, ρ_{fluid} the density of the fluid (in this case air) through which the droplets falls and F_d the drag force. According to Çengel et al. (2011), the general equation for the relationship between the drag coefficient (C_D) and the drag force (F_d) is given by:

$$F_d = \frac{1}{2} C_D \rho_{fluid} v_d^2 A \qquad (20)$$

where A represents the spray area, ρ_{fluid} the fluid density and v_d the droplet velocity. Together with the relationship between the Reynolds number and the drag coefficient for spherical droplets, this can be used to calculate the terminal velocity for each of the assumed droplet sizes. The effectiveness of the spray system can then be found from:

$$\eta_{overall} = \frac{c_0 - c}{c_0} \times 100 \qquad (21)$$

Summary

To be able to design a more efficient extraction system, it is necessary to know the extent of fume formation and how energy contributors (heat of formation, radiation and natural convection) will effect these fumes. The Mn formation is suspected to be controlled by the diffusion of the Mn through the gas-liquid interface, although Lee et al. (2005) indicates that no Mn vapour will form if the oxygen partial pressure is below 17kPa.

Various attempts to reduce the secondary fumes have led to the conclusion that the fumes will be captured by increasing the extraction volume or designing a closed structure in which casting operations will occur. These solutions are however expensive. In an attempt to find a more efficient and cost effective solution, Eramet Sauda installed water sprayers. As a result fume emissions were visibly reduced.

Two possible mechanisms were investigated which may be influential in the water curtain's ability to visibly reduce secondary fumes over the casting beds. The first involved the heat transfer between the fumes and the water spray, which were affected by the energy content of the fumes, the humidity of the air and the change in the droplet's diameter. The second possible mechanism was the mass transfer of Mn_3O_4 particulates from the fume to the droplets, which were affected by the particle's and droplet's characteristics as well as the forces acting on the droplet. CFD simulations were used to model the effects of the reduction mechanisms on the air flow patterns and compare the obtained results with the base case where the water spray system was not used.

Results and Discussion

Plant Specific Information

The parameters shown in Table 1 are the characteristics of the spray system and environment.

Table 1: Parameters at Eramet Sauda

Description	Value	Units
Total pressure	101.3	*kPa*
Hot air temperature	45	*°C*
Ambient temperature of the air	20	*°C*
Water temperature	12	*°C*
Relative humidity at ambient air temperature	70	*%*
Droplet diameter	250	*μm*
Height of shed roof	6	*m*
Sprayer spacing along the roof	2	*m*
Water flowrate out of the sprayers	15	*L/min*

Figure 5 provides a schematic representation of the shed and sprayer system.

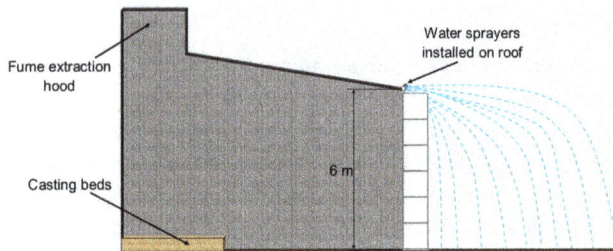

Figure 5: *Schematic of shed and sprayers*

Note the 1m by 1m control volume blocks extending vertically beneath the sprayers. These will be discussed further in *Heat Transfer* on page 32.

Fume Formation

Previous CFD Modeling

Els et al. (2013) used CFD modelling to simulate fumes rising from the casting beds. Initially they only modelled the heat transfer from the metal and structural surfaces due to convection and radiation. They found that the amount of transferred heat was significantly lower than the energy measured on-site as discussed in the next paragraph. They adjusted the model to include additional heat transfer. Figure 6 (Els et al. 2013) illustrates the results from the verification model and Figure 6 depicts the velocity plot of the fumes from the outside of the shed that covers the entire casting bed area.

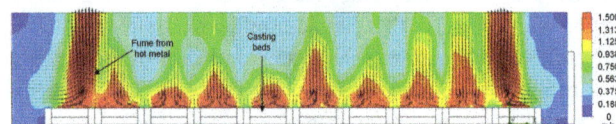

Figure 6: *Verification model velocity plot (Els et al. 2013)*

Previous Flowtests

Table 2 provides the data that Els et al. (2013) found from on-site measurements. They showed from the comparison of theoretical calculated values and the measured data that the energy released as a results of the formation of the Mn_3O_4 fumes produced about 20% of the total fume energy. Some additional fume energy may be attributed to radiation effects which may play an important role in the fumes' energy content.

Table 2: Previous test results (Els et al. 2013)

Test Point	MOR fan	LCFeMn pour point	Units
Velocity	36.5	32.6	*m/s*
Temperature	20.1	35.8	*°C*
Static pressure	-3.57	-2.57	*kPa*
Volume flow	48.5	43.3	*Am³/s*
	156861	134513	*Nm³/h*
Mass flow	58.1	49.3	*kg/s*
Energy	-	2.5	*MW*

Previous Conclusions

By using the model described above, Els et al. (2013) proved that a 100% fume removal efficiency will require either the extraction volume to be doubled or that the beds should be entirely enclosed. This would have been expensive and the spray system mechanisms (as explained in the following sections) were therefore developed. The spray system's role in visible fume reduction was investigated for the purpose of using this knowledge in future spray model designs. This investigation included air flow patterns, heat transfer and mass transfer of the particulates to the water droplets from the spray.

Air Flow Over the Casting Beds

Water disperses up to 8m outwards from the shed, and as a CFD input, this was broken down into 1m sections with temperature decreasing linearly over the 8m distance. The CFD is simulated with a length along the roof edge of 3m, exposing a 3m² surface area for air flow from under the roof edge into the first control volume as seen in Figure 7.

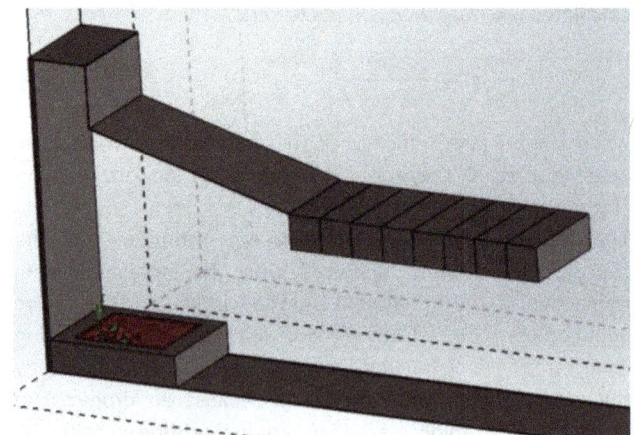

Figure 7: *3D representation of the model*

The heat sink was assumed to occur in this control volume. Since the heat sink will have an influence on the circulation pattern (i.e. the flow rate outward from under the roof edge and hence the outward velocity of the air), a trial and error approach was applied to find the air flowrate into the heat sink. Firstly, the heat sink values for a number of assumed air flow rates were determined using the energy balance described in *Heat Transfer Effects* on page 28. These heat sink values were used as input parameters for the CFD simulations and a velocity profile was determined. By comparing the assumed air flow rates to the resulting CFD output, the approximate velocity of the air could be determined. The calculated heat sink values based on different air inlet flowrates are displayed in Table 3.

Table 3: Heat sink results from the energy balance over the control volume

Description	Values			Units
Air flow rate	3.00	4.00	5.00	(m^3/s)
Air velocity	1.00	1.33	1.67	(m/s)
Block 1	8.90	11.7	14.5	(kW)
Block 2	7.99	10.5	13.0	(kW)
Block 3	7.05	9.25	11.5	(kW)
Block 4	6.10	7.99	9.87	(kW)
Block 5	5.12	6.70	8.27	(kW)
Block 6	4.12	5.38	6.64	(kW)
Block 7	3.10	4.04	4.97	(kW)
Block 8	2.06	2.66	3.26	(kW)

The CFD simulations using the heat sinks in Table 3 are shown in Figure 12, Appendix A. Comparing the CFD results and the calculations explained in Table 3, the air flow rate outwards from under the shed roof was determined to be roughly 4m³/s for the 3m length of roof modelled.

Heat Transfer

Overview
The heat transfer was modelled in two stages: the first involved using a control volume approach for the first (top) section shown in Figure 5, whilst the second involved using an incremental time approach for the rest of the sections in Figure 5.

First Section
Using the air flow rate as determined in *Air Flow Over the Casting Beds* on page 31, the amount of water that evaporated was calculated and the results appear in Table 4.

Table 4: Results after top meter heat transfer

Description	Value	Units
Hot air temperature	45	$°C$
Relative humidity at hot air temperature	24.4	%
Ambient temperature of air	20	$°C$
Water temperature	12	$°C$
Total heat sink	58.2	kW
Total water mass in	0.375	kg

Total mass evaporated	0.020357	kg
Mass of water out	0.0204	kg
Initial droplet diameter	250	$μm$
Droplet diameter leaving control volume	245	$μm$
Average temperature out (T_f)	19.73	$°C$

Lower Section
The air temperature varies slightly from the second meter to the floor, with the air mainly at a temperature of around 20°C and the water temperature at 19.73°C. At 20°C the air is approximately 70% humid and the total driving force for heat transfer is small. This implies that the air will not necessarily reach saturation and the control volume approach would be inaccurate, therefore an incremental time approach would be preferred.

Using *Equation* (12) at incremental steps of 0.01s, the droplet depletion rate is calculated for the remaining 5m (lower) section. This gives the droplet size as well as the mass of water that evaporated after each meter that the droplet falls vertically. The results are displayed in Table 5. From the change in the droplet diameter per meter shown in Table 5, one realises that the change in droplet diameter can be regarded as negligible. Therefore a constant diameter of 245.33μm was used for the mass transfer calculations in the control volumes below the first (upper) one.

Using the approach described above, it was shown (see Appendix A) that the inward air velocity under the edge of the roof increased somewhat with the introduction of the sprays, and that the air flowing over the casting beds will originate from the area where the moisture content is being increased by the water sprays.

Table 5: Results for the lower section

Sections	Mass evaporated (kg)	New drop diameter $(μm)$
2nd meter	8.0973E-06	245.327
3rd meter	8.2443E-06	245.324
4th meter	8.2441E-06	245.321
5th meter	8.2439E-06	245.319
6th meter	8.2437E-06	245.316

Transfer of Particles to Droplets
The capture efficiency was calculated by considering the relationship in Figure 4 between the capture efficiency and the separation number. *Equation* (16) was used to determine the separation number. The capture efficiency was calculated to be zero. This calculation is validated by de Nevers (2010, 302), who states that for particles less than 1μm the efficiency tends to be zero.

Calculations were done to determine the efficiency of droplet sizes equal to 250μm for different particle sizes given that the inlet flow rate of 15L.min⁻¹ was used. The graph in Figure 8 (in effect Figure 4 on an extended scale) shows that for particles between 0.9μm and 3μm, which are easily inhaled, the capture is negligible.

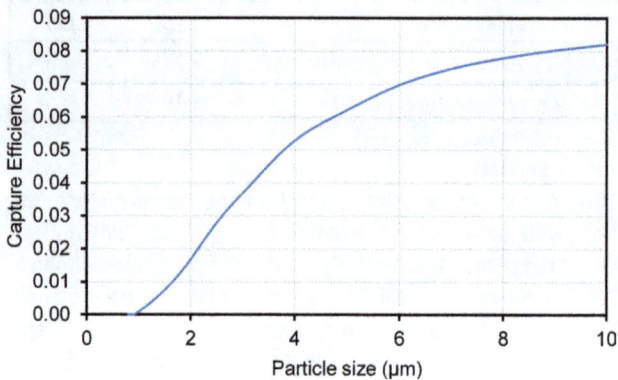

Figure 8: Capture efficiency for small particles

Since the PSD (Figure 3) shows a large number of very fine particles, an investigation was done on smaller droplet sizes and their capture efficiency. The graph in Figure 9 shows the capture efficiency of the small particle sizes at different droplet sizes.

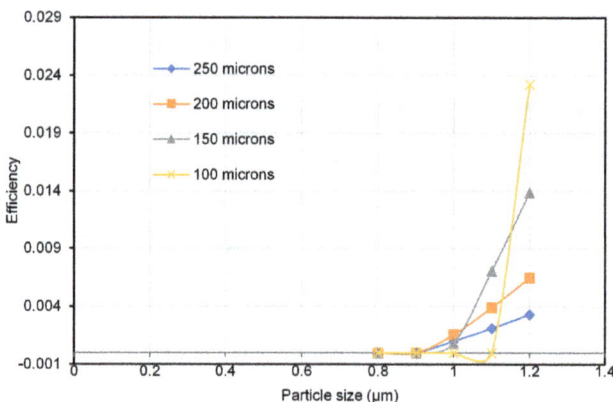

Figure 9: Captured efficiency variation with droplet size

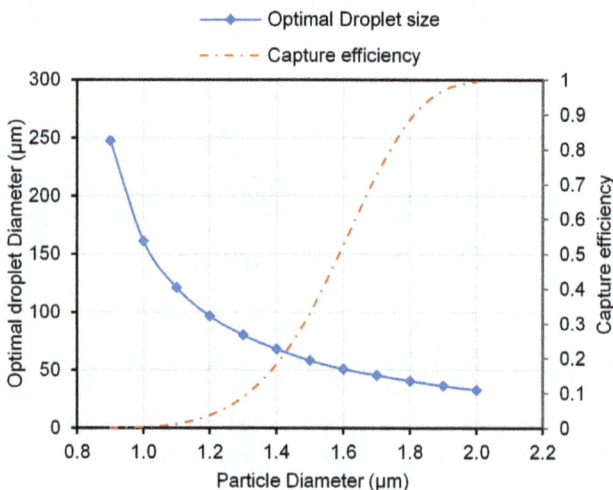

Figure 10: Capture efficiency for optimal droplet diameter

For particles less than 0.8μm, no droplet size showed any capture efficiency. Droplet sizes of 200μm and 250μm captured a small quantity of particles between 0.9μm and 1μm. This is verified by de Nevers (2010, 292), who also suggests that small

particles have a tendency to follow a stream line around the object, thereby not easily captured by small droplets.

The calculation procedure was repeated in an attempt to find an optimum droplet size for capture of the critical particle size. The optimal diameter that would result in the largest capture efficiency is shown in Figure 10.

Conclusions

To fully understand the fine water spray system's role in suppressing secondary emission fumes, it is necessary to understand the mechanism involved in the formation of these fumes, as well as possible energy contributors to the fumes' energy content. Considering this, a previous investigation was reflected upon, which involved a combination of heat transfer relationships and CFD modelling to determine possible energy contributors as well as possibilities for better extraction. It was shown that the oxidation reaction involved during the fumes' formation contributed only 20% of the fumes' energy. Based on these plume energy determinations it was found that a 100% fume removal efficiency from the casting shed was possible if either the extraction volume was increased or if the shed was completely enclosed. Unfortunately these solutions are expensive.

A fine water spray system was considered as an alternative solution. After the installation of the spray system a reduction in the amount of visible fume was noticed. Before the installation of the spray system, a slight circulation pattern was observed, which was further enhanced by the spray system. This phenomenon was investigated by considering the heat transfer between the water and the fume-containing air escaping from under the roof. The investigation showed that the heat transfer is most significant near the shed roof and that the main effect was a slight increase in the inward velocity of the air flowing over the casting bed.

Another possible reduction mechanism that was investigated was the transfer of particulates to the water droplets. However, it was found that the capture efficiency tends to zero as a result of the small Mn_3O_4 particulates. Droplet sizes that are about 200μm are able to capture 1μm particles better than smaller droplets, however still at a very low efficiency. Particles that are smaller than 0.8μm will most likely remain suspended in the air.

It is believed that a spray system will work very well for particles larger than 1.4μm. Operations that have fumes containing particles of this magnitude will be able to use a sprayer system as an effective and cheap fume suppressing technique. For operations that contain a large majority of particles between 1μm and 1.5μm, it is recommended that two separate sprayers should be used. The one sprayer should disperse droplets between 200μm and 300μm, while the second sprayer should disperse smaller droplets.

A fine mist sprayer was installed directly over the casting beds as shown in Figure 11 ensuring a more intense dispersion of moisture. It is clear that the fume concentration is dramatically decreased over the spray area. This leads the authors to believe that the increase in moisture content of the air over the beds – such as when a sprayer system is used – is instrumental in the formation of an oxide layer on the liquid metal, reducing metal vaporisation and thus fume formation. To optimise the sprayer system, further work is necessary to determine the mechanism of fume formation as well as the role of the water spray system to reduce the visible fume emissions.

Figure 11: Fine mist sprayers installed at bed top surface level

References

Çengel Y.A., Ghajar A.J. & Ma H. 2011. Heat and Mass Transfer: *Fundamentals & Applications*, 4th ed., McGraw-Hill.

de Nevers N. 2010. *Air pollution control engineering*. Waveland Press.

Dennis J.H., Hewitt P.J., Redding C.A. & Workman A.D. 2001. A model for prediction of fume formation rate in gas metal arc welding (GMAW), globular and spray modes, DC electrode positive. *Annals of Occupational Hygiene*, 45, 105-113.

Dushman S., Lafferty J.M. & Brown S.C. 1962. Scientific foundations of vacuum technique. *American Journal of Physics*, 30, 612-612.

Els L., Coetzee C. & Vorster O. 'Design of tapping fume extraction systems for ferroalloy furnaces', *Twelfth International Ferroalloy Congress, Helsinki, Finland*, 2010. 6-9.

Els L., Cowx P., Kadkhodabeigi M., Kornelius G., Andrew N., Smith P. & Rencken S. 'Analysis of a ferromanganese secondary fume extraction system to improve design methodologies', *Thirteenth International Ferroalloy Congress, Almaty, Kazakhstan*, 2013. 9-13.

Els L., Cowx P., Smith P. & Nordhagen R. 2014. Analysis and optimization of fume extraction from a ferromanganese furnace tapping operation.

Goering C., Bode L. & Gebhardt M. 1972. Mathematical modeling of spray droplet deceleration and evaporation.

Gonser M. & Hogan T. 2011. *Arc welding health effects, fume formation mechanisms, and characterization methods*. INTECH Open Access Publisher.

Goodfellow H.D. & Tähti E. 2001. *Industrial ventilation design guidebook*. Academic press.

Grant G., Brenton J. & Drysdale D. 2000. Fire suppression by water sprays. *Progress in energy and combustion science*, 26, 79-130.

Green D. & Perry R. 2007. *Perry's Chemical Engineers' Handbook, Eighth Edition*. McGraw-Hill Education.

Guézennec A.-G., Huber J.-C., Patisson F., Sessiecq P., Birat J.-P. & Ablitzer D. 2004. Dust formation by bubble-burst phenomenon at the surface of a liquid steel bath. *ISIJ international*, 44, 1328-1333.

Holterman H. 2003. *Kinetics and evaporation of water drops in air*. IMAG Wageningen.

Huber J., Rocabois P., Faral M., Birat J., Patisson F. & Ablitzer D. 'The formation of EAF dust', *58th Electric Furnace Conference and 17th Process Technology Conference*, 2000. 171-181.

Kopita R. 1955. The Use of an Impingement Baffle Scrubber in Gas Cleaning and Absorption. *Air Repair*, 4, 219-232.

Lee Y.E. & Kolbeinsen L. 2005. Kinetics of oxygen refining process for ferromanganese alloys. *ISIJ international*, 45, 1282-1290.

Mentor Graphics 2015. FloEFD FE 14.1.0.

Modest M.F. 2003. *Radiative Heat Transfer, 2nd Edition*, 2nd ed. Burlington, Academic Press.

Nuyttens D., Baetens K., De Schampheleire M. & Sonck B. 2007. Effect of nozzle type, size and pressure on spray droplet characteristics. *Biosystems Engineering*, 97, 333-345.

Ranz W. & Marshall W. 1952. Evaporation from drops. *Chem. Eng. Prog*, 48, 141-146.

Turkdogan E., Grieveson P. & Darken L. 1963. Enhancement of diffusion-limited rates of vaporization of metals. *The Journal of Physical Chemistry*, 67, 1647-1654.

Welty J.R., Wicks C.E., Rorrer G. & Wilson R.E. 2009. *Fundamentals of momentum, heat, and mass transfer*. John Wiley & Sons.

Williamson R.E. & Threadgill E. 1974. simulation for dynamics of evaporating spray droplets in a gricultural spraying. *Trans ASAE Gen Ed Am Soc Agric Eng*.

Witt P., Solnordal C., Mittoni L., Finn S. & Pluta J. 2006. Optimising the design of fume extraction hoods using a combination of engineering and CFD modelling. *Applied mathematical modelling*, 30, 1167-1179.

Appendix A

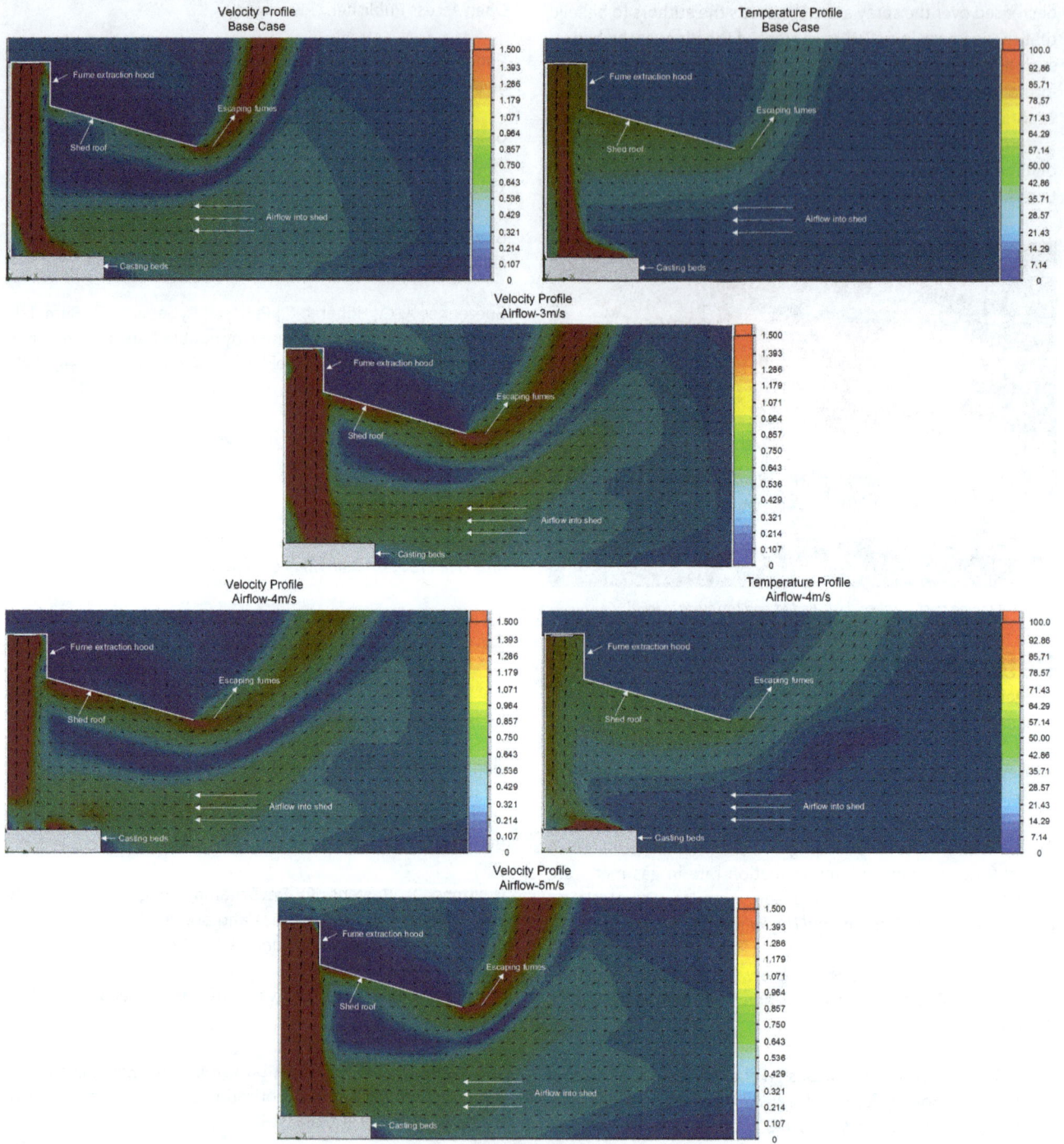

Figure 12: *CFD results showing the airflow in the casting shed*

7

Correcting respirable photometric particulate measurements using a gravimetric sampling method

Brigitte Language[*1], Stuart J. Piketh[1] and Roelof P. Burger[1]

[1]Unit for Environmental Sciences and Management, North West University, Potchefstroom, 2520, South Africa, bl23034149@gmail.com, Stuart.Piketh@nwu.ac.za, Roelof.Burger@nwu.ac.za

Abstract

According to the National Environmental Management: Air Quality Act of 2004 people have the right to clean air and a healthy environment. Particulate matter (PM) emissions pose a significant health threat. Both indoor and ambient air pollution contribute to the burden of disease associated with poor air quality. This is particularly true within the South African setting where low income households make use of different solid fuels for heating and cooking purposes resulting in high levels of PM emissions. This paper focuses on the evaluation mass concentration measurements recorded by continuous photometric PM instruments within KwaDela, a low income settlement in Mpumalanga located on the South African Highveld. Thus, obtaining a photometric calibration factor for both the DustTrak Model 8530 and the SidePak AM510. Sampling took place during August 2014 for a period of seven days. The photometric and gravimetric instruments were collocated within the indoor environment of selected households. These instruments were all fitted with 10mm Dorr-Oliver Cyclone inlets to obtain the respirable (PM_4) cut-point. The study found that both instruments tend to overestimate the indoor particulate mass concentrations when compared to the reference gravimetric method. The estimated photometric calibration factors for the DustTrak Model 8530 and SidePak AM510 are 0.14 (95%CI: 0.09, 0.15) and 0.24 (95%CI: 0.16, 0.30) respectively. The overestimation of the photometric measurements is rather significant. It is therefore important that the correction factors are applied to data collected in indoor environments prone to the combustion of solid fuels. The correction factors obtained from this and other studies vary as a result of the environment (ambient, indoor etc.) as well as the aerosol size fraction and the origin thereof. Thus, it is important to considered site specific calibration factors when implementing these photometric light-scattering instruments.

Keywords

respirable particulate matter, gravimetric analysis, light scattering photometer, photometric calibration factor; indoor air quality

Introduction

According to the World Health Organisation (2006) having access to air of good quality is necessary for a healthy life, this statement is supported by the South African National environmental management: Air Quality Act 39 of 2004. However, air pollution continues to be a major health problem globally, causing an estimated seven million premature deaths a year. The majority of these deaths are associated with the populations of developing countries (WHO, 2014).

The 2010 Global burden of disease study indicated that an estimated 3.5 million (uncertainty level: 2.7, 4.5) premature deaths are caused by household solid fuel use, an additional 0.5 million deaths can be attributed to ambient air pollution resulting from household emissions (Bruce et al., 2015). It is thus no surprise that the study rated ambient air pollution as the ninth and indoor air pollution as the third leading risk factors associated with the global burden of disease (Lim et al., 2012).

Recently focus has been drawn to the significance of indoor exposure to PM as most people tend to spend more than 85% of their time indoors (Yassin et al., 2012; Funk et al., 2014). Most data collected on PM concentrations are based on ambient measurements, which is not a reliable indicator of the particulate levels associated with indoor and personal exposures (Huang et al., 2007).

Measuring of particulate mass concentrations, which is the most widely reported parameter, is conducted mainly for scientific and regulatory reasons (McMurry, 2000).

The WHO and South African National Ambient Air Quality Standards (NAAQS) have set guidelines for exposure to PM_{10} and $PM_{2.5}$. The exposure guidelines for the annual and 24-hour averaging period for both PM_{10} and $PM_{2.5}$ are represented in Table 1. There is a significant difference between the WHO and NAAQS as the guidelines set by the WHO are much lower than those set by the NAAQS. It is important to note that there are no set guidelines for indoor PM exposure as the South African guidelines focus on ambient exposure. There is, however, not a set guideline for the respirable particulate fraction (PM_4) investigated in this study.

Table 1: WHO and NAAQS

		24 Hour (µg/m³)	1 Year (µg/m³)
PM$_{10}$	WHO	50	20
	NAAQS	75	40
PM$_{2.5}$	WHO	25	10
	NAAQS	65	25

Source: South Africa (2009 & 2012) and WHO (2006).

Ground-based PM monitoring is usually performed by using either continuous measurements collected by real-time PM monitoring instruments or filter-based manual sampling methods (Engel-Cox et al., 2013). The filter-based sampling method is a time integrated method obtaining PM mass concentrations through direct measurements, whereas the continuous instruments are based on various technologies and considered as an indirect measurement method. Continuous monitoring measurements make it possible to gain insight into levels of PM during shorter time intervals (Tasić et al., 2012).

Light-scattering photometers, such as the DustTrak Model II 8530 and SidePak AM510 (TSI Inc., Shoreview, MN, USA), are commonly used to measure PM mass concentrations (TSI Inc., 2002; Kim et al., 2004; TSI Inc., 2014). Previous studies done relating to the DustTrak (Tung et al., 1999; ; Heal, et al, 2000; Ramachandran, et al, 2000; Chung, et al 2001; Yanosky et al., 2002; Braniš and Hovorka, 2005; Kingham, et a., 2006; McNamara et al., 2011; Wallace, et al, 2011) and SidePak (Thorpe, 2007; Zhu et al., 2007; Jiang, et al., 2011; TSI Inc., 2013) photometric aerosol monitoring instruments have indicated a significant overestimation of the particulate concentrations when compared to a reference gravimetric method. These studies were all conducted in various settings and compared to different reference methods. It is therefore critical to estimate a calibration factor for each monitor within the specific sampling environment.

The aim of this study was to evaluate and obtain a photometric correction factor for two indoor photometric monitoring instruments, DustTrak II Model 8530 and SidePak AM510, situated within a South African low-income settlement prone to the indoor combustion of low-grade solid fuels, such as coal and wood.

Material and Methods

Experimental Design
The results presented in this article are part of a larger study on the measurements of ambient and indoor exposures experienced in a typical low-income settlement in South Africa. KwaDela (26°27'47.53"S; 29°39'51.73"E) is such a low-income settlement located in the Mpumalanga Highveld, part of the Highveld Priority Area, approximately 200 km South-East of Johannesburg.

The settlement is somewhat isolated, the closest town being Bethal, which is ~25 km West of KwaDela. A significant proportion of the settlement relies on the burning of solid fuels as their primary energy source used for everyday activities such as space heating and cooking. An evaluation of indoor PM$_4$ has been done for a one week period in August 2014. During the sampling period ambient air temperatures averaged around 12°C (low 3°C, high 25°C) while an average relative humidity of 64% was experienced.

From these twenty sampling houses two were randomly selected for the comparison study. The PM$_4$ measurements were collected by making use of both photometric direct-reading instruments and a gravimetric sampling method.

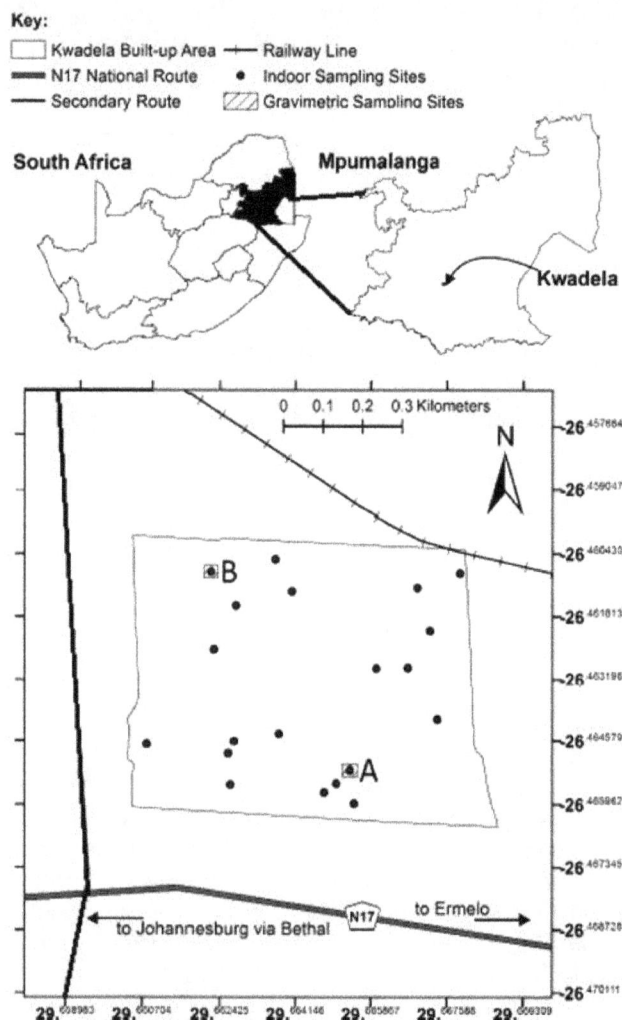

Figure 1: KwaDela low-income settlement in Mpumalanga, South Africa. The spatial distribution of the indoor and gravimetric sampling sites are also represented.

Continuous Monitoring Instruments
PM$_4$ concentrations in indoor air has been measured using two photometric light scatting monitors, namely the **DustTrak II Model 8530** and **SidePak AM 510** (TSI Inc., Shoreview, MN, USA). These instruments do not have a built in PM$_4$ impactor, thus a **10-mm Nylon Dorr Oliver Cyclone** inlet (TSI Inc., Shoreview, MN,

USA) was used with each instrument. The instruments were sampled at a flow rate of 1.7 L.min⁻¹, to acquire the required 50% cut size at 4 µm (the cyclone removes 100% of 10 µm particles and 50% of 4 µm particles, this in turn resembles the 0% of 10 µm particles and 50% of 4 µm particles which enter the lung (Sensidyne, 1999)). The DustTrak operated for 24 hours a day, in 12 hour intervals from 10h00 to 22h00 and again from 22h00 to 10h00. The SidePak operated for 12 hours a day, in 6 hour intervals from 10h00 to 16h00 and again from 16h00 to 22h00.

These specific sampling times were chosen as to avoid a sample collecting PM over both peak burning period found in the settlement; which could result in filter overloading. PM_4 concentrations were logged in five minute intervals. The output for the particulate mass concentration was given in milligram per cubic meter (mg.m⁻³) (TSI Inc., 2002; Kim et al., 2004; TSI Inc., 2014). By averaging the five minute interval concentrations over the sampling duration, the time-integrated measurement were calculated (Kim et al., 2004). Data was downloaded from the instruments at the end of each sampling event, by connecting the instrument via a USB connection to a computer, using the TSI TrackPro Software. Table 2 gives the manufacturer's specification for both the DustTrak and the SidePak photometric monitoring instruments.

Table 2: *Manufacturer specifications for the photometric instruments*

	DustTrak II 8530	SidePak AM510
Flow Rate (L/min)	1.7-2.4 (1.7)	0.7-1.8 (1.7)
Particle Size Range (µm)	0.1 - ±10	0.1 - ±10
Mass Concentration Range (mg/m³)	0.001-100	0.001-20
Laser Beam Wavelength (nm)	780	670
Operating Temperature (°C)	0 - 50	0 - 50
Temp. Coefficient (mg/m³ per °C)	+0.001	+0.0005
Zero Stability (mg/m³) over 24-hr at 10 second time-constant	±0.001	±0.001
Calibration	Arizona Test Dust	Arizona Test Dust

Source: TSI Inc. (2002 & 2014).

Gravimetric Sampling Method

The gravimetric sampling was done by exposing 37mm cassettes, at a constant flow rate of 1.7 L.min⁻¹, using Gilian GilAir 3 (Sensidyne, Clearwater, FL, USA) pumps. The pumps were fitted with 10-mm Nylon Dorr Oliver Cyclone inlets to obtain the 50% cut size at 4 µm. The gravimetric sampling occurred in line with the photometric monitors. Thirty-seven millimetre Borosilicate Microfiber Filters (ADVANTEC MFS Inc., Pleasanton, CA, USA), used in the 37 mm cassettes, were weighed prior to and after sampling. Weighing was done by making use of a XP26 DeltaRange Microbalance (Mettler-Toledo AG, Greifensee, CH) having a sensitivity of 1µg.

Photometric Calibration Factor

The PM_4 photometric measurements could be adjusted by making use of a calibration factor to approximate the actual PM_4 mass concentration. By doing a comparison between the PM_4 mass concentrations obtained from the photometric monitors and the reference gravimetric method a specific calibration factor was developed for each instrument. The calibration factor was calculated by the following equation (4):

$$Cal.Factor = \frac{Grav.Conc.}{Inst.Conc.}(Cur.Cal.Fac.) \qquad (1)$$

The DustTrak and SidePak measurements were then corrected by multiplying the five minute averages with the specific photometric calibration factor assigned to each instrument.

Quality Assurance and Quality Control

Various procedures were integrated into the sampling to ensure the quality of the photometric measurements. Preceding the start of the sampling campaign the monitors were sent for factory calibration using the respirable fraction of standard ISO 12103-1, A1 Arizona test dust. Before each sampling event the instruments were zero-calibrated by attaching the zero-filter as per the manufacturer's instructions. Flow rates were checked prior to and after each sampling event to ensure that the target flow rate was maintained. Filters were handled with care during weighing and loading activities as to prevent contamination and loss of gained PM and insure filter weight accuracies. The micro-balance was situated in a clean-lab, having controlled access, to limit external interference during weighing. The balance was levelled and calibrated, with the weights provided by the manufacturer, prior to each weighing session. It also has an internal function that removes any static that might influence the mass measurements.

Statistical Analysis

Basic statistical analyses were performed by using STATISTICA version 12 (StatSoft Inc.). All statistical analyses were performed with a 0.95 confidence and a 0.05 significance. The correlation coefficient analyses was performed to indicate the direction and strength of the linear relationship between the concentrations obtained from the real-time photometric instruments and gravimetric sampling. Furthermore, comparisons were made between initial and corrected PM_4 mass concentrations (one day case study) as well as cumulative distributions for a one week period.

Results and Discussion

Photometric Calibration Factors

Twenty-eight sets of comparison samples were collected during a week sampling in August 2014. A total of seventeen sets were valid, eight sets contributing to the evaluation of the DustTrak and nine to the evaluation of the SidePak. The other eleven sets were voided due to the loss of filter mass during gravimetric sampling (8), SidePak monitor experiencing a battery failure (1), and incorrect flow rates (2).

The linear regression for the 12-hour integrated DustTrak concentrations against the 12-hour gravimetric concentration resulted in a correlation coefficient (r^2) of 0.79, which gives an indication that the DustTrak measurements have a strong positive correlation when compared to the gravimetric concentrations. The linear regression for the 6-hour integrated SidePak concentrations against the 6-hour gravimetric concentration data resulted in a correlation coefficient (r^2) of 0.64, indicating a moderate positive correlation. In addition, an analysis was done of the ratio of the 12-hour integrated DustTrak concentrations and 6-hour integrated SidePak concentrations against the 12- and 6-hour gravimetric concentrations. The median ratio value for the DustTrak is 11.54 (low 3.76, high 31.25) with a standard deviation of ±9.23, while the median ration value for the SidePak is 3.83 (low 1.11, high 19.80) with a standard deviation of ±6.49. An average ratio of 7.32 and 4.16 existed between the DustTrak and SidePak and their respective gravimetric concentration. The ratios for the both these instruments vary dramatically from one day to the next. This may indicate that there is a significant variation in the day-to-day variability within a single household.

The estimated photometric calibration factor for the DustTrak is 0.14 (**95%Cl**: 0.09, 0.15) whereas the SidePak has an estimated calibration factor of 0.24 (95%Cl: 0.16, 0.30). The DustTrak calibration factor is significantly lower than those produced by previous studies. The SidePak calibration factor, while not identical to previous studies is slightly lower. The differences could be due to various aspects such as (1) having a reduced sensitivity when measuring lower PM concentrations (Jimenez et al., 2011), (2) the variations in chemical composition of aerosols and the type of aerosol (Jiang et al., 2011), (3) the difference in density between Arizona test dust and the type of aerosol measured, combustion aerosol tend to have a lower density than the test dust (TIS Inc., 2013), (4) the effect of temperature and relative humidity (McNamara et al., 2011), and (5) the different size fractions associated with aerosols (Yanosky et al., 2002).

Cumulative Distribution

The initial and corrected PM_4 data shows the cumulative exceedances (Table 3) of all WHO and NAAQS standards are similar for both the DustTrak and SidePak instruments. Initial (DustTrak and SidePak) measurements exceed the 75 μm/m³ NAAQS PM_{10} level for 35% and 28% of observed measurements, while the corrected measurements exceed the level for 5% of the observed measurements. The highest level of cumulative exceedances are of the 25 μm/m³ WHO $PM_{2.5}$ level. This level is exceeded for 92% and 100% of the observed measurements, while the corrected measurements exceed the level for 10% and 18% of the observed measurements.

Table 3: Summary of the cumulative exceedances of PM_4 observed measurement concentrations for a one week period

24 Hour (μg/m³)	DustTrak Model 8530		SidePak AM510	
	Initial	**Corrected**	**Initial**	**Corrected**
WHO $PM_{2.5}$	92%	15%	**100%**	18%
WHO $PM_{2.5}$	65%	8%	**68%**	10%
NAAQS $PM_{2.5}$	42%	6%	35%	6%
NAAQS PM_{10}	35%	5%	28%	5%

Conclusion

Due to the linear relationship between negative health effects and increased PM concentrations, this observation indicates that the residents within KwaDela are chronically exposed to high levels of PM_4.

The development of a PM_4 calibration factor for an indoor environment prone to the combustion of solid fuels, such as coal and wood, has implications for both scientific and regulatory studies especially with regard to epidemiological and exposure assessments.

Historically, researchers have made use of averaged 24-hour values to characterise and estimate exposures to PM levels. It is, however, possible to measure exposures over short-term periods by making use of real-time PM monitors. Light-scattering photometer instruments are advantageous to use for monitoring indoor PM_4 concentrations due to the fact that they provide real-time data giving us insight into short-term changes in exposure levels. These instruments are also portable and require that minor maintenance be done periodically, making it easy to deploy within an indoor monitoring network, especially one within a low-income settlement such as KwaDela. The estimation of calibration factors for indoor solid fuel combustion reinforces certainty in studies that utilise these real-time monitoring instruments intended for this purpose.

A specific calibration factor was estimated for the DustTrak (0.14) and SidePak (0.24). These calibration factors should primarily be utilised where DustTrak and SidePak monitoring is conducted to quantify PM_4 exposure within an indoor environment where solid fuel combustion takes place.

References

Braniš M. and J. Hovorka, 2005, 'Performance of a photometer DustTrak in various indoor and outdoor environments', Abstracts of the 2005 Evaluations and Assessment Conference (EAC 2005), Ghent, 2005, Sep. 28–Oct. 10, p. 535.

Bruce N., et al, 2015, 'WHO indoor air quality guidelines on household fuel combustion: Strategy implications of new evidence on interventions and exposure – risk functions', Atmospheric Environment, 106:451-457.

Chung, et al, 2001, 'Comparison of real-time instruments used to monitor airborne particulate matter', Journal of the Air & Waste Management Association, 51:09-120.

Engel-Cox, J., Oanh, N.T.K., van Donkelaar, A., Martin, R.V. & Zell, E., 2013, 'Towards the next generation of air quality monitoring: particulate matter', Atmospheric Environment 80:584-590.

Funk, W.E., Pleil, J.D., Pedit, J.A., Boundy, M.G., Yeatts, K.B., Nash, D.G., Trent, C.B., Sadig, M.E., Davidson, C.A. & Leith, D., 2014, 'Indoor air quality in the United Arab Emirates', Journal of Environmental Protection 5:709-722.

Heal M.R., et al, 2000, 'Intercomparison of five PM_{10} monitoring devices and the implications for exposure measurement in epidemiological research', *Journal of Environmental Monitoring*, 2:455-461.

Jiang, R.T., Acevedo-Bolton, V., Cheng, K.C., Klepeis, N.E., Ott, W.R. & Hildeman, L.M., 2011, 'Determination of response of real-time SidePak AM510 monitor to secondhand smoke, other common indoor aerosols, and outdoor aerosols', Journal of Environmental Monitoring 13:1695-1702.

Jiménez, A.S., van Tongeren, M., Glea, K.S., Steinsvag, K., MacCalman, L. & Cherries, J.W., 2011, 'Comparison of the SidePak personal monitor with the Aerosol Particle Sizer (APS)', *Journal of Environmental Monitoring* 13:1841-1846.

Kim, J.Y., Magari, S.R., Herrick, R.F., Smith, T.J., Christiani, D.C., 2004, 'Comparison of fine particulate measurements from a direct-reading instrument and a gravimetric sampling method', *Journal of Occupational and Environmental Hygiene* 1:707-715.

Kingham S., et al, 2006, 'Winter comparison of TEOM, MiniVol and DustTrak PM_{10} monitors in a woodsmoke environment', *Atmospheric Environment*, 40:338-347.

Lim, S.S. et al. 2012, 'A comparative risk assessment of burden of disease and injury attributes to 67 risk factors and risk factor clusters 13 in 21 regions, 1990-2010: a systematic analysis fir the Global Burden of Disease Study 2010', *Lancet* 380 (9859):2224-2260.

McMurry, P.H., 2000, 'A review of atmospheric aerosol measurements', *Atmospheric Environment* 34:1959-1999.

McNamara, M.L., Noonan, C. & Ward, T.J., 2011, 'Correction factor for continuous monitoring of wood smoke fine particulate matter', *Aerosol and Air Quality Research* 11:315-322.

Ramachandran G. et al, 2000, 'Comparison of short-term variation (15-min averages) in outdoor and indoor $PM_{2.5}$ concentrations', *Journal of the Air & Waste Management Association*, 50: 1157 to 1166.

Sensidyne, 1999, 'Personal Cyclone Sampler' – GIL 2320 Rev B 9901.

South Africa., 2009, 'National Environmental Management: Air Quality Act, 2004 (Act No. 39 of 2004) – National Ambient Air Quality Standards', (Notice 1210). *Government Gazette*, 32816:6, 24 Des.

South Africa., 2012, 'National Environmental Management: Air Quality Act, 2004 (Act No. 39 of 2004) – National Ambient Air Quality Standards for particulate matter with aerodynamic diameter less than 2.5 micron metres', (Notice 1210). *Government Gazette*, 35463, 7, 29 Des.

Tasić, V., Jovašević-Stojanović, M., Vardoulakis, S., Milošević, N.,

Kovačevicć, R. & Petrović, J., 2012, 'Comparative assessment of a real-time particle monitor against the reference gravimetric method of PM_{10} and $PM_{2.5}$ in indoor air', *Atmospheric Environment* 54:358-364.

Thorpe, 2007, 'Assessment of personal direct-reading dust monitors for the measurement of airborne inhalable dust', *Ann. Occup. Hy*, 51(1):97 to 112.

TSI Inc., 2002, 'SidePak Personal Aerosol Monitor Model AM510: User Guide'

TSI Inc., 2013, 'Rationale for programming a photometer calibration factor (PCF) of 0.38 for ambient monitoring', Application note EXPMN-007.

TSI Inc., 2014, 'DustTrak II Aerosol Monitor Model 8530/8531/8532/8530EP: Operation Manual'

Tung T.C.W., Chao C.Y.H, and Burnett J., 1999, 'A methodology to investigate the particulate penetration coefficient through building shell', *Atmospheric Environment*, 33:881 to 893.

Wallace L.A., et al, 2011, 'Validation of continuous particulate monitors for personal, indoor, and outdoor exposures', *Journal of Exposure Science and Environmental Epidemiology*, 21: 49 to 64.

World Health Organization, 2006, 'WHO air quality guidelines for particulate matter, ozone, nitrogen dioxide and sulfur dioxide: Global update 2005. 15'.

World Health Organisation, 2014, 'Burden of disease from the joint effects of Household and Ambient Air Pollution for 2012'.

Yanosky, J.D., Williams, P.L. & MacIntosh, D.L., 2002, 'A comparison of two-direct reading aerosol monitors with the federal reference method for $PM_{2.5}$ in indoor air', *Atmospheric Environment* 36:107-113.

Yassin, M.F., AlThaqeb, B.E.Y. & Al-Matiri, E.A.A., 2012, 'Assessment of indoor $PM_{2.5}$ in different residential environments', *Atmospheric Environment* 56:65-68.

Zhu K., Zhang J. and Lioy P.J., 2007, 'Evaluation and comparison of continuous fine particulate matter monitors for measurement of ambient aerosols', *Journal of the Air & Waste Management Association*, 57: 1499 to 1506.

8

Influence of coal-particle size on emissions using the top-lit updraft ignition method

Lethukuthula Masondo*[1], Daniel Masekameni[1,2,3], Tafadzwa Makonese[1,2], Harold J Annegarn[4], Kenneth Mohapi[1]

[1]Department of Environmental Health: University of Johannesburg
[2]SeTAR Centre, Faculty of Engineering and the Built Environment
[3]Department of Geography, Environmental Management and Energy Studies, Faculty of Science, University of Johannesburg, PO Box 524, Auckland Park 2006, Johannesburg
[4]Energy Institute, Cape Peninsula University of Technology, PO Box 652, Cape Town, 8001

Abstract

Despite the Government's intervention of an intensive electrification program in South Africa, which has resulted in more than 87% of households being connected to the grid, a majority of low-income households still depend on solid fuel (coal and wood) as a primary source of energy, especially on the central Highveld. In informal settlements, combustion of coal is done in inefficient self-fabricated braziers, colloquially known as imbaulas. Emissions from domestic coal combustion result in elevated household and ambient air pollution levels that often exceed national air quality limits. Continued dependence on coal combustion exposes households to copious amounts of health-damaging pollutants. Despite the health significance of coal-burning emissions from informal braziers, there is still a dearth of emissions data from these devices. Consequently, evaluating the emission characteristics of these devices and to determine the resultant emission factors is needed. The effects of ignition methods and ventilation rates on particulate and gaseous emission from coal-burning braziers are reported in literature. However, to date there are no studies carried out to investigate the influence of the size of coal pieces on brazier emission performance. In this paper, we report on controlled combustion experiments carried out to investigate systematically, influences of coal particle size on gaseous and condensed matter (smoke) emissions from informal residential coal combustion braziers. Results presented are averages of three identical burn-cycles of duration three hours or fuel burn-out, whichever was the soonest.

Keywords

coal particle size; brazier; imbaula, emission factor, smoke

Introduction

Energy is an important factor for economic growth, community development and sustenance of life in South Africa (Masekameni et al., 2014). Globally, more than 3 billion people rely on solid fuels combusted in open fires or traditional stoves, for purposes of cooking and space heating (Smith et al., 2012). Emissions from solid fuels account for 4.3 million deaths per year globally (Gordon et al., 2014). These deaths are more common in developing countries with South Africa being no exception (WHO, 2012). It is argued that these deaths can be reduced by use of cleaner burning cook stoves and the introduction of efficient ignition methods that lead to low emissions (Masekameni et al., 2015).

More than half of the South African population still depend on coal and wood for cooking and space heating needs (Kimemia et al., 2011). In the low to medium economic stratum, these fuels are burnt in inefficient stoves and open fires that do not allow for complete combustion, thus impacting on human and environmental health (Kimemia et al., 2011). In the central Highveld of South Africa, a majority of low-income households burn coal and wood in self-fabricated braziers known colloquially as imbaulas, which are constructed from discarded steel drums (Kimemia, 2010). The braziers have holes punched around the sides to provide primary air needed for combustion. These devices are used extensively in winter resulting in severe indoor and ambient air pollution (Makonese et al., 2015). The use of poor quality coal in these devices, results in high emissions of gaseous and particulate matter (Makonese et al., 2014a). Coal fuel commonly used in informal settlements of South Africa is the untreated bituminous coal with ash content of up to 40% and with energy content between 15 and 25 MJ/kg (Annegarn & Sithole, 1999; Pemberton-Pigott et al., 2009).

Smith et al (2014) contends that the continued use of coal in poorly ventilated and inefficient stoves leads to increased exposure to health damaging pollutants and the development of respiratory diseases. Bruce et al (2000) found that 1300 lives, amongst children below the age of five are lost each year due to excessive inhalation of particulate matter from domestic coal combustion between the years 1995-2000 in developing countries. Annegarn and Sithole (1999) reported that these stoves lack performance improvements resulting in increased emissions of particulates and gases.

To reduce emissions of noxious gases and particulate matter, existing or new braziers should be optimised. Optimisation of solid fuel burning appliances should consider stove design parameters (i.e. number and distribution of holes above and below the fire grate, the position of the grate in the bucket), fuel characteristics (such as fuel particle size and fuel quality), and operational practices (including ignition methods and fire tending practices) (Makonese, 2015; Masekameni et al., 2014). For example, the top-lit updraft (TLUD) ignition method has been reported to be an effective way of igniting a fire in a coal brazier. A coal fire ignited using the top-lit updraft (TLUD) method produces less visible smoke compared to a fire ignited using the conventional/traditional method. This TLUD ignition method has become a national priority energy intervention programme due to its estimated 80% reduction in ambient particulate air pollution and 20% reduction in coal use at no additional cost (Makonese et al., 2014b).

The effects of ventilation rates (as a function of the size and distribution of holes around the brazier), coal quality, and ignition methods on emissions of gases and particulates from coal burning braziers are reported in literature (Makonese, 2015; Makonese et al., 2014b). However, there are still limited studies carried out to investigate the influence of the size of coal pieces on brazier emission performance.

In light of the above, this study aims to investigate the influence of coal particle size on the emissions performance of coal-burning braziers using the top-lit updraft method. In this paper, three coal particle sizes are evaluated for emission factors of carbon monoxide (CO_{EF}), CO/CO_2 ratio and $PM_{2.5}$ emission factors, using a high ventilation laboratory designed brazier.

Methodology

Fuel Preparation

D-grade coal from Slater mine in Mpumalanga was chosen for our experiments. The fuel is preferred by local coal merchants. Three different coal particle sizes ranges 20 – 40 mm (small), 40 – 60 mm (medium) and 60 – 80 mm (large) were used to investigate emissions performance in a high ventilation rate (i.e. measured as a function of the number and size of air holes on the sides of the brazier) brazier. Large coal nuggets were crushed into small pieces before sieving them through a 20 x 40 mm wire sieve for the 20 – 40 mm size range for small coal size.

For medium size, coal was sieved through a 40 x 60 mm wire mesh, while for large coal size a 60 x 80 mm sieve was used. In order to ensure that the correct sizes were obtained in each category, the technicians checked the dimensions of a sample of already sieved individual coal particles using a ruler. 4 000 g of selected coal fuel were used for each size category.

Coal analysis

The coal was characterized for thermal content, major elemental (proximate) analysis, moisture and ash content by an independent laboratory (Bureau Veritas Inspectorate Laboratories (Pty) Ltd). The fuel samples were analysed on an air-dried basis. Experimental results presented in this paper are based on the proximate and ultimate analysis results for the D-grade coals used in making the fires. Fuel specifications used during the experiments are provided in Table 1.

Table 1: *Fuel analysis*

Parameter (Air Dried Basis)	Standard Method	Slater Coal D-Grade
Moisture content (%)	ISO 5925	3.5
Volatiles (%)	ISO 562	20.3
Ash (%)	ISO 1171	24.2
Fixed carbon (%)	By difference	52.0
Calorific value (MJ kg^{-1})	ISO 1928	23.4
Calorific value (Kcal kg^{-1})	ISO 1928	5590
Total sulphur (%)	ASTM D4239	0.63
Carbon (%)	ASTM D5373	62. 6
Hydrogen (%)	ASTM D5373	2.72
Nitrogen (%)	ASTM D5373	1.43
Oxygen (%)	By difference	4.96

Choice of fire-ignition methods

The TLUD ignition method was used to investigate the influence of coal particle size on emissions of carbon monoxide and particulate matter, and the CO/CO_2 ratio.

The order of laying a fire during a top-lit ignition fire entailed the following: first, placing the major portion of the coal load on the support grid in the brazier, then paper and wood kindling, with a few lumps of coal added at an appropriate time after the fire has been lit. In our experiments, 3 000 g of coal was added to the bottom of the brazier onto the fuel grate. 36 g of paper and 450 g of kindling were added. After igniting the kindling, ~1 000 g of coal was added to the brazier above the kindling.

Tests were conducted under controlled laboratory conditions, keeping parameters such as ignition method and ventilation rates constant.

Stove characterisation

The brazier used in our experiments is shown in Figure 1. The brazier has a fuel support grate, made of wire although it is common to have some braziers operated without a fire grate. With a fire grate in place the rate of burning is increased. It should be noted that there is no standard brazier, as the devices vary greatly in terms of the number and sizes of the side holes (i.e. affecting ventilation rates), the presence of a grate and its position in the metal drum (Kimemia et al., 2011).

Ventilation rates affect the overall performance of the stove and these rates differ significantly from one device to the other. To evaluate realistically and compare the performance of two or more braziers, ventilation rates need to be specified. Ventilation rates for the experimental brazier used in the study are given in Table 2.

Figure 1: Brazier ignited using the TLUD ignition method

Table 2: Stove description

Brazier type	Height (mm)	Dia. (mm)	Grate height (mm)	Area of holes below grate (cm²)	Area of holes above grate (cm²)
High ventilation	370	290	185 (50%)	248 (61%)	159 (39%)

Test apparatus

The hood method was used for evaluating emissions (Ahuja et al., 1987). The gas samples were analysed using two Testo® flue gas analysers model (Testo® 350XL/454). The sampling configuration for the undiluted flue gases included, in sequence, a stainless steel tube, a filter holder, and a flue gas analyser. For the diluted channel, the sampling configuration included, in sequence, the dilution system, a Teflon tube channel, and a second Testo® flue gas analyser. Traditional coal stoves (*Imbaulas*) emit high levels of particulate emissions; therefore, the dilution system was used to maintain aerosols emissions within the detection limit of the instrument (150 mg/m³). The Testo® measures Carbon dioxide (CO_2), Carbon monoxide (CO), Nitrogen oxides (NO_x), Nitrogen dioxide (NO_2), Hydrogen (H_2), Sulphur (S), Sulphur dioxide (SO_2) and Oxygen (O_2).

Particle mass concentrations and size segregated mass fraction concentrations were monitored using a DustTrak DRX 8533 aerosol monitor. The DustTrak DRX Model 8533 is a laser-based instrument that measures size fractions of the sampled aerosol, from which mass fractions are deduced. The instrument

simultaneously measures size segregated mass fraction concentrations (i.e. PM_1, $PM_{2.5}$, PM_4, PM_{10}, and Total Particle Mass - TPM) over a wide concentration range (0.001–150 mg/m³) in real time. Data points were recorded at 10 s intervals.

Testing apparatus

A schematic description of the sampling train is shown in Figure 2. Section A in the schematic show the mixing point, where raw exhaust sample mixes with compressed air. The diluted sample was drawn at point B. A raw exhaust sample was collected at point C.

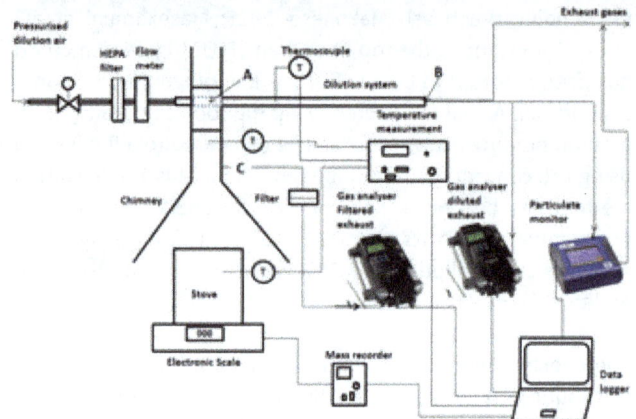

Figure 2: Schematic illustration of the experimental dilution set-up for the SeTAR dilution system, showing the mixing point (A) and the sampling diluted exhaust gas point (B) and sampling point (C) is for raw exhaust gases.

Quality control

For each fuel/stove combination, a series of preliminary burn cycles were carried out to standardise procedures and to minimize the natural variability due to differences in operator behaviour. To familiarise the operators with the testing procedure and with the characteristics of the stove, these trial runs were conducted repeatedly until a stable mode of operation was established. Thereafter three definitive tests were conducted for each fuel/stove combination.

After each fuel/stove combination was tested, the probes were cleaned and the pumps and machines checked and zeroed (Makonese, 2015). Before tests were conducted, the sampling dilution system components were cleaned, assembled, and tested for leaks to prevent contamination from the surrounding air.

A calibration exercise was performed to check the accuracy of the flow rates through each of the critical flow orifices. The sampling dilution system was cleaned prior to testing to minimize pre-existing organic and metal compounds, including the use of high power compressed air and water to remove large particles.

The collection trains, including the stainless steel piping and sampling nozzles, were cleaned with soap, water-rinsed and then air-dried with compressed air. The dilution system was then assembled and connected to the testing rig for a trial run of the tests.

Results

CO emission factors

Carbon monoxide emission factors (CO_{EF}) of the three coal particle sizes results are presented pairwise to compare between coal particle sizes in Table 3). Differences between CO_{EF} pairs are tested for significance using a student T-test, to indicate whether changes in coal particle size result in a significant difference in the emission factor.

A change in coal particle size did not cause a significant difference on the CO_{EF} for the small (20 – 40 mm and medium (40 – 60 mm) particle sizes. However, there is a significant difference in CO_{EF} between medium (40 – 60 mm) and large (60 – 80 mm) coal particles sizes. Results show that there also are significant differences in CO_{EF}, at the 95% confidence level, between small (20 – 40 mm) and large (60 – 80 mm) coal particle sizes. The CO_{EF} for large coal pieces is about three fold higher than for medium and low pieces.

Table 3: CO_{EF} of three coal particle sizes in a high ventilation stove using TLUD ignition method

Coal particle size	Ignition method	CO_{EF} (g/MJ) (n = 3)	Statistical analysis		
			F-Test	P-Value	Sig @ 95%
Medium vs Small	TLUD	1.6 ± 0.09 1.5 ± 0.04	0.31	0.07	No
Large vs Medium	TLUD	4.3 ± 0.22 1.6 ± 0.09	0.31	<0.01	Yes
Large vs Small	TLUD	4.3 ± 0.22 1.5 ± 0.04	0.07	<0.01	Yes

CO/CO$_2$ ratio over the Burn Cycle

Results of the influence of coal particle sizes on CO/CO_2 ratio are shown in Table 4, for the three different coal particle sizes. There are no significant differences in the CO/CO_2 ratio between small and medium coal particle sizes. However, pairwise comparison between medium and large coal particle sizes indicated a significant difference at the 95% confidence level.

Increasing coal particle size ranges from 20 – 40 mm to 60 – 80 mm, leads to an increase in the CO/CO_2 ratio by ~ 65%. The nominal combustion efficiency is reduced from 97.5% to 92.6%. These results are expected – larger coal particle sizes burn poorly relative to small and medium coal particle sizes.

Table 4: Pairwise comparison by coal particle size of average CO/CO_2 ratio over the burn cycle

Coal particle size	Ignition method	CO/CO_2 ratio [%] (n = 3)	Statistical analysis		
			F-Test	P-Value	Sig @ 95%
Medium vs Small	TLUD	2.5 ± 0.51 2.8 ± 0.22	0.29	0.39	No
Large vs Medium	TLUD	7.4 ± 0.66 2.5 ± 0.51	0.75	<0.01	Yes
Large vs Small	TLUD	7.4 ± 0.66 2.8 ± 0.22	0.19	<0.01	Yes

PM$_{2.5}$ Emission Factors

Results of pairwise comparison of average PM$_{2.5}$ emission factors over the burn cycle of the three coal particle sizes are presented in Table 5. There are no significant differences in PM$_{2.5}$ emission factors between medium and small coal particle sizes. Pairwise comparison between large and medium coal particle sizes resulted in a significant difference in PM$_{2.5}$ emission factors. A similar result is observed between large and small coal particle sizes. Reducing coal particle size ranges from 60–80 mm to 20–40 mm leads to a 50% reduction in PM$_{2.5}$ emission factors.

Table 5: Pairwise comparison by coal particle size of average $PM_{2.5EF}$ over the burn cycle

Coal particle size	Ignition method	Avg. PM$_{2.5}$ (g/MJ) (n = 3)	Statistical analysis		
			F-Test	P-Value	Sig @ 95%
Medium vs Small	TLUD	0.31 ± 0.02 0.42 ± 0.06	0.29	0.39	No
Large vs Medium	TLUD	0.75 ± 0.04 0.31 ± 0.02	0.75	<0.01	Yes
Large vs Small	TLUD	0.75 ± 0.04 0.42 ± 0.06	0.19	<0.01	Yes

PM$_{2.5}$ concentration time series plots over the burn cycle

Time series plots of PM$_{2.5}$ concentrations are shown in Figure 2, for the three coal particle sizes. All three-coal particle sizes experienced high peaks at ignition as the kindling burned and consequently ignited the coal. The PM$_{2.5}$ concentration drops sharply within a few minutes, and then peaks again during pyrolysis phase.

The small and medium coal particle size indicates an earlier ignition of the main fuel bed relative to larger coal size range, indicated by lower PM$_{2.5}$ emissions 30 minutes after ignition (Figure 3). The largest coal size bed (with 80 mm coal particles) takes over 90 minutes to drive off most of the PM and is characterised by an unsteady burn rate. This result is similar to that of Yang et al (2005) who reported that a larger particle-size bed tends to burn more transiently compared to a smaller particle-size bed, which tends to quickly build up to a steady burn pattern. This suggests that the control of a brazier burning larger coal particle sizes needs to be more carefully planned because of the constant variation of the burn pattern as a function of the fuel size (Yang et al., 2005).

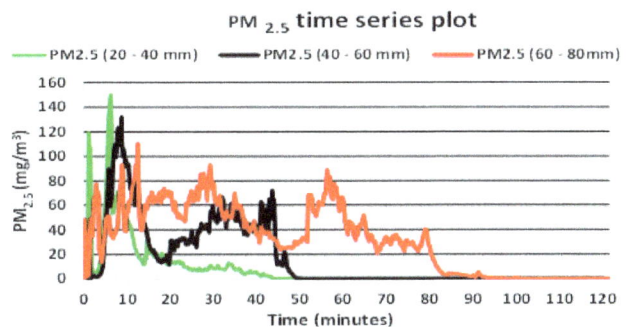

Figure 3: PM$_{2.5}$ time series plots for the three coal particle sizes in a high ventilation stove

In summary, evidence presented shows that coal particle size ranges have an influence on gaseous and particulate emissions from coal braziers. Large coal particle sizes result in poor combustion efficiencies and increased CO and $PM_{2.5}$ emissions. It is recommended that coal particle size ranges of between 20–40 mm be used for optimal brazier performance.

Significance and conclusion

In general, the following conclusions can be drawn from this study. Particulate and CO emission factors increase with an increase in the mean size range of the fuel. Small and medium coal particle sizes produced comparable emissions (CO and $PM_{2.5}$) and CO/CO_2 ratio. The ignition time and the time to reach full pyrolysis are shorter with a bed of smaller particles compared to a bed of larger particles, when the devices are operated under the same conditions. Small coal particle size ranges presented a uniform flame propagation speed for most parts of the combustion process, while large particles showed a less stable transient features where the burning rate, although lower compared to small coal particle sizes, fluctuates throughout the combustion process.

If these results are validated by further testing using stoves with medium and low ventilation rates, as found within the range of artisan manufactured braziers, it would imply that pollution reductions can be achieved by supplying a regulated graded coal size, in the range 20–40 mm, to the domestic coal market. However, small coal particles burn quicker and therefore more of the fuel will be required to complete any given task, leading to an increased financial burden on the user.

Acknowledgements

This work was supported financially by the University of Johannesburg through a URC/Faculty of Science grant to the SeTAR Centre, and in part from a grant from the Global Alliance for Clean Cookstoves (GACC) to the SeTAR Centre as a Regional Stove Testing and Development Centre. The authors thank Gumede K, Gxabuza N, Shabangu M and Maseki J, for their assistance with the combustion experiments.

References

Ahuja, D.R., Veena, J., Smith, K.R., Venkataraman, C. 1987. Thermal performance and emission characteristics of unvented biomass-burning cook-stoves. A proposed standard method for evaluation, *Biomass*, 12, 247-270.

Annegarn, H.J, Sithole, S. 1999. *"Soweto Air Monitoring Project (SAM), Quarterly Report to the Department of Minerals and Energy"*, Report No. AER 20.001 Q-SAM, University of Witwatersrand, Johannesburg.

Balmer, M. 2007. Household coal use in an urban township in South Africa, *Journal of Energy in Southern Africa*, 18(3), 27-32.
Bruce, N., Perez-Padilla, R., Albalak, R. 2000. Indoor air pollution in developing countries: A major environmental and public health challenge, *Bulletin of the World Health Organization*, 78, 1078-1092.

Gordon, S. B., Bruce, N. G., Grigg, J. et al., 2014. Respiratory risks from household air pollution in low and middle-income countries, *The Lancet Respiratory Medicine Commission*, 2(10). P 832-860.

Kimemia, D., Annegarn, H., Robinson, J., & Pemberton-Pigott, C. 2011. Optimising the Imbaula stove. *Proceedings of the Domestic Use of Energy (DUE) Conference*, 1 April 2011, Cape Town.

Kimemia, K.D., 2010. Biomass Alternative Urban Energy Economy, The ventilation of Setswetla, Alexandra Township, Gauteng. MSc thesis, University of Johannesburg, Johannesburg.

Makonese, T. 2015. Systematic investigation of smoke emissions from packed bed residential coal combustion devices, PhD Thesis, University of Johannesburg, South Africa.

Makonese, T., Masekameni, D., Annegarn H., & Forbes, P. 2015. Influence of Fuel-Bed temperatures on CO and condensed matter emissions from packed-bed residential coal combustion. *Proceedings of the Domestic Use of Energy (DUE) Conference*, 31 March – 2 April 2015, Cape Town.

Makonese, T., Forbes, P., Mudau, L., & Annegarn, .H. J. 2014a. Aerosol particle morphology of residential coal combustion smoke, *The Clean Air Journal*, 24(2), 24–28.

Makonese, T., Masekameni, D., Annegarn, H., Forbes, P., Robinson, J. & Pemberton-Pigott, C. 2014b. Domestic lump coal combustion: Characterization of performance emissions from selected braziers. *Proceedings of the Twenty-Second Domestic Use of Energy (DUE) Conference*, 30 March–2 April 2014, Cape Town. IEEE Conference Publications, DOI: 10.1109/DUE.2014.6827755, 2014, pp. 1-7.

Masekameni, M.D., Makonese, T., Annegarn H.J. 2014. Optimisation of ventilation and ignition method for reducing emissions from coal-burning imbaulas. *Proceedings of the Twenty-Second Domestic Use of Energy (DUE) Conference*, 30 March–2 April 2014, Cape Town. IEEE Conference Publications, DOI: 10.1109/DUE.2014.6827755, 2014, pp. 1-7.

Masekameni, M.D., Makonese, T., Annegarn H.J. 2015. A Comparison of emissions and thermal efficiency of three improved liquid fuel stoves. *Proceedings of the Twenty-Third Domestic Use of Energy (DUE) Conference*, 31 March–3 April 2015, Cape Town. IEEE Conference Publications, DOI: 10.1109/DUE.2015.7102965, 2015, pp. 71-76.

Pemberton-Pigott, C, Annegarn, H. & Cook, C. 2009. Emissions reductions from domestic coal burning: Practical application of combustion principles. *Proceedings of the Domestic Use of Energy (DUE) Conference*, 2 April 2009, Cape Town.

Smith, K.R., Balakrishnan, K., Butler, C., Chafe, Z., Fairlie, I., & Kinney, P. 2012. Energy and health. In: *Global Energy Assessment: toward a Sustainable Future*, (Johansson T.B., Patwardhan A, Nakicenovic N, Gomez-Echeverri L, eds). New York: Cambridge University Press, 255–324.

Smith, K.R., Bruce, N., Balakrishnan, K., Adair-Rohani, H., Balmes, J., Chafe, Z. 2014. Millions dead, how do we know and what does it mean? Methods used in the Comparative Risk Assessment of household air pollution, *Annual Review of Public Health*, 35,185–206.

WHO. 2012. Indoor air pollution: Household indoor air pollution (online). Available from: http://www.who.int/indoorair/en/ [Accessed 15 May 2015].

Yang, B.Y, Ryu, C., Khor, A., Sharifi, V.N., Swithenbank, J. 2005. Fuel size effect onpinewood combustion in a packed bed. *Fuel*, 84: 2026-2038.

9

Assessment of ambient air pollution in the Waterberg Priority Area 2012-2015

Gregor T. Feig[1], Seneca Naidoo[2], Nokulunga Ncgukana[3]

South African Weather Service, 442 Rigel Ave South, Erasmusrand, Pretoria, South Africa
[1]Now at the Council for Scientific and Industrial Research, gfeig@csir.co.za
[2]Now at the Council for Scientific and Industrial Research, SNaidoo8@csir.co.za
[3]lunga.ncgukana@weathersa.co.za

Abstract

The Waterberg Priority Area ambient air quality monitoring network was established in 2012 to monitor the ambient air quality in the Waterberg Air Quality Priority Area. Three monitoring stations were established in Lephalale, Thabazimbi and Mokopane. The monitoring stations measure the concentrations of PM_{10}, $PM_{2.5}$, SO_2, NO_x, CO, O_3, BTEX and meteorological parameters. Hourly data for a 31 month period (October 2012-April 2015) was obtained from the South African Air Quality Information System (SAAQIS) and analysed to assess patterns in atmospheric concentrations, including seasonal and diurnal patterns of the ambient concentrations and to assess the impacts that such reported pollution concentration may have. Local source regions for SO_2, PM_{10}, $PM_{2.5}$ and O_3 were identified and trends in the recorded concentrations are discussed.

Keywords

Waterberg Priority Area, Air Pollution, PM_{10}, $PM_{2.5}$, SO_2, O_3

Introduction

The Waterberg coal fields extend across the border between South Africa and Botswana. These coal fields are estimated to hold a reserve of approximately 6Gt (6×10^9 Mg). This coal reserve is regarded as the last remaining large coal resource in the country (Hartnady 2010). As such, in the National Development Plan (National Planning Commission 2010), the Waterberg coal fields have been earmarked for further industrial development, related to exploitation of the coal resource for *inter-alia* power generation.

Due to the expected development within the Waterberg Area and the existing mining and metallurgical activities in the western arm of the bushveld igneous complex (Venter et al. 2012), there is concern regarding the future and current air quality in these regions. As a result the Waterberg District Municipality (Limpopo Province) and the Bojanala Platinum District Municipality (North West Province) were declared as the Waterberg Priority Area in 2012 since the Minister of Environmental Affairs expected the levels of pollutants in the Waterberg District to exceed the National Ambient Air Quality Standards (NAAQS) (DEA 2009) in the near future and that a significant trans-boundary situation exists between the Waterberg District Municipality in Limpopo Province and the Bojanala Platinum District Municipality in the North-West Province (Department of Environmental Affairs 2012) (Figure 1). A draft air quality management plan has been developed which details the major pollutant sources and

receptors (Department of Environmental Affairs 2014).

A number of studies in ambient air pollution have been conducted in the past. A passive sampling network was operated over the northern parts of South Africa between 2005 and 2007 (Josipovic et al. 2010). This study reported occasional high concentrations of SO_2 at Thabazimbi and Mokopane, The source of the SO_2 at Thabazimbi was attributed to the Thabazimbi iron ore mine and smelter, the Matimba power station and the Grootgeluk coal mine in Lephalale. The elevated levels of SO_2 at Mokopane were attributed to the Mokopane platinum mine and the Matimba power station.

A long term air quality measurement campaign was conducted in the Bojanala District of the Waterberg Priority area (Venter et al. 2012). This campaign took place at Marikana from February 2008 to May 2010. In the Venter et al. 2012 study it was reported that the concentrations of NO_x, SO_2 and CO were within the NAAQS but there were significant exceedances of the standards for ozone and particulate matter. It was also within this region that Hirsikko et al. 2012 reported a high frequency of new particle formation events and an average particle number concentration of 10^4/cm³. They further postulated that SO_2 was the likely feed material for the new particle formation.

Due to the known and expected future exceedances of the NAAQS in the Waterberg Priority Area, it was necessary to establish an ambient air quality monitoring network, to monitor the ambient concentrations of criteria pollutants in the area. This study examines the first two and a half years of the monitoring results for the three ambient air quality stations in the Waterberg Priority area. This study intends to characterise the spatial and temporal patterns in criteria pollutant concentrations, and to identify potential pollutant sources.

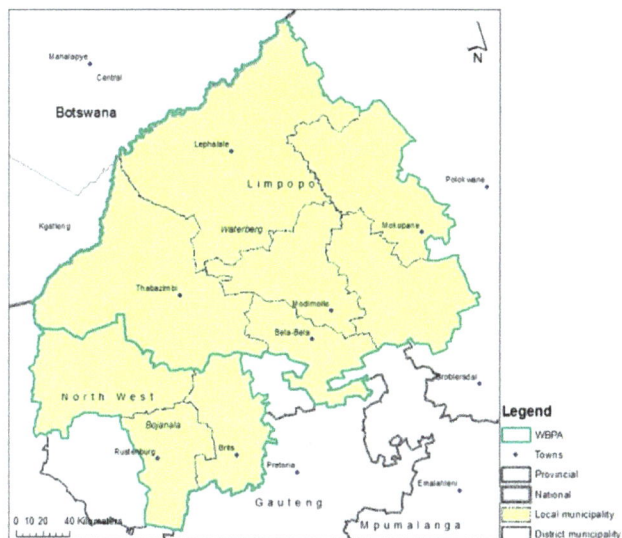

Figure 1: *Waterberg Priority Area (Department of Environmental Affairs 2014)*

Methods

Three ambient air quality monitoring stations were established in the Waterberg Priority Area in October 2012. The stations are located at Thabazimbi, Lephalale and Mokopane (Figure 1).

Each of the stations is fully equipped to monitor the following parameters at a temporal resolution of 1 minute:

- Sulphur Dioxide (SO_2)
- Particulate matter of aerodynamic diameter >10 um (PM_{10})
- Particulate matter of aerodynamic diameter > 2.5um ($PM_{2.5}$)
- Oxides of Nitrogen ($NO_x = NO + NO_2$)
- Ozone (O_3)
- Carbon Monoxide (CO)
- VOCs (Benzene, Toluene, Ethyl benzene, Xylene)
- Meteorological Parameters
  - Wind Speed
  - Wind direction
  - Pressure
  - Temperature
  - Relative Humidity
  - Solar Radiation
  - Rainfall

Data from the monitoring stations reports in real time to a server located at the South African Weather Service. On a monthly basis the data was assessed and validated to remove spikes and calibration data and to adjust drifts in the data.

Daily checks were performed on the stations by remotely accessing the data logging system, if problems with the station were identified non-routine maintenance was carried out. On a bi-weekly basis routine site visits were conducted to ensure that the instrumentation was functioning and to perform a zero and span (80% of instrument maximum) check using NMISA certified calibration gases to ensure that the drifts and deviations of the instrument were within the specified ranges, if the instrumentation was found not to respond within the data quality criteria the data was flagged and corrected or removed from the dataset. On a quarterly basis multipoint calibration verifications (zero, 80% of instrument range and three intermediate points) were conducted utilizing NMISA certified reference gases. A full calibration of all the instruments was conducted on an annual basis by a South African National Accreditation System (SANAS) accredited calibration laboratory.

Table 1: *Station location and monitoring purpose*

Latitude	Longitude	Data Recovery October 2012-April 2015	Monitoring purpose
Lephalale			
S23.681	E27.722	85.0%	Impact on human health in a low income residential community impacted by domestic combustion, vehicular emissions, biomass burning, the Grootgeluk coal mine and the Matimba power station. Future impacts from Medupi coal fired power station are expected.
Thabazimbi			
S24.591	E27.391	80.8%	Located in a low income residential community, domestic combustion, biomass burning, vehicular and road emissions and iron ore mining activity.
Mokopane			
S24.155	S24.155	85.9%	Human health impacts in a low income residential area, impacted by domestic combustion, vehicular and road emissions and small scale industry as well as biomass burning.

For this study hourly averaged data for SO_2, PM_{10}, $PM_{2.5}$ and O_3 were extracted from the South African Air Quality Information System (SAAQIS) and revalidated to remove negative concentrations and data spikes that were not removed during the original validation. A data completeness rule of 80% was used for data averaging to the hourly average that was utilised and any subsequent averaging. The data was assessed using the Open Air Package in R (Uria-Tellaetxe and Carslaw 2014; Carslaw 2014).

Results and Discussion

The monitoring data from the three monitoring stations were assessed to characterise the atmospheric dynamics of the site, including the diurnal and seasonal cycles and the potential pollutant sources. The results are presented for the four criteria pollutants of greatest concern in South Africa, namely sulphur dioxide (SO_2), particulate matter (PM_{10} and $PM_{2.5}$) and ozone (O_3) (Thompson et al. 2011; Lourens et al. 2011; Venter et al. 2012).

Sulphur Dioxide

The concentration of SO_2 at the monitoring stations is presented in Table 2. The ambient SO_2 concentrations over the period are low with mean values in the range of 1-1.5ppb. The 90 percentile at all the stations does not exceed 5ppb. In comparison the annual National standard is 19ppb. Compliance with the national standards is presented in the National State of the Air Report.

Table 2: SO_2 measurement summary

	Lephalale	Makopane	Thabazimbi
N	21280	20053	21250
% recovery	95.1%	86.8%	95.0%
Mean	2.19	1.68	2.14
Median	0.82	0.97	1.11
10 percentile	0.278	0.45	0.38
25 percentile	0.48	0.64	0.63
75 percentile	1.7	1.82	2.25
90 percentile	3.30	3.56	4.52

The relation between the concentration of SO_2 and wind speed and direction is presented in Figure 3. The Lephalale plot shows a hotspot of SO_2 associated with low to medium speed winds from the westerly and north westerly sectors. This corresponds to the location of the Matimba power station and the Grootgeluk coal mine which are likely the source of the SO_2.

The monthly trend of the SO_2 concentrations at the monitoring stations is presented in Figure 4. Of the three stations only Mokopane showed a statistically significant trend in the SO_2 concentrations. At Mokopane the SO_2 concentration decreased at a rate of 0.3ppb per year (P> 0.001). The decrease at the Mokopane station appears to have occurred from late 2013 or

early 2014. The SO_2 concentrations at Lephalale and Thabazimbi did not show any statistically significant trends.

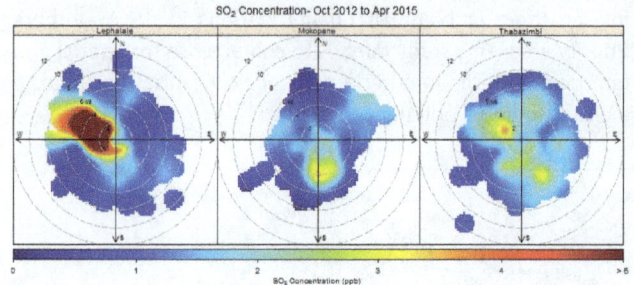

Figure 2: Polar Plot of the SO_2 concentration at the Waterberg monitoring stations

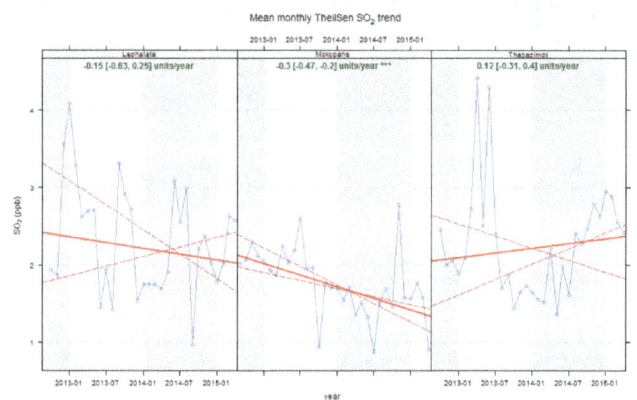

Figure 3: Monthly Mean Deseasonalised SO_2 Trend, for the period Oct 2012-April 2015. The solid trend line represents the mean slope of the trend, while the dashed trend lines represent the 95% confidence interval of the slope

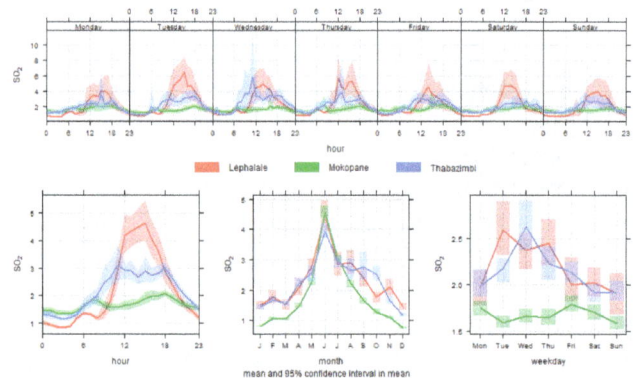

Figure 4: SO_2 Time variation plot for the period October 2012 to April 2015

The time variation in the hourly SO_2 concentrations is presented in Figure 5. It is shown that there is a strong diurnal profile in Lephalale and Thabazimbi, where there is a peak in the SO_2 concentrations during the day; this is typical of sites influenced by pollution from industrial stack emissions, which is brought to the surface during periods of high convection (Venter et al. 2012; Zhou et al. 2012) For the Lephalale site, the mid-day peak occurs throughout the week, which would correspond to a high level source that is emitting continuously. For the Thabazimbi site the largest SO_2 peak occurs on the Wednesday morning and may indicate a specific process that occurs on a weekly basis at one of the facilities in the area. All three sites show a strong seasonal profile in the SO_2 concentrations, where the highest values are recorded during the winter months.

PM_{10}

The average PM_{10} concentration at the three Waterberg stations for the period October 2012- April 2015 was 52µg/m³ for Thabazimbi, 40.6µg/m³ for Mokopane and 26µg/m³ for Lephalale (Table 3). The 90th percentile for the Thabazimbi and Mokopane stations is higher than 100µg/m³. The average PM_{10} concentration over the 2.5 year measurement period (October 2012 to April 2015) is greater than the annual PM_{10} standard (40µg/m³).

Table 3: PM_{10} measurement summary

	Lephalale	Makopane	Thabazimbi
Number of measurements	20739	19608	15027
% recovery	92.68%	84.8%	67.2%
Mean	26.04	40.64	52.31
Median	19.11	26.93	30.93
10 percentile	5.94	8.60	8.01
25 percentile	11.03	14.88	16.52
75 percentile	32.95	50.64	59.46
90 percentile	53.37	87.48	115.62

When the relation between the PM_{10} concentrations and the wind speed and wind direction are considered (Figure 7) the periods of highest PM_{10} concentration correspond to periods of high wind speed, typically greater than 6m/s. These are periods when the high PM_{10} concentrations are likely to be attributable to the generation of windblown dust. Marticorena & Bergametti (1995) modelled the generation of aeolian dust based on threshold friction velocities, or the point at which the wind speed is high enough to entrain soil particles from the surface. This approach has successfully been used to model dust generation in other regions (Kocha et al. 2011; Schmechtig et al. 2011).

The trend in the PM_{10} concentrations seems to be decreasing in all three sites (Figure 8) where there is a decrease in the PM_{10} concentration of between 4.5 and 6.5 µg/m³/year. This, however, is only statistically significant at the Lephalale site (p<0.01).

The seasonal, diurnal and day of week time variation plots for PM_{10} are presented in Figure 9. The PM_{10} concentrations show a distinct diurnal pattern at all stations, with peaks occurring in the morning (06:00) and in the evening (18:00). The evening peak is greater than the morning peak. This pattern holds for all the stations and is likely linked to domestic combustion or traffic sources. The seasonal cycle shows a strong peak during the winter months from April to October, which corresponds to the periods where there is increased atmospheric stability over

the interior of the country (Preston-Whyte et al. 1976; Scott & Diab 2000; Lourens et al. 2011) increased biomass burning (Korontzi 2005) and domestic combustion for space heating (Wernecke et al. 2015) .

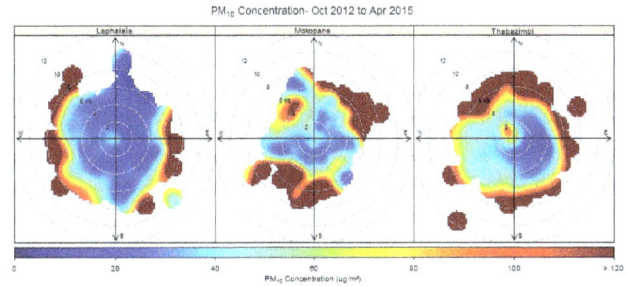

Figure 5: Polar Plot of the PM_{10} concentration at the Waterberg monitoring stations

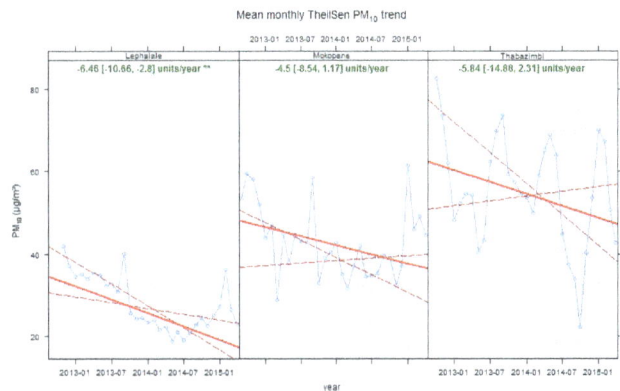

Figure 6: Monthly Mean Deseasonalised PM_{10} Trend for the period Oct 2012-April 2015 the solid trend line represents the mean slope of the trend, while the dashed trend lines represent the 95% confidence interval of the slope

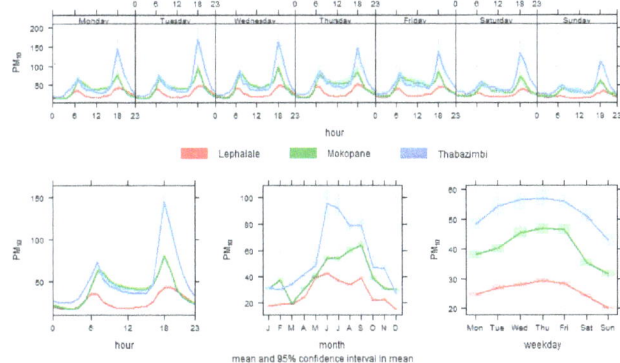

Figure 7: PM_{10} Time variation plot for the period October 2012 to April 2015

A strong diurnal pattern in the PM_{10} concentrations was observed occurring in the early morning and in the evening. This is strongest at the Thabazimbi and Mokopane stations where the evening peak PM_{10} concentration is considerably greater than the morning peak. To further allude to the domestic combustion component of the PM_{10} source a day of week pattern was also observed at all the stations, with higher average PM_{10} concentrations being observed between Monday and Friday, followed by decreases on Saturday and Sunday, especially in the morning peak over the weekend. This could indicate that the behaviour patterns of the people in these areas changes over the weekends and there is less vehicular traffic and the timing of activities may be more staggered than during the work week. For all the sites there is a strong contrast in the temporal profiles of PM_{10} and SO_2, indicating that these pollutants are generated at different sources.

$PM_{2.5}$

The mean $PM_{2.5}$ concentration at the three Waterberg monitoring stations ranges from 12.3µg/m³ for Lephalale to 20 µg/m³ and 20.3µg/m³ for Thabazimbi and Makopane respectively (Table 4). Thabazimbi and Mokopane showed the highest $PM_{2.5}$ values. For the 2.5 year monitoring period considered here the average $PM_{2.5}$ concentration at all the sites is below the national standard for the period of the measurement (25µg/m³), however for Mokopane and Thabazimbi it exceeds the stricter standard (20µg/m³) that came into effect at the beginning of 2016.

Table 4: *$PM_{2.5}$ measurement summary*

	Lephalale	Makopane	Thabazimbi
Number of measure-ments	20743	19633	14687
% recovery	92.70%	85.0%	65.6%
Mean	12.34	20.29	19.98
Median	9.49	12.88	10.74
10 percentile	2.50	3.90	1.92
25 percentile	4.93	7.29	5.35
75 percentile	16.54	22.86	20.85
90 percentile	25.79	43.95	43.56

The polar plot figures, which show the relationship between the wind speed, direction and the $PM_{2.5}$ concentration show that similarly to PM_{10} the periods of highest $PM_{2.5}$ concentration are associated with periods of high wind speed (> 6m/s). At the Mokopane station there is a hotspot of high $PM_{2.5}$ concentration associated with moderate wind speeds (4-6m/s) from the south-west and north-west. This wind direction corresponds to the location of the low income residential areas and associated agricultural areas of Sekgakgapeng and Masodi, respectively.

The trend analysis for $PM_{2.5}$ (Figure 12) shows no significant trends in the $PM_{2.5}$ concentrations at any of the sites. This is in contrast to the PM_{10} concentrations which show a mean decrease at all sites and which is statistically significant at the Lephalale site. This could indicate that the sources of PM_{10} and $PM_{2.5}$ at Waterberg sites are different and therefore different management interventions are needed to address them.

The time variation plots for $PM_{2.5}$ are similar to those of PM_{10}; there is a clear seasonal pattern, with the highest $PM_{2.5}$ concentrations recorded during the winter period. A strong diurnal pattern exists with morning and evening peaks and there is a weekly cycle where an increase in the $PM_{2.5}$ concentrations is observed

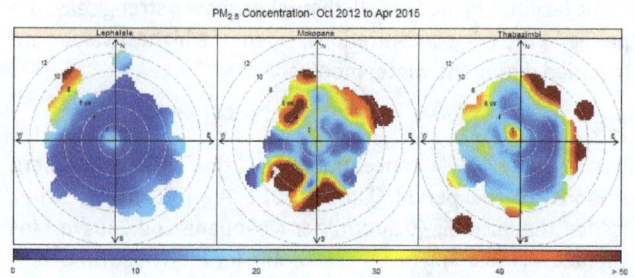

Figure 8: *Polar Plot of the $PM_{2.5}$ concentration at the Waterberg monitoring stations*

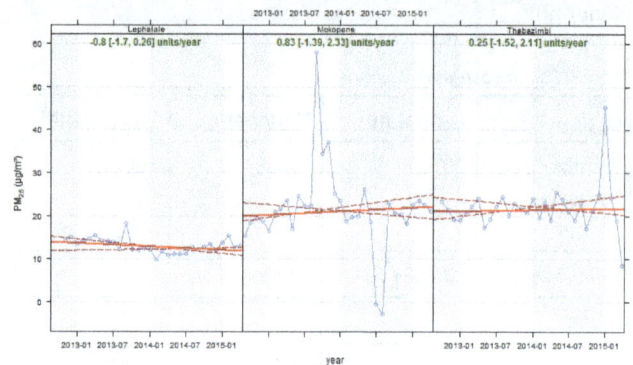

Figure 9: *Monthly Mean Deseasonalised $PM_{2.5}$ Trend for the period Oct 2012-April 2015. The solid trend line represents the mean slope of the trend, while the dashed trend lines represent the 95% confidence interval of the slope*

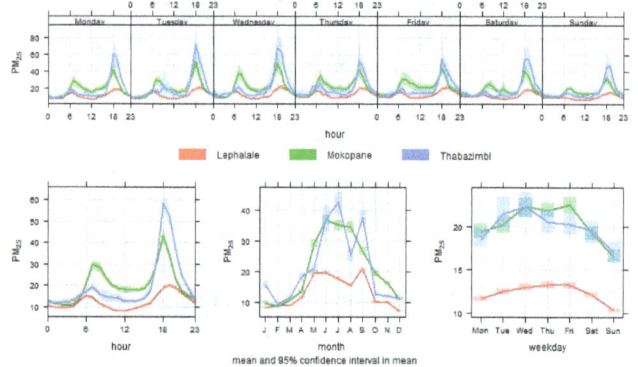

Figure 10: *$PM_{2.5}$ Time variation plot for the period October 2012 to April 2015*

between Monday and Friday followed by a large decrease over the weekend, with an especially large reduction in the morning peak over the weekends.

Ozone

The mean ozone concentrations recorded over the period ranges between 24.2 ppb (Lephalale) to 28.2 ppb (Mokopane and Thabazimbi) (Table 5).

The periods of high ozone concentration are typically associated with periods of relatively strong winds, specifically from the north westerly sectors for Lephalale and Mokopane. In a study of ambient air quality in the Vaal Triangle high ozone concentrations were observed under similar conditions during the approach of a cold front over the South African interior (Feig et al. 2014). For all the sites during periods of very low wind (as is typical of night time conditions) the ozone concentrations are very low The periods of high ozone concentration at the Thabazimbi site are associated with winds from the north east and easterly directions.

Table 5: *Ozone measurement summary*

	Lephalale	**Makopane**	**Thabazimbi**
Number of measure-ments	21061	20708	20255
% recovery	94.12%	89.6%	90.5%
Mean	24.27	28.20	28.21
Median	23.52	27.23	27.88
10 percentile	4.81	11.88	7.44
25 percentile	12.65	18.62	16.18
75 percentile	34.07	36.65	38.52
90 percentile	43.57	45.41	48.78

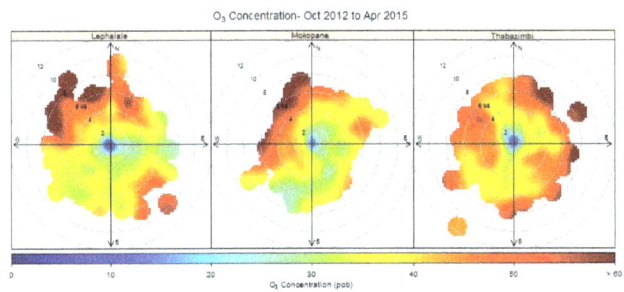

Figure 11: *Polar Plot of the O_3 concentration at the Waterberg monitoring stations*

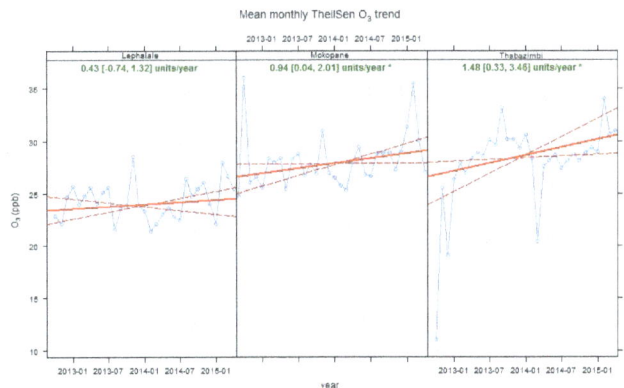

Figure 12: *Monthly Mean Deseasonalised O_3 Trend for the period Oct 2012-April 2015 the solid trend line represents the mean slope of the trend, while the dashed trend lines represent the 95% confidence interval of the slope*

The trend analysis (Figure 16) indicates that over the monitoring time period there is a statistically significant increase in the monthly ozone concentration (P<0.05) at the Mokopane and Thabazimbi stations. This increase is 0.94ppb/year and 1.48ppb/year for the Mokopane and Thabazimbi station respectively.

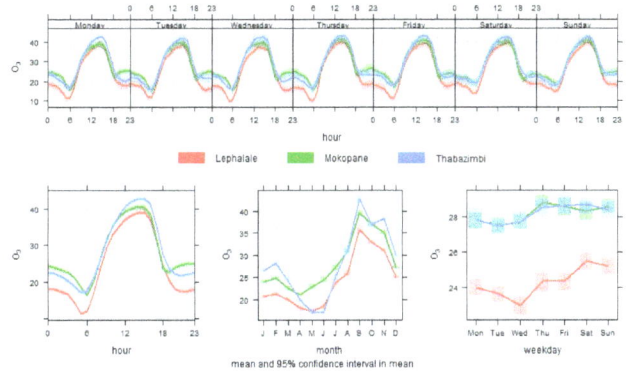

Figure 13: *O_3 Time variation plot for the period October 2012 to April 2015*

There is strong seasonal trend in the ozone concentrations observed at all the stations, with peaks in the ozone concentrations being observed in the September/October periods, which has been frequently observed and reported (Thompson et al. 2003; Thompson et al. 2011; Thompson et al. 2014; Scholes & Scholes 1998; Zunckel et al. 2004)

Conclusion

This study aims to provide an assessment of the ambient air quality monitoring that has occurred in the Waterberg priority area between October 2012 and April 2015. Data recovery from the three monitoring stations has been good with a valid data capture percentage of greater than 80 % for all the stations.

The recorded SO_2 concentrations are generally low. At the Lephalale station there is a strong source of SO_2 located to the west and north-west of the station that is likely a high level industrial source. The Thabazimbi station shows a distinct peak in SO_2 concentrations on Wednesday mornings.

The PM_{10} and $PM_{2.5}$ concentrations show a strong seasonal pattern with the highest values occurring during the winter months. There has been a statistically significant decrease in the PM_{10} concentrations at Lephalale over the monitoring time period, which is not seen in the $PM_{2.5}$ concentrations. The periods of high PM_{10} concentration are associated with high wind speeds. In addition to the temporal patterns in the PM_{10} and $PM_{2.5}$ concentrations are indicative of local domestic combustion or traffic sources in that there are strong peaks in the early morning and evening, and a strong weekend effect is seen especially with regards to a reduction in the morning peak during on Saturdays and Sundays.

The concentrations of ozone are highest in the spring period, similarly to what has been found in the Vaal Triangle high ozone events may occur during the advance of a frontal system across the country (Feig et al. 2014).

The Waterberg priority area was declared in anticipation of the development of air quality problems associated with the development of the Waterberg coal fields. The initial analysis indicates that the area may already be facing air quality problems, prior to the initiation of the major planned developments in the area. The continued operation of these ambient air quality monitoring stations will be vital in assessing the pollutant concentrations in the area and monitoring how the pollutant levels change with the implementation of the planned developments. This paper also demonstrates the value of utilizing advanced data analysis methods in order to identify potential pollution sources and to track trends in the ambient air quality over the region.

Acknowledgements

The authors would like to thank SAAQIS for the provision of the data used in this study. Installation and maintenance of the monitoring stations during the period of this study was done by C and M Consulting Engineers. Funding for the monitoring network was provided by the South African Department of Environmental Affairs.

References

Carslaw, D.C., 2014. Editorial Modern tools for air quality data analysis – openair. *Clean Air Journal*, 24(2), pp.4–5.

DEA, 2009. National Ambient Air Quality Standards. *Government Gazette*, p.4.

Department of Environmental Affairs, S.A., 2014. *The Waterberg-Bojanala Priority Area Draft Air Quality Management Plan*,

Department of Environmental Affairs, S.A., 2012. *Waterberg Priority Area (WPA) Declaration_15-6-2012.pdf*,

Feig, G. et al., 2014. Analysis of a period of elevated ozone concentration reported over the Vaal Triangle on 2 June 2013. *Clean Air Journal*, 24(1), pp.10–16.

Hartnady, C.J.H., 2010. South Africa's diminishing coal reserves. *South African Journal of Science*, 106(9-10), pp.1–5.

Hirsikko, A. et al., 2012. Characterisation of sub-micron particle number concentrations and formation events in the western Bushveld Igneous Complex, South Africa. *Atmospheric Chemistry and Physics*, 12, pp.3951–3967.

Josipovic, M. et al., 2010. Concentrations, distributions and critical level exceedance assessment of SO_2, NO_2 and O_3 in South Africa. *Environmental Monitoring and Assessment*, 171(1-4), pp.181–196.

Kocha, C. et al., 2011. High-resolution simulation of a major West African dust-storm : *Quarterly Journal of the Royal Meteorological Society*.

Korontzi, S., 2005. Seasonal patterns in biomass burning emissions from southern African vegetation fires for the year 2000. *Global Change Biology*, 11(10), pp.1680–1700. Available at: http://onlinelibrary.wiley.com/doi/10.1111/j.1365-2486.2005.001024.x/full [Accessed July 14, 2011].

Lourens, A.S. et al., 2011. Spatial and temporal assessment of gaseous pollutants in the Highveld of South Africa. *South African Journal of Science*, 107, pp.1–8.

Marticorena, B. & Bergametti, G., 1995. Modeling the atmospheric dust cycle 1 Design of a soil-derived dust emission scheme. *Journal of Geophysical Research: Atmospheres*.

National Planning Commission, 2010. *National Development Plan (2030)*,

Preston-Whyte, R.A.R., Diab, R. & Tyson, P., 1976. Towards an inversion climatology if southern Africa: Part II, non-surface inversions in the lower atmosphere. *South African Geographical Journal*, 58(2), pp.151–163. Available at: http://scholar.google.com/scholar?hl=en&btnG=Search&q=intitle:Towards+an+inversion+climatology+of+southern+Africa:+Part+1,+surface+inversions#0 [Accessed July 18, 2011].

Schmechtig, C. et al., 2011. Simulation of the mineral dust content over Western Africa from the event to the annual scale with the CHIMERE-DUST model. *Atmospheric Chemistry and Physics*, 11(14), pp.7185–7207. Available at: http://www.atmos-chem-phys.net/11/7185/2011/ [Accessed July 25, 2011].

Scholes, R. & Scholes, M., 1998. Natural and human-related sources of ozone-forming trace gases in southern Africa. *South African Journal of Science*, 94, pp.422–427. Available at: http://researchspace.csir.co.za/dspace/handle/10204/774 [Accessed July 18, 2011].

Scott, G.M. & Diab, R.D., 2000. Forecasting Air Pollution Potential : A Synoptic Climatological Approach. *Journal of the Air & Waste Management Association*, 50, pp.1831–3289. Available at: http://dx.doi.org/10.1080/10473289.2000.10464216.

Thompson, A.M. et al., 2003. Southern Hemisphere Additional Ozonesondes (SHADOZ) 1998–2000 tropical ozone climatology 2. Tropospheric variability and the zonal wave-one. *Journal of Geophysical Research*, 108, pp.1998–2000.

Thompson, A.M. et al., 2011. Strategic ozone sounding networks: Review of design and accomplishments. *Atmospheric Environment*, 45(13), pp.2145–2163. Available at: http://linkinghub.elsevier.com/retrieve/pii/S135223101000364X [Accessed April 11, 2012].

Thompson, A.M. et al., 2014. Tropospheric ozone increases over the southern Africa region: bellwether for rapid growth in Southern Hemisphere pollution? *Atmospheric Chemistry and Physics*, 14, pp.9855–9869. Available at: http://www.atmos-chem-phys.net/14/9855/2014/.

Uria-Tellaetxe, I. & Carslaw, D.C., 2014. Conditional bivariate probability function for source identification. *Environmental Modelling & Software*, 59, pp.1–9. Available at: http://linkinghub.elsevier.com/retrieve/pii/S1364815214001339 [Accessed July 16, 2014].

Venter, A.D. et al., 2012. An air quality assessment in the industrialised western Bushveld Igneous Complex , *South Africa. S Afr J Sci*, 108, pp.1–10.

Wernecke, B. et al., 2015. Indoor and outdoor particulate matter concentrations on the Mpumalanga highveld – A case study. , 25(15), pp.12–16.

Zhou, W. et al., 2012. Observation and modeling of the evolution of Texas power plant plumes. *Atmospheric Chemistry and Physics*, 12(1), pp.455–468. Available at: http://www.atmos-chem-phys.net/12/455/2012/ [Accessed August 7, 2013].

Zunckel, M. et al., 2004. Surface ozone over southern Africa: synthesis of monitoring results during the Cross border Air Pollution Impact Assessment project. *Atmospheric Environment*, 38(36), pp.6139–6147. Available at: http://www.sciencedirect.com/science/article/pii/S1352231004007216 [Accessed April 22, 2015].

10

Mercury emissions from South Africa's coal-fired power stations

Belinda L. Garnham[*1] and Kristy E. Langerman[1]

[1]Eskom Holdings SOC Limited, Megawatt Park, [1]Maxwell Drive, Sunninghill, Sandton, RoosBe@eskom.co.za, RossKe@eskom.co.za

Abstract

Mercury is a persistent and toxic substance that can be bio-accumulated in the food chain. Natural and anthropogenic sources contribute to the mercury emitted in the atmosphere. Eskom's coal-fired power stations in South Africa contributed just under 93% of the total electricity produced in 2015 (Eskom 2016). Trace amounts of mercury can be found in coal, mostly combined with sulphur, and can be released into the atmosphere upon combustion. Coal-fired electricity generation plants are the highest contributors to mercury emissions in South Africa. A major factor affecting the amount of mercury emitted into the atmosphere is the type and efficiency of emission abatement equipment at a power station. Eskom employs particulate emission control technology at all its coal-fired power stations, and new power stations will also have sulphur dioxide abatement technology. A co-beneficial reduction of mercury emissions exists as a result of emission control technology. The amount of mercury emitted from each of Eskom's coal-fired power stations is calculated, based on the amount of coal burnt and the mercury content in the coal. Emission Reduction Factors (ERF's) from two sources are taken into consideration to reflect the co-benefit received from the emission control technologies at the stations. Between 17 and 23 tons of mercury is calculated to have been emitted from Eskom's coal-fired power stations in 2015. On completion of Eskom's emission reduction plan, which includes fabric filter plant retrofits at two and a half stations and a flue gas desulphurisation retrofit at one power station, total mercury emissions from the fleet will potentially be reduced by 6-13% by 2026 relative to the baseline. Mercury emission reduction is perhaps currently not the most pressing air quality problem in South Africa. While the focus should then be on reducing emissions of other pollutants which have a greater impact on human health, mercury emission reduction can be achieved as a co-benefit of installing other emission abatement technologies. At the very least, more accurate calculations of mercury emissions per power station should be obtained by measuring the mercury content of more recent coal samples, and developing power station-specific ERF's before mercury emission regulations are established or an investment into targeted mercury emission reduction technology is made.

Keywords

mercury emissions, coal-fired power stations emissions, Eskom

Introduction

Mercury is a persistent and toxic substance that accumulates in the food chain, and even though mercury is present in trace amounts, exposure increases as it accumulates. Coal contains mostly ash, carbon and small amounts of sulphur, and the trace amounts of mercury that are mostly combined with the sulphur can be released into the atmosphere upon combustion (Miller 2007a). Coal is the primary and most widely used fuel in the electricity generation industry and constitutes 43% of the total fuel used globally (Pirrone et al. 2009), thus the importance of controlling or limiting the emissions in this industry is evident.

Due to the combustion of coal in coal-fired power stations, mercury is present in the immediate exhaust gas as vapour phase Hg^0 (elemental mercury) (Srivastava et al. 2006; EPA 2011a). As a result of oxidation reactions, oxidised mercury (Hg^{2+}) or particulate bound mercury (Hg^p) may be formed. Oxidised mercury, Hg^{2+}, can form in the presence of chlorine, Cl, (forming $HgCl_2$), and particulate-bound mercury, Hgp, can form in the presence of fly ash or unburnt carbon remnants. The formation of Hgp with fly ash and unburned carbon occurs as a result of chlorination before the conversion of elemental mercury to $HgCl_2$. The particulate-bound mercury can then be captured by downstream particulate abatement technology such as Fabric Filter Plants (FFP's) or Electrostatic Precipitators (ESP's).

Very limited information on the status of mercury emissions in African countries is available, although mercury emissions in Africa are increasing due to the rapid economic development in these countries. South Africa signed the Minamata Convention on Mercury on 10 October 2013; however, there is no legislation regarding mercury emissions at present. South Africa, being the most industrialised country in the continent of Africa, also has limited information on levels of mercury in resources, mercury in products and mercury in emissions (Pirrone et al. 2009). It has previously been estimated that emissions of mercury as a result of power generation account for 77% of the total mercury emitted in South Africa (Pirrone et al. 2009). This figure, unless mitigated, is unlikely to decrease significantly, as coal is the main source of energy and the demand for base energy will increase with the increase in population.

Uncertainties in the mercury emission inventories can hamper the development of policies, but do not negate the benefits in establishing a baseline of control that can be further developed or improved. Calculations presented in this paper are intended to give an indication of the magnitude of mercury emissions from coal-fired power stations in South Africa. The mercury emission inventory needs to be refined by updating the measurements of mercury in coal, and determining South Africa-specific emission reduction factors based on mercury emission measurements at South African power stations.

This paper presents a baseline of the emitted mercury from 2011 to 2015, as a result of coal combustion in the electricity generation process, from the individual Eskom Holdings SOC Ltd ("Eskom") coal-fired power stations, utilising two sets of emission reduction factors. An estimate of expected mercury emission reduction in future is also presented taking into consideration Eskom's emission reduction plan. This paper did not include the three other non-Eskom owned smaller coal-fired power stations currently operating in South Africa: the Rooiwal Power Station and the Pretoria West Power Station in Pretoria, and the Kelvin Power Station in Johannesburg as well as the Sasol Secunda boilers. These three stations have a total generating capacity of 1 080 MW (compared to Eskom's installed 38 548 MW coal-fired power stations capacity – Eskom, 2016). The inclusion of these stations would not significantly change the values reflected in this paper. The results of these estimates can be used to compare co-benefits of emission control technologies implemented at Eskom's power stations at present or potentially in the future.

Methodology and Data

Eskom's coal-fired power stations

Eskom currently operates 14 coal-fired power stations located across South Africa. Medupi and Kusile Power Station are two new build power stations. Medupi's first unit came online in 2015, with all units expected to come online by 2021. Kusile Power Station is expected to commence with electricity production from 2017.

These power stations burn bituminous type coal, and employ particulate matter (PM) control technology – either Fabric Filter Plants (FFP's) or cold-sided Electrostatic Precipitators (ESP's) with SO3 flue gas conditioning (FGC). Kusile and Medupi Power Stations will also employ Flue Gas Desulphurisation (FGD), from commissioning at Kusile and six years after commissioning at Medupi.

Eskom's Emission Reduction Plan is focused on the reduction of SO_2 and PM through the retrofit of some ESP fitted stations with FFPs, as well as the addition of Flue Gas Desulphurisation (FGD) units at one or two stations. Eskom's current and future emission control technology is indicated in Table 1.

Table 1: *Current and future emission control technology at Eskom's coal-fired power stations*

Power Station	Emission control technology	
	Status Quo *	Emission reduction plan
Arnot	FFP's	No change
Camden	FFP's	No change
Duvha	FFP's & ESP's+FGC	HFTs onto ESP units
Grootvlei	FFP's & ESP's+FGC	All units with FFP
Hendrina	FFP's	No change
Kendal	ESP's + FGC	HFTs
Komati	ESP's + FGC	No change
Kriel	ESP's + FGC	FFP retrofit
Kusile*	FFP's & wet FGD	No change
Lethabo	ESP's + FGC	HFT upgrades
Majuba	FFP's	No change
Matimba	ESP's + FGC	HFT upgrades
Matla	ESP's + FGC	HFT upgrades
Medupi	FFP's	Wet FGD retrofit
Tutuka	ESP's	FFP retrofit

FFP's = Fabric Filter Plants, ESP's = Electrostatic Precipitators, HFT's = High frequency transformers, FGC = Flue Gas Conditioning
*Kusile is only expected to come into operation from 2017

Data sources

The data utilised to calculate the emitted mercury from the power stations includes the following:

- the amount of coal burnt per annum (tons) per power station (actuals from 2010 to 2015 and projections from 2016 on);
- the mercury content (ppm) of the coal used at each power station; and
- mercury emission reduction factors sourced from available literature.

Amount of coal burnt

The current and historic amount of coal burnt is provided by Eskom and is calculated by measuring the amount of coal coming into the power stations (to the coal stock yards, from the coal mines) and comparing this to the height, width, length as well as density dimensions of the coal stock yards. The amount of coal coming into the station is determined and weighed as it arrives. Surveys of the coal stockpiles are conducted quarterly. The data used for historic coal burnt is presented in Table 2.

Table 2: *Annual coal burnt at Eskom's coal power stations from 2011 to 2015 (Mt)*

	2011	2012	2013	2014	2015
Arnot	6.98	6.82	5.97	6.03	5.45
Camden	4.23	4.90	5.25	5.23	4.53
Duvha	9.18	8.77	10.14	7.47	7.55
Grootvlei	3.58	3.48	4.15	4.28	3.25
Hendrina	6.96	6.02	5.13	6.52	5.36
Kendal	15.85	15.73	15.85	13.70	15.03
Komati	1.14	2.26	2.96	2.61	2.52
Kriel	9.21	8.30	8.43	8.13	8.19
Lethabo	17.52	16.32	16.17	16.22	15.37
Majuba	13.38	13.50	13.27	13.09	11.41
Matimba	14.93	14.63	13.95	13.70	13.44
Matla	11.31	11.70	10.75	11.41	11.66
Tutuka	10.74	11.18	10.61	11.15	10.78
Medupi	0.00	0.00	0.00	0.00	0.89
Total	125	124	123	120	115

The projected coal burnt data utilised for the estimates from the 2016 Eskom financial year onwards, are from Eskom's 10-year production plan of December 2015, which was the most recent available at the time of drafting this paper. This plan forecasts the amount of coal that will be burnt at each power station based on past trends of usage, expected maintenance, and prediction of total electricity demand. A financial year runs from the beginning of April to the end of March in the following year.

Mercury content of coal

The mercury content of coal data used is provided by Eskom (Delport, 2007) for six monthly samples collected from the end of 2004 until mid-2005, and from UNEP (2014) for monthly and annual coal samples collected between 2010 and 2012. The analysis of the coal samples by Delport (2007)was performed using a gold amalgamated spectroscopy technique. This sampling was conducted at the Consulting Research and Development Department (now known as the Research, Testing and Development Department, Group Sustainability Division). This data is reflected in Table 3. No recent samples have been measured by Eskom.

The average mercury in coal content for the coal used at Eskom's power stations ranges between 0.17 ppm (Arnot Power Station) and 0.38 ppm (Matimba Power Station; Table 3). At many power stations, the variability in mercury content is high from month to month. Kriel, Lethabo and Matimba Power Station have the highest variability in the mercury in coal content (Figure 1).

Table 3: *Average mercury in coal content (ppm) of monthly and annual composite samples of coal used at Eskom's power stations. Annual samples are shown in bold.*

Power Stations	2004 - 2005 samples (Delport, 2007)	2010-2012 samples (UNEP, 2014)	Average (ppm)
Arnot	0.17, 0.17, 0.13, 0.17, 0.15, 0.26	0.16, 0.12	0.17
Duvha	0.25, 0.26, 0.20, 0.19, 0.27, 0.23	0.21, 0.18, 0.19	0.22
Hendrina	0.21, 0.23, 0.17, 0.20, 0.20, 0.24	0.28, 0.23, 0.22	0.22
Kendal	0.25, 0.22, 0.29, 0.46, 0.31, 0.34	0.22, 0.21, 0.18, 0.21	0.27
Kriel	0.39, 0.35, 0.23, 0.23, 0.26, 0.55	0.12, 0.13, 0.14	0.27
Lethabo	0.25, 0.23, 0.29, 0.33, 0.46,0.64	0.40, 0.46, 0.15	0.35
Majuba	0.36, 0.33, 0.25, 0.29, 0.21, 0.28	0.29, 0.23, 0.28, 0.22	0.27
Matimba	0.61, 0.70, 0.33, 0.37, 0.22, 049	0.23, 0.25, 0.19	0.38
Matla	0.18, 0.20, 0.45, 0.22, 0.25, 0.49	0.28,0.23, 0.20	0.28
Tutuka	0.36, 0.34, 0.33, 0.23, 0.26, 0.24	0.18, 0.30, 0.22	0.27
Grootvlei		0.37, 0.28, 0.34, 0.32	0.33
Camden		0.20, 0.25, 0.22, 0.19	0.22
Komati		0.23, 0.20, 0.24	0.22
Medupi**			0.38
Kusile***			0.27

** Matimba's data was assumed for Medupi (due to proximity and source)

***Kendal's data is assumed for Kusile (due to proximity and source)

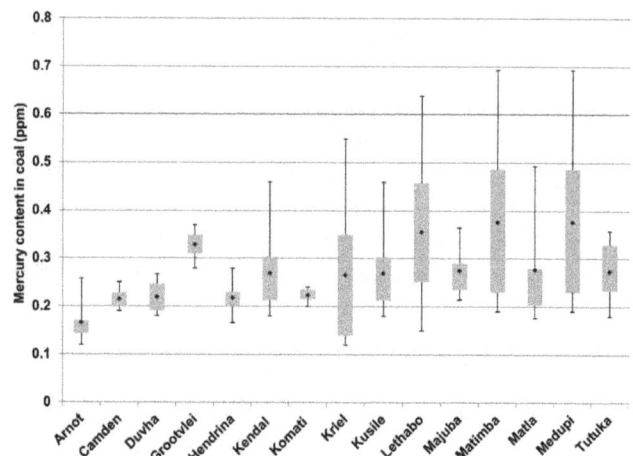

Figure 1: *The mercury in coal content, based on the measurements presented in Table 3. Diamonds show average mercury content (ppm); the boxes show the range between the first quartile and the third quartile, and the whiskers show the minimum and maximum values*

Although the mercury in coal measurements are extremely limited, it may be tentatively suggested that the mercury in coal content has declined at Kendal and Matimba between 2004/2005 and 2010-2012.

Emission Reduction Factors

The Emission Reduction Factors (ERF's) utilised to calculate the amount of mercury emitted from a coal-fired power station should take into consideration the type of emission control technology at a power station and the type of coal burnt as both of these can significantly affect the amount of mercury emitted. Two sets of emission factors are utilised in this paper, those from the EPA (2013) as well as those in the latest UNEP toolkit (2015), as represented in the Table 44. These factors are for bituminous coal.

Table 4: ERF's to calculate mercury emissions from bituminous coal

Emission control technology	ERF (%)	
	EPA 2013	UNEP 2015
FFP's	89	50
CS-ESP's	36	25
FFP's+CS-ESP's*	62.5	37.5
FFP+wet FGD	90	65
CS-ESP+wet FGD	66	65***

* This is determined by averaging the factors for FFPs & CS-ESPs
** The ESP factors consider general ESPs (not specifically CS-ESPs)
*** No factor available for an ESP & wet FGD combination in particular (only particulate matter filter & wet FGD)

The United Nations Environment Programme (UNEP) produced a toolkit with the intention to assist countries with developing mercury inventories. The first toolkit was developed in 2005, updated in 2011, and again in 2015. Default emission reduction factors are presented in the toolkit, those being 25% reduction of mercury from 'general ESPs' and a 50% mercury reduction from FFPs. The toolkit indicates that these default factors are based on "a limited database, and are expert judgments based on summarised data only with no considered systematic quantitative approach (i.e. consumption-weighted concentration and distribution factors derivation)".

In the United States, the Environmental Protection Agency (EPA) initiated an Information Collection Request (ICR) from 'Electric Utility Steam Generating Units', in 1998. One of the three main elements of the 1999 ICR was the acquisition of data by coal sampling and stack testing in order to determine mercury reductions from different 'representative unit configurations'. The data from the ICR 1999 indicated that bituminous coal-fired stations, with emission control technology, have higher levels of mercury capture than sub-bituminous or lignite coal-fired power stations with the same emission control technology. Currently the US EPA utilises emission factors based on the US ICR of 1999 and these are captured in their base case (v5.13). These factors are incorporated into the Integrated Planning

Model (IPM) of the US EPA which presents "forecasts of least cost capacity expansion, electricity dispatch, and emission control strategies" (EPA 2010, EPA 2013). The emission reduction factors from the UNEP 2015 toolkit as well as the EPA 2013 are used for the calculation of mercury emissions in this paper, and reflected in Table 44.

Data analysis method

The method used for the calculation of the amount of mercury emissions from each of the coal-fired power stations is a mass-balance formula, as also presented in the 'Toolkit for Identification and Quantification of Mercury Releases' (UNEP 2015).

Results and discussion

Mercury emissions from 2011 to 2015

The few studies that have estimated the amount of mercury emissions as a result of coal-fired power stations in South Africa reflect results that vary greatly. These results range from 9.8 tons of mercury emitted in 2004 (Dabrowski et al. 2008) to just over 83 tons in 2000 (Pacyna et al. 2006). According to our calculations, the amount of mercury emitted from the Eskom coal fleet in 2015 is between 16.8 tons (12.2-20.1 tons for the first-third quartile range) and 22.6 tons (16.9-26.6 tons for the first-third quartile range), depending on the selection of ERFs (EPA 2013 and UNEP 2015). The annual mercury emitted from each power station in 2015 is shown in Figure 2.

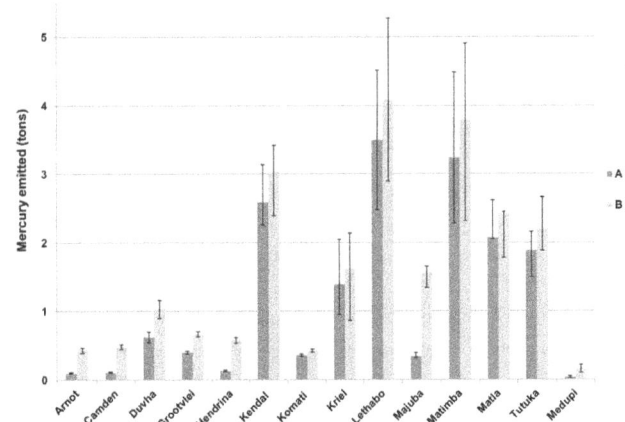

Figure 2: *The mercury emitted per power station in 2015 using the two ERF's (A=EPA 2013 and B = UNEP 2015). The error bars reflect the first and third quartiles of mercury emissions per station, per ERF.*

Matimba, Lethabo, Kendal, Matla, Kriel and Tutuka emitted the highest amount of mercury in 2015. These six stations produced on average just below 82% of the total mercury emitted by the fleet of coal power stations. Matimba and Lethabo alone contribute to just below 38% of the total mercury emitted by the fleet in 2015.

A slight decrease in mercury emissions is noticed from 2011 to 2015 (Figure 3). This is a direct link to the quantity of coal burnt since 2011 (Table 2), which has declined from an average of 17.9 tons in 2011 to an average of 16.8 tons in 2015, using the

EPA (2013) emission reduction factors, or from an average of 24.3 tons in 2011 to 22.6 tons in 2015 using the UNEP (2015) emission reduction factors. With the implementation of Eskom's emission reduction plan (Table 1), a further decrease in mercury emissions is expected due to the higher co-beneficial reduction of mercury emissions from the selected PM and SO_2 technology selected to be retrofitted (further discussed in section 3.2).

Mercury emission monitoring in power stations' stacks is required to validate mercury emission reduction factors and mercury emission estimates. The results in this paper show that estimates from previous reports could have been over estimates of the current mercury emissions from the power stations (Pacyna et al. 2006, Pirrone et al. 2009).

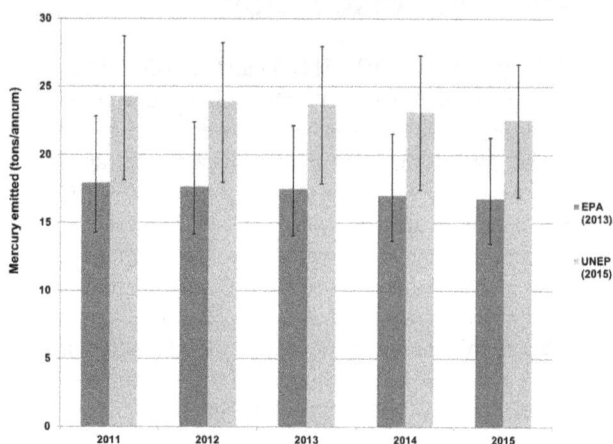

Figure 3: Mercury emissions from Eskom's coal-fired power stations from 2011-2015 using the two ERFs. The error bars reflect the first and third quartiles of mercury emissions.

Power stations with FFPs as emission control technology (Arnot, Camden, Hendrina, Majuba and Medupi) produce on average 0.01-0.07kg of mercury per GWh sent out (Figure 4 and Figure 5). This is significantly lower than the stations with ESPs (Kendal, Kriel, Lethabo, Matimba, Matla, Komati and Tutuka), emitting just under 0.11-0.13kg of mercury per GWh sent out. These differences are a direct reflection of the differing mercury emission factors from the EPA 2013 and UNEP 2015. The two stations, Grootvlei and Duvha, which had half ESPs and half FFPs implemented at the stations in 2015 emit just over 0.07-0.1kg mercury per GWh energy sent out.

The highest mercury-emitting coal power stations in 2015 are Lethabo (3.5 and 4.1 tons, with the EPA (2013) and UNEP (2015) emission reduction factors, respectively), Matimba (3.2 and 3.8 tons) and Kendal (2.6 and 3.0 tons). These figures are a function of the amount of coal burnt and the mercury content in coal. When comparing the amount of average mercury emitted per GWh of energy sent out, Matimba and Lethabo are the top two 'mercury per GWh' emitting stations, then followed by Kendal, Matla and Tutuka.

Currently in Eskom, five coal-fired power stations pre-wash coal prior to combustion (these are Arnot, Duvha, Hendrina, Lethabo and Matimba). Arnot and Hendrina have FFP's as an emission

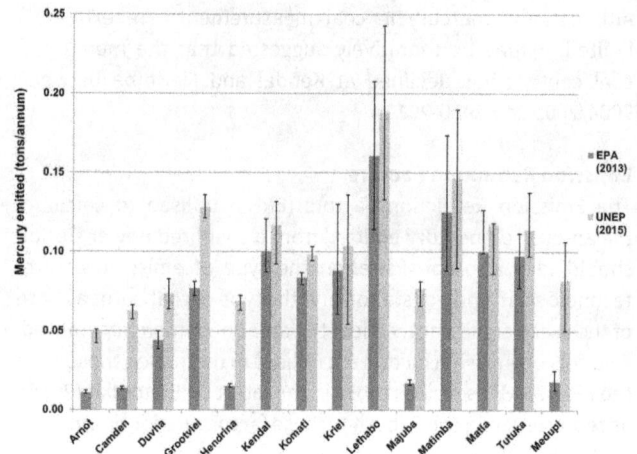

Figure 4: Mercury emitted per energy unit sent out per station using the two ERF's

control technology. Camden and Majuba, also have FFPs, and no coal washing. Arnot and Hendrina emit a lower amount of mercury per GWh sent out than Camden and Majuba (Figure 4). This can also be seen when comparing Duvha's mercury emitted per GWh sent out with Grootvlei which does not pre-wash coal, and which also has half ESPs and half FFPs.

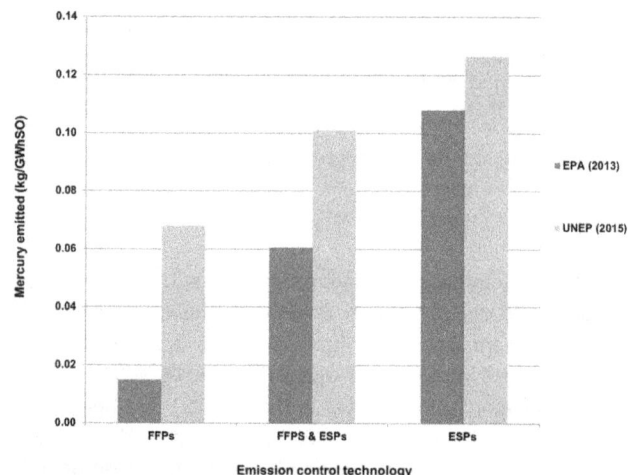

Figure 5: Relative mercury emitted per GWhSO per emission control technology in 2015

Projected mercury emissions from 2016 to 2026

In the absence of any emission abatement retrofits or upgrades, it is inconclusive whether total mercury emissions from Eskom's coal-fired power stations will increase or decrease over the next 10 years. Figure 6 and Figure 7 show future mercury emission trends with and without the implementation of Eskom's emissions reduction plan. Variations from year-to-year are mostly due to the projected total quantity of coal burnt per power station as well as the decommissioning of Arnot, Camden and Komati Power Stations within this time frame. The additions of the Medupi units as well as the commissioning of Kusile Power Station units are considered in the total mercury emissions per year.

The trends in mercury emissions over time in Figures 6 and 7 are influenced both by trends in coal burnt (which increases towards 2024/25), and also by the changing load factors of the different power stations over time. Since the different power stations have different particulate abatement technologies installed, the different emission reduction factors for the abatement technology installed result in the trends plotted.

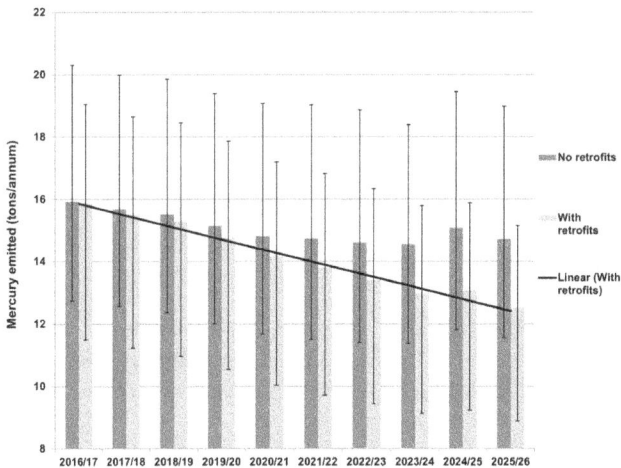

Figure 6: *Projected annual mercury emissions from 2016/17 to 2025/26 assuming no retrofits and the implementation of Eskom's emission reduction plan, using the EPA 2013 ERFs*

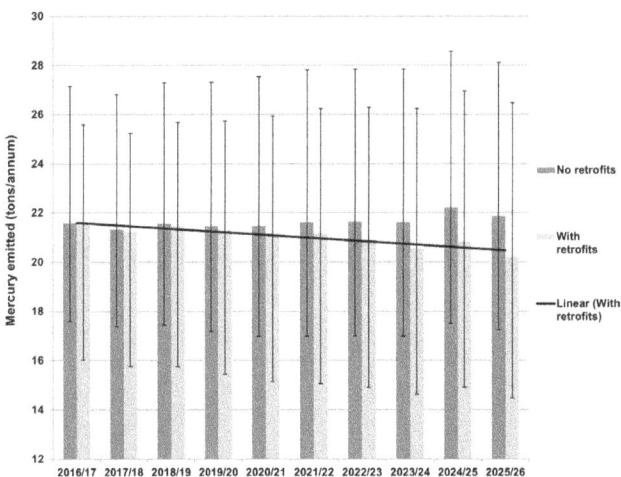

Figure 7: *Projected annual mercury emissions from 2016/17 to 2025/26 assuming no retrofits and the implementation of Eskom's emission reduction plan, using the UNEP 2015 ERFs*

Total mercury emissions are expected to be reduced by between 6% and 13% over the next 10 years as a result of the implementation of Eskom emission reduction/retrofit plan (Figure 6 and Figure 7). The retrofits in this time period (i.e. up to end of March 2026) include FFP retrofits on the ESP units at Grootvlei Power Station (affecting the emissions from 2016 onwards), FFP installations at Tutuka and Kriel Power Stations (affecting the emission from the power stations from 2019 and 2020, respectively), and lastly the installation of FGD at Medupi (affecting the emissions from 2022). The retrofits are assumed to occur at a pace of one unit per financial year, with the co-beneficial reduction of mercury emissions being realised the

year following retrofit. It is assumed that ESP upgrades and burner modifications (including the addition of high frequency transformers to an ESP) have no effect on mercury emissions, which is a conservative assumption.

Mercury emissions will be reduced even further with the further decommissioning of older coal units and the retrofitting of FGD outside of this 10 year window.

Conclusion

Factors that should be considered when selecting the most appropriate mechanism for mercury emission reduction from a power station include:

- the need to reduce or control specific pollutants. Focus should be on the need to control priority pollutants (those that have an effect on human health and are emitted in high quantities), meeting current emission limits and the subsequent co-beneficial reduction of mercury;
- the economic feasibility of the control technologies (for the power stations as well as the country). Additional to the cost of emission control technology, and the resultant impact on the economy, meeting the electricity demand in South Africa should be considered; and
- the life of the stations. This factor ties in with economic feasibility. Implementing costly technology on an old power station is not necessarily sustainable.

As most of South Africa's energy is derived from the combustion of coal, focus on energy saving initiatives, the introduction of renewables into the energy mix as well as improving the power stations' efficiency, would contribute to the reduction of mercury.

Setting and subsequent meeting of mercury control requirements at a power station and for a country begins with an assessment of baseline mercury removals achieved by emission control technologies already in place. This baseline assessment of the mercury emissions from the coal power stations in South Africa are a stepping stone to the drafting of regulations for the country.

The main driver for the establishment of mercury regulations in South Africa is the global concern of bio-accumulation in the food chain. Due to the goal of lowering the mercury emissions globally, the persistent nature of mercury, as well as the developing economic state of South Africa, a cap limit for the total mercury emissions from the generating coal-fired power station sector is recommended. This would allow for economic feasibility and flexibility in terms of retrofitting or implementing new technology in those power stations with a longer life, and thus a reduction could be established on the total mercury emissions from the fleet. A national cap limit on the mercury emission thus seems more appropriate than a specific limit for the individual coal stations.

Emission monitoring in power stations' stacks is required to develop power station-specific ERF's and validate mercury emission reduction estimates before mercury emission regulations are established or an investment into targeted mercury emission reduction technology is made.

Acknowledgements

Eskom coal burnt data was provided by Eskom Holdings SOC Ltd. This paper was developed based on a minor dissertation completed for partial fulfilment of the requirements for the author, Belinda Garnham (nèe Roos), MSc degree at the University of Johannesburg, under the supervision of Mrs Thea Schoeman (Roos 2011).

References

Dabrowski, J.M., Ashton, P.J., Murray, K., Leaner, J.J. and R.P. Mason. (2008). Anthropogenic mercury emissions in South Africa: Coal combustion in power plants. *Atmospheric Environment* 42: 6620–6626.

Delport, W. (2006). The fate of Mercury in coal after combustion with and without SO_3 injection in the gaseous emission stream, report number TSI/MECH/L023

EPA (Environmental Protection Agency). (2010). Documentation for EPA Base Case v.4.10 Using the Integrated Planning Model. *https://www.epa.gov/airmarkets/documentation-integrated-planning-model-ipm-base-case-v410*, Accessed June 2016.

EPA (Environmental Protection Agency). (2013). Documentation for EPA Base Case v.5.13 Using the Integrated Planning Model. *https://www.epa.gov/sites/production/files/2015-07/documents/documentation_for_epa_base_case_v.5.13_using_the_integrated_planning_model.pdf*, Accessed June 2016.

EPA (Environmental Protection Agency). (2011a). Emission Control Technologies. *https://www.epa.gov/sites/production/files/2015-07/documents/chapter_5_emission_control_technologies.pdf*, Accessed June 2016

EPA (Environmental Protection Agency). (2011b). Air Toxics standards for utilities: MACT Floor Analysis-Coal HG – Revised (18/05/2011). *http://www.epa.gov/ttn/atw/utility/utilitypg.html*, Accessed June 2011.

Eskom integrated report, 2015/2016, *http://www.eskom.co.za/IR2016/Documents/Eskom_integrated_report_2016.pdf*, accessed June 2016

Miller G.T., (2007a). *Sustaining the earth: An integrated approach*. Eighth edition. Canada. Thomson.

Pacyna, E.G., Pacyna, J.M., Steenhuisen and F., S. Wilson. (2006). Global anthropogenic mercury emission inventory for 2000. *Atmospheric Environment* 40: 4048–4063.

Pirrone, N., Cinnirella, S., Feng, X., Finkelman, R.B., Friedli, H.R., Leaner, J., Mason, R and A.B. Mukherjee. (2009). Global Mercury Emissions to the atmosphere from Natural and Anthropogenic Sources. In: Mercury Fate and Transport in the Global Atmosphere: Measurements, models and policy implications (Pirrone N. and Mason R. Eds.), Springer, New York.

Roos B L, 2011, Mercury Emissions from coal-fired power stations in South Africa, *http://ujdigispace.uj.ac.za/bitstream/handle/10210/6296/Roos.pdf?sequence=1*, accessed June 2016

Srivastava, R.K., Hutson, N., Martin, B., Princiotta, F. and J. Staudt. (2006). Control of mercury emissions from coal-fired electric utility boilers. *Environmental Science and Technology*. 40: 1385 – 1393.

UNEP (United Nations Environmental Programme). (2014). Collaborative studies for mercury characterisation in coal and coal combustion products, republic of South Africca. *http://www.unep.org*, Accessed October 2016.

UNEP (United Nations Environmental Programme). (2015). Toolkit for identification and quantification of mercury releases, version 1.3 April 2015. *http://www.chem.unep.ch,* Accessed June 2016.

11

The health benefits of attaining and strengthening air quality standards in Cape Town

Samantha Keen[1], Katye Altieri[2]

[1]Energy Research Centre, University of Cape Town, Private Bag X3, Rondebosch, 7700, South Africa, samantha.keen@uct.ac.za
[2]Energy Research Centre, University of Cape Town, Private Bag X3, Rondebosch, 7700, South Africa, katye.altieri@uct.ac.za

Abstract

The link between pollution and poor health and mortality has been established globally. Developing countries carry most of the burden of ill health from air pollution, and urban centres like the City of Cape Town even more so. Effective air quality management to protect human health relies on the attainment of air quality standards. This study uses the Benefits Mapping and Analysis Program (BenMAP) along with a locally derived exposure-response function and air quality monitor data to investigate whether the consistent attainment of current or more stringent air quality standards would avoid loss of life. The results show that attaining the PM_{10} 24-hour mean South Africa National Standard limit and the PM_{10} and SO_2 24-hour mean World Health Organisation guidelines in Cape Town reduces levels of pollutants and does reduce excess risk of mortality in Cape Town.

Keywords

health impact assessment, air pollution, mortality, BenMAP, Cape Town, PM_{10}, SO_2, South African National Standard (SANS) limit, World Health Organisation (WHO) guideline

Introduction

The link between ill health and mortality and air pollution has long been established, for both long- and short-term exposure (Brunekreef and Holgate 2002). This association is almost certainly greater in developing than in developed countries (Shah et al. 2015). One of the reasons for this increased vulnerability to the ill effects of poor air quality is the cumulative impact of poor environmental health (Norman et al. 2010).

In Cape Town, concerns about air quality have motivated investigation of the constituents and causes of poor air quality, for example the occurrence of a brown haze over Cape Town in the colder months of May to September (Wicking-Baird et al. 1997) and the unusually high concentrations of particulate matter (PM) in the township of Khayelitsha (City of Cape Town 2008). Studies of the association of local pollutant exposure and morbidity include health surveys near the local crude oil refinery and an audit of hospital admissions during periods of atmospheric temperature inversions (Truluck 1993; White et al. 2009), and note that parts of the City suffer a disproportionate burden of air pollution-related ill health (Scorgie and Watson 2004).

Air pollution in Cape Town is mainly attributed to traffic vehicles, industry and the oil refinery, as well as to the domestic burning of fuels and the burning of waste (including car-tyres). Air quality interventions by the City of Cape Town include random vehicle emissions testing and the requirement for industry to apply for air emissions licences. The main goal of the National Environmental Management Air Quality Act (NEM:AQA) is to attain air quality objectives, and to this end the City monitors ambient air quality.

A health impact assessment to quantify the impact of attaining air quality objectives is based on the exposure-response relationship and population exposure data, which is derived from air quality and population data. A shortage of local epidemiological research to quantify the exposure-response relationship has encouraged the use of functions from the international literature in order to estimate the impact of pollution on human health (Scorgie et al. 2004), albeit with the caveat to do so with caution (Wichmann 2005).

The World Health Organisation (WHO) recommends that local health impact assessments use multi-site study or meta-analytic summary health impact functions except where the target population differs from the aggregate in its response to air pollution (WHO 2001). Globally, South Africa bears a sixth of the burden of HIV/AIDS infection and is second only to Lesotho in incidence rates (WHO 2015a); co-infection of HIV/AIDS and tuberculosis profoundly compromises the immune system (Powlowski et al. 2012). In Cape Town, both HIV/AIDS and TB are recorded as being in the top ten causes of death (Lehohla 2014). These facts suggest that the health of the local population may be significantly more vulnerable than the aggregate in available metaanalyses and multi-site studies, which are predominantly based on developed country data (Wichmann 2005).

The only locally proposed exposure-response functions are based on City of Cape Town exposure to elevated 24-hour mean levels of air pollutants and excess risk of mortality (with 95% confidence interval) for 2001–2006. The association was calculated for particulate matter of less than 10 μm (PM_{10}), nitrogen dioxide (NO_2) and sulphur dioxide (SO_2) with cerebrovascular, cardio-vascular and respiratory diseases (Wichmann & Voyi 2012). The study used a case-crossover study design with a time-stratified approach (using matching days of the week) to quantify associations of cerebrovascular disease mortality with inter-quartile range (IQR) increases of 12 $μg/m^3$ in PM_{10} 24-hour mean values; of cardiovascular and cerebrovascular disease mortality associated with elevated IQR levels of 12 $μg/m^3$ of NO_2; and of cardiovascular disease mortality and an IQR increase of 12 $μg/m^3$ of SO_2.

The case-crossover study design and use of impact lags of the same day to 5 days control for confounding factors that are considered constant in the short-term, for example seasonality or vulnerability attributable to smoking. The use of conditional regression analysis controlled for confounding as a result of temperature, humidity and day of week effects. The study conducted statistical analyses for the warmer months (September – April), cooler months (May – August), and the entire period to control for seasonality.

The study design employed by Wichmann and Voyi (2012) is not without limitation. Case-crossover design studies, as a specific type of time series studies, characteristically fail to identify displaced mortality, or 'harvesting', of sufferers of chronic exposure (Bateson & Schwartz 1999). This displaced mortality is identified by a sharp and temporary decrease in mortality immediately after a period of pollutant-associated excess mortality. This creates uncertainty as to whether the cause of excess mortality is acute exposure, longer-term lower level chronic exposure, or the incidence of both. However, the long-term data (over a period of decades) that would be needed to separate out and quantify the impact of chronic exposure (Rabl et al. 2011) are not available.

It is generally accepted that it is difficult to isolate the human health impact of any single pollutant. This is especially so for pollutants which tend to co-vary. SO_2 and PM are a good example of this, in part because SO_2 oxidises to sulphate, which coalesces with, and is then measured as part of the PM mass concentration (WHO 2006). In addition, vulnerability is driven by multiple factors (Norman et al. 2010) and few diseases are attributable to only one pollutant.

Notwithstanding the need to bear these limitations in mind when interpreting study results, the need for local data studies is well recognised (Wichmann 2005, Norman et al. 2010). This study uses the locally derived exposure-response function to investigate whether the consistent attainment of SANS limits and WHO guidelines would reduce associated excess mortality risk in Cape Town.

Materials and methods

The Benefits Mapping and Analysis Program (BenMAP) is open source software developed by the United States Environmental Protection Agency to calculate human health impacts and the economic costs thereof as a result of changes in ambient air pollution. Inputs to BenMAP include air quality morbidity or

mortality incidence, population data, GIS shapefile(s) for the area of interest and a health-impact or exposure-response function. The program has been successfully applied for analyses globally and is effective in integrating data available at different scales (Hubbell et al. 2009).

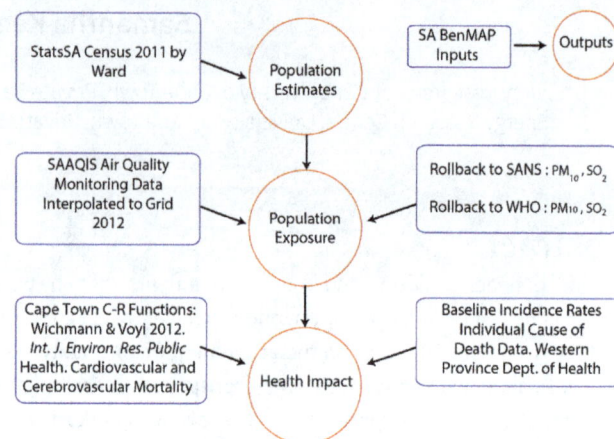

Figure 1: Schematic of stepwise inputs and outputs for BenMAP calculations of health impacts.

Data inputs

The health impact function is adopted from Wichmann and Voyi's (2012) epidemiological case-crossover study, which was based on 2001–2006 mortality data and 24-hour mean pollutant monitor data. The function is based on the 24-hour mean, so only the 24-hour mean SANS limits and WHO guidelines are applied in this study. The South African National Standard (SANS) limit for ambient concentrations of PM_{10} is 75 $μg/m^3$ and for SO_2 it is 125 $μg/m^3$ (SANS 1929: 2004). The more stringent WHO guidelines are 50 $μg/m^3$ and 20 $μg/m^3$, respectively.

Air quality monitoring data was downloaded from the City of Cape Town Open Data Portal (City of Cape Town 2014). The data was processed to remove all non-number entries and formatted for input to BenMAP. Analysis of data availability reveals considerable data gaps for some stations (Appendix: Table 1). Monitor stations are located where risk of exposure to pollutants is high because of proximity to pollutant sources and risk of population exposure. For areas that lack air quality data, either because there is no monitor station or because of gaps in data, BenMAP conducts user-specified interpolation. This study used the Voronoi Neighbour Averaging method, which uses data from more than one nearby monitor (the algorithm is more fully described in RTI International (2015)).

Individual cause of death data for the Western Cape was provided by the Western Cape Health Department. The WHO international statistical classification of diseases and related health problems, 10th revision (ICD-10) (WHO 2015b) was used to aggregate the data to broader categories including death as a result of cardiovascular and cerebrovascular disease and to calculate all age and gender mortality rates. This is in keeping with the method used to derive the exposure-response function (Wichmann and Voyi 2012).

Cape Town municipality and ward GIS shapefiles provided the grid on which BenMAP mapped the air quality, population and incidence data. BenMAP calculated the incidence values using the mortality data and input Census 2011 ward population data (Statistics South Africa 2012).

Figure 2: *Map of air quality monitor stations in the Cape Town Metropolitan area. (City of Cape Town. State of Environment Report. 2009)*

Sensitivity analysis

To estimate the impact of a change in air quality related health endpoints, BenMAP runs the health impact function linking the monitor data and incidence data (the Baseline scenario) and a Rollback to a hypothetical case in which levels of air quality data are reduced or limited (the Control scenario). The health impact function form used in this study is:

$$\Delta Y = (1-(1 / EXP(Beta * \Delta AQ))) * Y_0 * Pop \qquad (1)$$

where Y_0 is the incidence function in the 'Baseline' scenario, Pop is the population, ΔAQ is the change in air quality between the Baseline and Control scenarios, and Beta is the impact function for an IQR increase in pollutant, given as the 95th percentile range in order to take account of the uncertainty in the effect estimates.

The incidence function is calculated by:

$$Mortality\ rate * Pop \qquad (2)$$

A Monte Carlo approach is used to create a normal distribution of incidence. The approach is well described elsewhere (RTI International 2015).

The sensitivity analyses for PM_{10} are set to the SANS limit of 75 µg/m³ and the WHO guideline of 50 µg/m³. The analysis for SO_2 uses the SANS limit of 125 µg/m³ and the WHO guideline of 20 µg/m³.

Results and discussion

BenMAP generated Baseline and Control air quality surface maps to overlay the GIS shapefile layers (municipality and ward boundaries). A third air quality surface map shows the reduction in pollution exposure (ΔAQ). Assuming uniform distribution of each ward population, the change in incidence (ΔY) was calculated on a ward by ward basis (each ward representing a shapefile grid cell). The wards near the monitors known for high concentration readings and with high population density benefitted most from attaining the WHO guidelines and SANS limit. The wards near the Khayelitsha monitor station (Figure 2) benefit most from reduced excess mortality risk as a result of attaining the PM_{10} SANS limit and WHO guideline. The wards with the Bellville South and with the Tableview monitor stations benefit notably from attainment of the SO_2 WHO guideline. Bellville South is an industrial area and Tableview is home to the oil refinery.

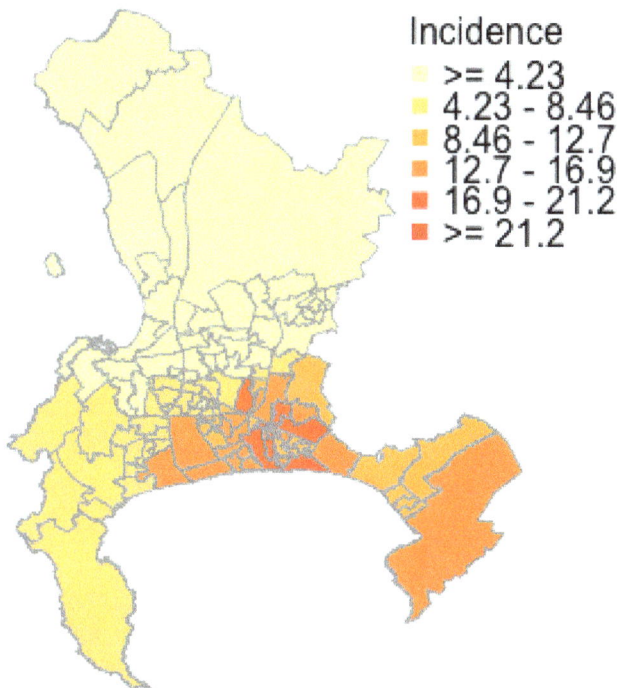

Figure 3: *Reduction in excess mortality as a result of attaining the PM_{10} SANS 24-hour mean limit.*

The inherent uncertainty in this impact assessment (as a result of incomplete input data-sets and health impact function limitations) should be considered in any interpretation of the results below. To minimise the risk of double counting, the

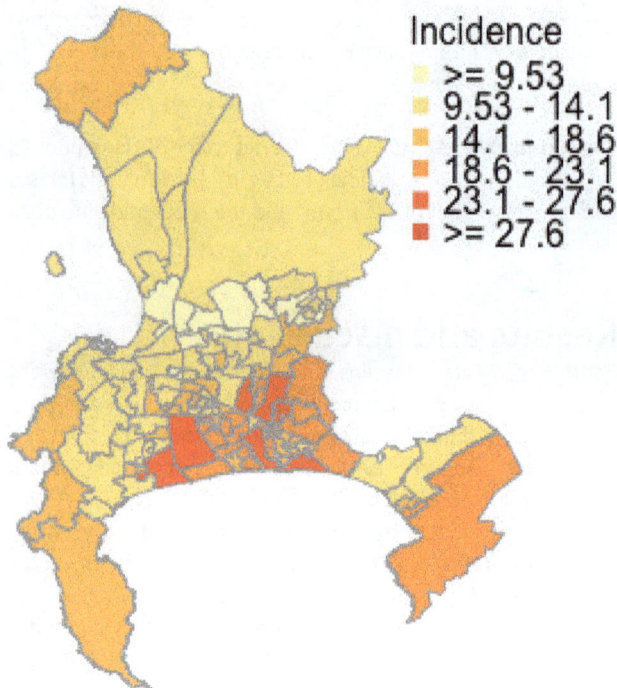

Figure 4: *Reduction in excess mortality as a result of attaining the PM_{10} WHO 24-hour mean guideline.*

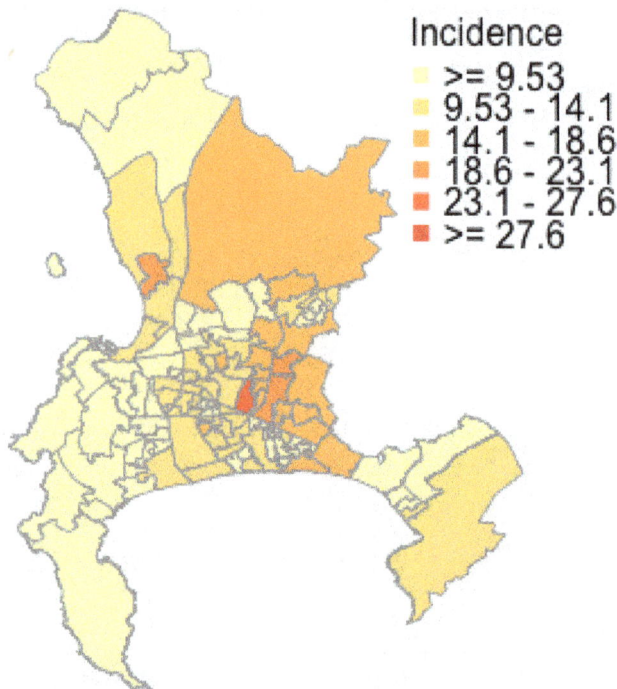

Figure 5: *Reduction in excess mortality as a result of attaining the SO_2 WHO 24-hour mean guideline.*

impact of reducing the two pollutants cannot be summed. The use of the ICD-10 classification reduces, but does not eliminate misattribution of pollutant with health endpoints. Acute exposure impacts might be under-estimated by any displaced mortality effect.

PM_{10}-related mortality reductions

Attaining the PM_{10} SANS limit
Attaining the PM_{10} SANS limit of 75 µg/m³ is estimated to reduce the excess mortality by a mean of up to 21 cases of excess mortality in the worst affected wards, and 857 cases across the City for the year.

Attaining the PM_{10} WHO guideline
Attaining the PM_{10} WHO guideline of 50 µg/m³ is estimated to reduce excess mortality by a mean of up to 28 cases of cerebrovascular-related deaths annually in some wards (Figure 4). Over the whole municipality the reduction is 1 690 cases of excess mortality for the year.

SO_2-related mortality reductions

Attaining the SO_2 SANS limit
There was no change in incidence brought about by limiting peak concentrations of SO_2 to no more than 125 µg/m³.

Attaining the SO_2 WHO guideline
Attaining the WHO guideline for SO_2 avoids a mean of 20 cases of cardiovascular-related excess daily mortality in some wards. Across the City, the estimated avoided excess mortality is 1 174 cases for the year (Figure 5).

Considerations, limitations and further work
The level of uncertainty in the results is driven in part by the level at which the input data is aggregated, largely by the absence of data, and in part by the use of interpolation for absent data. For example, in this study the population data is at ward level and the pollutant monitors, located in some wards only, report incomplete data sets. Loss of life data at ward level is not available in light of the need to protect anonymity. For this reason Health District level data was aggregated across the municipality.

The exposure-response function applied across the City is based on data from three air quality monitoring stations in relatively close proximity: City Hall, Goodwood and Tableview (Wichmann and Voyi 2012). In the light of social inequality across the city of Cape Town, the difference in risk factors for poor health varies considerably (Norman et al. 2010). More localised health impact functions, for example at the health district level, would likely reduce uncertainty in results. Uncertainty in health effect functions might be further reduced using multi-pollutant models and by long-term cohort studies to quantify the chronic impacts of air pollution.

Conclusions
The attainment of the SANS 24-hour mean limit for PM10 reduces annual excess mortality by 857 cases. This benefit is nearly doubled by attaining the WHO 24-hour mean PM10 guideline. The attainment of the SO2 24-hour mean limit reduces annual excess mortality by 1 174 cases. The areas of Cape Town that benefit most from the air quality improvements are those near

the Khayelitsha monitoring station, the Tableview monitoring station and local oil refinery respectively, and the industrial area of Bellville South. The production of locally estimated exposure-response functions has the potential to support the prioritisation of attaining air quality standards.

Acknowledgements

Funding for this work came from the London School of Economics International Growth Centre Project No. 895430. The Western Cape Health Department provided cause of death data and related support. The City of Cape Town made air quality data available on their Open Portal web page.

References

Bateson TF, Schwartz J. Control for seasonal variation and time trend in case-crossover studies of acute effects of environmental exposures. Epidemiology. 1999 Sep 1;10(5):539-44.

Briggs D. Environmental pollution and the global burden of disease. British Medical Bulletin. 2003 Dec 1;68(1):1-24.

Brunekreef, B. & Holgate, S.T., 2002. Air pollution and health. *The Lancet*, 360(9341), pp.1233– 1242.

Burger, L., von Groenewaldt, R. and Bird, T., 2014. *Atmospheric Impact Report: Sasol Secunda Facility*, Halfway House. Available at: http://www.srk.co.za/files/File/South-Africa/publicDocuments/Sasol_Postponement/SOGS_/ANNEXURE_A_SYNFUELS_Atmos pheric_Impact_Report.pdf.

City of Cape Town, Air quality monitoring. Available at: https://www.capetown.gov.za/en/Water/Publis hingImages/AQM_2.jpg [Accessed June 12, 2016].

City of Cape Town, 2008. *Khayelitsha Air Pollution Strategy Project (KAPS) Report*, City of Cape Town.

City of Cape Town, 2014. Open data portal: Air quality. Available at: https://web1.capetown.gov.za/web1/OpenDat aPortal/DatasetDetail?DatasetName=Air quality [Accessed May 9, 2016].

Hubbell, B.J., Fann, N. & Levy, J.I., 2009. Methodological considerations in developing local-scale health impact assessments: Balancing national, regional, and local data. *Air Quality, Atmosphere and Health*, 2(2), pp.99–110.

Lehohla, P., 2014. Comparison of the demographic profiles of two cities: Cape Town and Tshwane, 1997–2011.

Karim, S.S.A., Churchyard, G.J., Karim, Q.A. and Lawn, S.D., 2009. HIV infection and tuberculosis in South Africa: an urgent need to escalate the public health response. *the Lancet*, 374(9693), pp.921-933.

Norman, R., Bradshaw, D., Lewin, S., Cairncross, E., Nannan, N., Vos, T. and Collaborating, S.A.C.R.A., 2010. Estimating the burden of disease attributable to four selected environmental risk factors in South Africa. *Reviews on environmental health*, 25(2), pp.87-120.

Pawlowski, A., Jansson, M., Sköld, M., Rottenberg, M.E. and Källenius, G., 2012. Tuberculosis and HIV co-infection. *PLoS Pathog*, 8(2), p.e1002464.

Rabl A, Thach TQ, Chau PY, Wong CM. How to determine life expectancy change of air pollution mortality: a time series study. Environmental Health. 2011 Mar 31;10(1):1.

RTI International, 2015. BenMAP: Environmental Benefits and Mapping Analysis Program – Community Editition. User's manual appendices. Available at: https://www.epa.gov/sites/production/files/2015-04/documents/benmap- ce_user_manual_appendices_march_2015.pdf.

Scorgie, Y., Paterson, G., Burger, L.W., Annegarn, H.J. and Kneen, M.A., 2004. Socio-Economic Impact of Air Pollution Reduction Measures – Task 4a Supplementary Report: Quantification of Health Risks and Associated Costs Due to Fuel Burning Source Groups. *Report compiled on behalf of NEDLAC*.

Scorgie, Y. & Watson, R., 2004. *Updated air quality situation assessment for the City of Cape Town. Report done on behalf of: City of Cape Town, Report No. APP/04/CCT- 02*, Cape Town.

Shah, A.S., Lee, K.K., McAllister, D.A., Hunter, A., Nair, H., Whiteley, W., Langrish, J.P., Newby, D.E. and Mills, N.L., 2015. Short term exposure to air pollution and stroke: systematic review and meta-analysis. *bmj, 350*, p.h1295.

Statistics South Africa, 2012. Census 2011, *Statistical release (Revised)* P0301. 4., Pretoria.

Truluck, T.F., 1993. *Hospital admission patterns of childhood respiratory illness in Cape Town and their association with air pollution and meteorological factors*. University of Cape Town.

White, N., van der Walt, A., Ravenscroft, G., Roberts, W. and Ehrlich, R., 2009. Meteorologically estimated exposure but not distance predicts asthma symptoms in schoolchildren in the environs of a petrochemical refinery: a cross-sectional study. *Environmental Health*, 8(1), p.1.

World Health Organization, 2001. Quantification of the health effects of exposure to air pollution. *Copenhagen, WHO, Regional Office for Europe*.

World Health Organization. Air quality guidelines: global update 2005: particulate matter, ozone, nitrogen dioxide, and sulfur dioxide. World Health Organization; 2006.

World Health Organization, 2015a. *Global tuberculosis report 2015*. World Health Organization.

World Health Organization, 2015b. International Statistical Classification of Diseases and Related Health Problems 10th Revision. *http. apps.who.int/classifications/icd10/browse/2010/en*.

Wichmann, J., 2005. Air pollution epidemiological studies in South Africa: Need for freshening up. Reviews on Environmental Health, 20(4), pp.265-301.

Wichmann, J. & Voyi, K. 2012. Ambient air pollution exposure and respiratory, cardiovascular and cerebrovascular mortality in Cape Town, South Africa: 2001-2006.

Appendix A. Air quality monitor data availability and air quality targets

Table 1: Percentage of hourly mean air quality monitor data available for 2013.

Monitoring station	SO$_2$ (%)	PM$_{10}$
Atlantis	58%	-
Bellville South	88%	97%
Bothasig	77%	-
City Hall	98%	-
Foreshore	-	97%
Goodwood	19%	48%
Khayelitsha	68%	69%
Killarney	15%	29%
Plattekloof	53%	52%
Tableviews	89%	80%
Wallacedene	47%	50%

12

Characterising the impact of rainfall on dustfall rates

Jared Lodder[*1], Martin A. van Nierop[1], Elanie van Staden[1], and Stuart J. Piketh[2]

[1]Gondwana Environmental Solutions, 562 Ontdekkers Road, Florida, Roodepoort, 1716, South Africa, info@gesza.co.za
[2]Unit for Environmental Sciences and Management, North-West University, Potchefstroom, 2520, South Africa, Stuart.Piketh@nwu.ac.za

Abstract

Soil moisture increased the cohesion potential between particles, reducing the ability of the particle to be entrained. Dust suppression techniques are designed to increase soil moisture and therefore soil cohesion through the application of water or water-based chemicals to surfaces that have known potential for dust entrainment. Rainfall has the ability to act as a natural dust suppression mechanism; however, there is a paucity of literature on the actual effectiveness of rainfall in this regard. The ASTM D1739 methods for dustfall monitoring, commonly used in South Africa, and the National Dust Control Regulations (2013), both state that rainfall should be recorded when conducting dustfall monitoring. The rationale is that rainfall or the absence thereof, results in lower or higher dustfall rates, respectively. A suitable study site was identified in Mpumalanga, South Africa. This site had eight non-directional dustfall samplers in the near vicinity of an air quality monitoring station. Dustfall results from the eight samplers were analysed based on four scenarios, two that considers the presence of rainfall and two that consider the absence of rainfall. This analysis was further combined with wind speed data. This study, over a 24-month period indicates that there is no substantial evidence that above average rainfall will result in below average dustfall. This occurred for one month out of 24 months. Conversely, there is no consensus that the absence of rainfall will result in higher dustfall rates, which occurred cumulatively 30% of the time. Additional environmental and / or operational information may have a greater influence on dustfall compared to rainfall. Careful consideration should be taken to prevent misrepresentation of causational effects of rainfall on dustfall results. Management of dust should be undertaken through dust mitigation measures irrespective of the natural rainfall regime.

Keywords

Dustfall, air quality, ASTM D1739, Rainfall, dust suppression, South Africa

Introduction

Soil moisture increases the cohesion between soil particles increasing the resistance of soil particles to wind-blown entrainment and erosion (Wiggs, Baird and Atherton 2004). Wind tunnel studies of moisture content and wind speeds have allowed for the development of critical moisture thresholds (Han et al. 2009; McKenna-Neuman and Nickling 1989). The critical threshold is the moisture content value whereby the potential for entrainment and sediment transport is suppressed (Wiggs, Baird and Atherton 2004) albeit with a wide variation of results (Namikas and Sherman 1995).

Fugitive dust suppression techniques aim reducing the materials erodibility by increasing soil moisture content and / or soil cohesion through the application of water and water-based chemicals to sources of dust (Thompson and Visser 2002; Tsai, Lee and Lin 2003). Rainfall has the ability to act as a natural dust suppressant through the same mechanism and anthropogenic dust control measures. However, there is a paucity of data on the effectiveness of rainfall as a natural dust suppressant mechanism.

South Africa promulgated the National Dust Control Regulations (NDCR) in 2013 (South Africa 2013), which specifies the acceptable dustfall rates within Residential and Non-Residential areas. Furthermore, the NDCR states the monitoring methodology and the reporting requirements. Of interest in this study, is the requirement for rainfall to be included in dustfall reports (ASTM 1970; South Africa 2013). The implication of requiring knowledge on rainfall during a dustfall monitoring survey suggests that rainfall has the ability to impact dustfall rates. The hypothesis is that rainfall should reduce dustfall rates through increased soil moisture, which increases soil cohesion thus reduces the potential for wind-blown entrainment (Wiggs and Holmes 2011).

This study aims to test this hypothesis and subsequently to improve our understanding of the importance of including and discussing rainfall in relation to dustfall results.

Study Site

The actual study site is undisclosed for confidentiality reasons. The study site is located within the Mpumalanga Province, South Africa. The broader area surrounding the study site (approximately 15 km radius) includes formal and informal settlements, mining, agriculture and power generation.

A dustfall network comprised of eight non-directional dustfall samplers without windshields was strategically located based on the objectives of the air quality monitoring plan. This resulted in the closest dustfall sampler being located approximately 1 km from an ambient air quality monitoring station (AQMS), while the furthest sampler was located approximately 10 km from the AQMS.

The samplers, in relation to each other, are on average 8 km apart, with the furthest distance 17 km. The primary and secondary sources of dustfall for each site is provided (Table 1). Primary sources of dustfall are agriculture and mining while two sites could be considered as baseline monitoring due to their location within either residential or smallholding areas. Secondary sources of dustfall include large-scale industrial development, mining and agricultural activities.

Table 1: Primary and secondary sources of dustfall for each site.

Site	Primary source	Secondary source
Site 01	Agriculture	Large-scale construction (~10 km2)
Site 02	Agriculture	Large-scale construction (~10 km2)
Site 03	Mine haulage road	Large-scale construction (~10 km2)
Site 04	None; residential	None
Site 05	Agriculture and waste rock stockpile	None
Site 06	None; smallholding	Coal mining
Site 07	Agriculture	None
Site 08	Sand mining	Agriculture

The AQMS monitors various meteorological and pollutant parameters. The parameters pertinent to this study include rainfall, wind speed and wind direction, which were monitored using a MetOne 300 series (rainfall) and MetOne 034B (winds).

Dustfall monitoring networks are typically developed to monitor the potential for wind-blown dust from specific sources, such as tailings dams, stockpiles and roads. As such, it is useful to analyse data from the network as a whole unit, instead of analysing each individual samplers. This process allows for improved understanding of regional impacts on the individual dustfall units. However, this study site provides challenges in conducting this type of analysis due to the multiple sources of dustfall.

Methods

Dustfall and Meteorological Monitoring
Dustfall monitoring commenced in November 2011 at all eight sites and continued until April 2015. The same type of sampler and collector (buckets) were used throughout the study. All collectors were positioned at above 2 m from ground-level. The sample preparation and analysis were conducted in accordance with the ASTM D1739 requirements (ASTM 1970).

Meteorological monitoring commenced in December 2012 and

continued until April 2015. The rain gauge was attached to the top of the AQMS, approximately 2 m above ground, while the wind sensor was located on a mast, approximately 5 m above ground.

Data from the dustfall monitoring was collected on monthly basis, while the rainfall and wind data was collected at 5-minute averages.

Rainfall and meteorological data was analysed for the period January 2013 to December 2014, where both datasets were concurrently operational. In addition, the time-period would allow for two full years to be analysed.

Dustfall
The monthly dustfall rates from each site, over the two-year period, were averaged to determine baseline dustfall data per month. Each month's dustfall data was compared to the monthly average to determine whether the observed dustfall rates were above or below the baseline average.

Rainfall
Rainfall data from four meteorological stations managed by the Department of Water and Sanitation (DWS) was collected to determine baseline rainfall for the dustfall network (Department of Water and Sanitation 2008). The four stations are located between 20 and 50 km from the dustfall network and had rainfall data extending back to 1962 (one station).

Observed rainfall data was for the dustfall network was collected from the AQMS. The observed rainfall for each month (January 2013 to December 2014) was compared to the baseline data to determine if the observed rainfall was above or below the baseline average.

Data analysis
Dustfall and rainfall for each month was compared to the baseline dustfall and rainfall. Based on this process, four scenarios were identified:
1. Rainfall reduces dustfall rates when monthly dustfall rates are below average and rainfall is above average;
2. Rainfall does not reduce dustfall rates when monthly dustfall rates are above average and monthly rainfall is above average;
3. The absence of rainfall may increase dustfall rates when the monthly dustfall rates are above average and monthly rainfall is below average; and
4. The absence of rain does not increase dustfall when monthly dustfall rates are below average and rainfall is below average.

This process was conducted under two conditions, 1) conservative and 2) stringent:
1. The conservative approach considered all valid data points for each site for each month. Each individual dustfall point was compared to the monthly rainfall and classified into one of the four scenarios. This enabled all data points (186 valid points) to be used in the analysis.
2. The stringent approach was applied where all eight sites exhibited the same trend (e.g., all eight sites had recorded

above-average dustfall rates). The premise for this approach is that the dustfall network should be considered as whole unit and that all sites should exhibit the same trend, which could indicate regional conditions. As such, only 80 valid points were used in this analysis.

Additional data was analysed in attempt to further understand the findings within each of the scenarios, including;

a)   The total rainfall [mm]; and

b)   The average monthly wind speed [m/s].

Results

Rainfall

The study site is located in a summer rainfall region of South Africa. The average seasonal rainfall (DWS Rainfall) and the observed seasonal rainfall follows the expected rainfall trends for a summer rainfall area (Table 2).

Table 2: *Average Seasonal Rainfall [mm] from the DWS Stations (Department of Water and Sanitation 2008) and the observed rainfall from site.*

Season	Months (abbr.)	Rainfall [mm]	
		Baseline	Observed
Summer	DJF	119	110
Autumn	MAM	50	62
Winter	JJA	10	1
Spring	SON	79	82

Further correlation was between observed rainfall per month as a function of the baseline data and average observed monthly rainfall as a function of the baseline data was conducted. The findings indicate a R2 of 0.65 and 071 respectively. The correlation between monthly-observed rainfall as a function of the baseline rainfall was further analysed in terms of seasonal trends. All winter months experienced below average rainfall whilst all other seasons contained a variation of above average and below average rainfall (Figure 1).

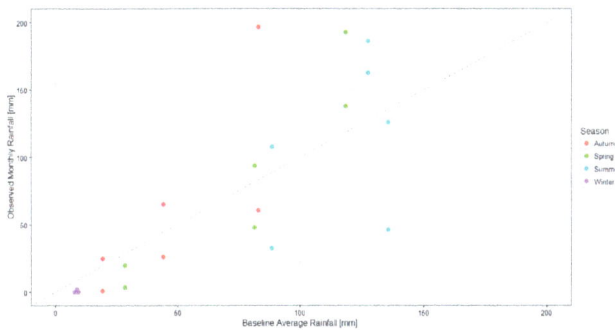

Figure 1: *Comparison of observed monthly rainfall (y-axis) with long-term monthly-averaged rainfall (x-axis) from the Department of Water and Sanitation (2008) with the colours representing the seasons.*

The correlation between the two datasets suggests that the rainfall experienced at site were representative of typical rainfall conditions in the area. Above average rainfall on a monthly basis occurred nine out of 24 months (Table 3).

Wind speeds

Average monthly wind speeds over the two-year period ranged between 1.6 and 3.2 m/s. Average seasonal winds range from Autumn (1.8 m/s) to Spring (2.9 m/s). The higher winds during Spring are consistent with typical conditions on the Highveld.

Table 3: *Above (blue) and below (red) average rainfall occurrences*

January 2013	February 2013	March 2013	April 2013
May 2013	June 2013	July 2013	August 2013
September 2013	October 2013	November 2013	December 2013
January 2014	February 2014	March 2014	April 2014
May 2014	June 2014	July 2014	August 2014
September 2014	October 2014	November 2014	December 2014

Dustfall

The total number of valid dustfall rates was 186 out of a maximum of 192. The 186-dustfall rates were analysed. The observed monthly rainfall at site for all months was plotted as a function of observed monthly dustfall rates (Figure 2). The graph indicates a wide range of rainfall and dustfall rates above and below the average rainfall and dustfall rates (blue).

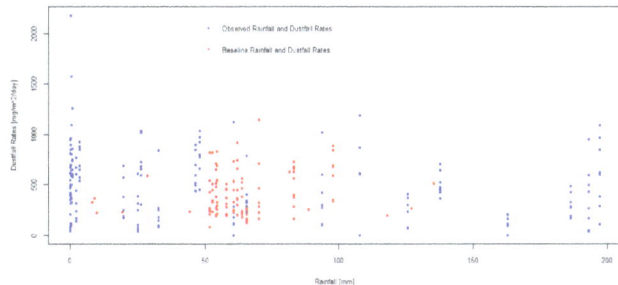

Figure 2: *Dustfall rates as a function of rainfall for observed monthly data (red) and averaged data dustfall and rainfall (blue).*

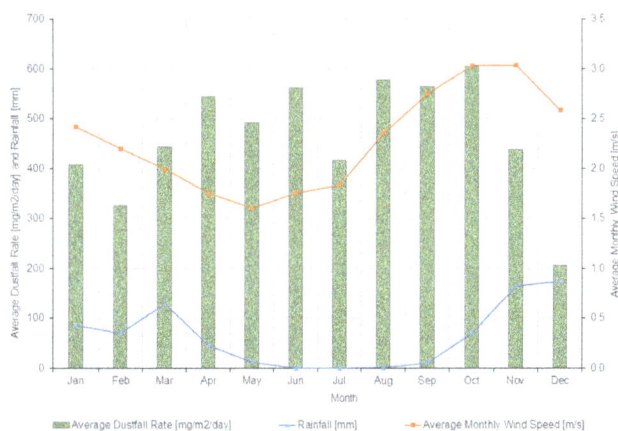

Figure 3: *Monthly averaged dustfall rates (green bars), total monthly rainfall (blue line) and averaged wind speeds (red line).*

The averaged dustfall rates, monthly wind speeds and observed rainfall was plotted (Figure 3). This graph clearly illustrates the expected meteorological trends characteristic of the Highveld. Wind speeds increased from the end of August, peaking in

October / November. Rainfall increases from Spring through to Summer and declines sharply during Autumn, with little to no rainfall occurring during the Winter periods. The average monthly dustfall indicates two peaks, April to June and August to October. Whilst the average rainfall for these two peaks is similar (20 versus 28 mm, respectively), these two peaks have contrasting average wind speeds (1.7 versus 2.7 m/s, respectively).

A comparison of each scenario was performed considering the conservative approach (Table 4), where all data was considered, and the stringent approach (Table 5), where only those periods where the entire network responded in the same way. The average wind speed, total rainfall and the frequency of occurrence to each scenario (valid counts) was analysed. The two approaches indicate similar trends in valid counts; however, there is a marked contrast in the average rainfall and wind speeds for each approach.

The original hypothesis, that increased rainfall should reduce dustfall, occurs the least amount of times in both approaches.

Table 4: *Statistics for each of the four scenarios under conservative approach.*

Scenario	Valid counts [%]	Average Rainfall [mm]	Average Monthly Wind speed [m/s]
1	15	142	2.4
2	17	143	2.5
3	35	20	2.1
4	33	28	2.2

Table 5: *Statistics for each of the four scenarios under stringent approach.*

Scenario	Valid counts [%]	Average Rainfall [mm]	Average Monthly Wind speed [m/s]
1	4	25	1.6
2	9	162	2.9
3	13	9	1.7
4	17	36	2.2

Discussion

Of the two approaches considered, the stringent approach provides, in the opinion of the author, more telling information on the potential effect of rainfall, or the absence of rainfall on dustfall rates. As such, the discussion will be focused on the results provided in Table 5. It is important to note that this approach excluded 57% of the total dataset, due to the stringent criteria of inclusion. In spite of this, the objective was to identify broad trends that could be used to guide further research into this topic.

The presence of rainfall on dustfall rates

The ability of rainfall to reduce dustfall rates is investigated based on Scenario 1 (above average rainfall and below average dustfall) and Scenario 2 (above average rainfall and above average dustfall).

Scenario 1 occurred during May 2013 only, which equates to 4% of the time. Typical rainfall during May is approximately 19 mm. Scenario 1 recorded 25 mm of rainfall, which is not substantially higher than the average. This rainfall occurred of a 2-day period. During the same period however, the average monthly wind speeds of 1.6 m/s is the lowest compared to all scenarios. It is unlikely that the above average rainfall that occurred during Scenario 1 would have resulted in reduced dustfall. There is a much greater likelihood that the reduced wind speeds during this period, prevented above average dustfall to occur.

Scenario 2 occurred during two months (November and December 2013). The hypothesis of Scenario 2 is that even when rainfall is above average, dustfall rates are above average. The highest rainfall (162 mm) and the highest wind speeds (2.9 m/s) occurred during these two months. More than half of the days during this period experienced rainfall. The potential for dust entrainment should have been the lowest during this period, based on the likelihood of increased soil / dust cohesion. It is likely that due to the high wind speeds (compared to the other Scenarios), the potential mitigating effects of rainfall dissipated in favour of wind speeds. This finding is supported in wind a wind erosion study in the Free State Province conducted by Wiggs and Holmes (2011). This study supported the initial hypothesis of this study, that increased rainfall should reduce dustfall, however, during a period of high rainfall, the dustfall rates were observed to be the highest, dispelling this hypothesis.

The absence of rainfall on dustfall rates

Scenarios 3 and 4, which considered the impact of the absence of rainfall on dustfall results accounted for 30% of the dataset. The key hypothesis in relation to the absence of rainfall is that under conditions of low rainfall, it is expected that dustfall rates would be higher. This is considered in Scenario 3, which occurred over three months (April, June and July 2014. It is known that the winter periods, whilst they are the driest, are also some of the calmest periods during the year. As such, this hypothesis is not found to be true in this study either.

Alternatively, Scenario 4 occurred the most during this study, a total of four months out of 24 (January, June, July and September 2013). The anomaly in this scenario is the relatively high wind speeds, coupled with the absence of rainfall, still resulted in below average dustfall rates. This is again, contrary to expectations that higher dustfall rates should occur under dry, windy conditions.

Whilst Scenarios 3 and 4 did not conform to expectations, it does indicate that the absence of rainfall may have a greater influence on observed dustfall rates, than the presence of rainfall would have.

Is rainfall an effective dust suppression mechanism

Soil moisture is a critical factor in determining whether soil particles can be entrained (Tsai, Lee and Lin 2003). Material that has a gravimetric moisture content above 0.2% is considered extremely resistant to wind entrainment (McKenna-Neuman and Nickling 1989). Rainfall, depending on the duration and intensity, should theoretically, provide sufficient moisture to reduce the potential for entrainment.

Based on the four scenarios, the rainfall does not suggest that it has an impact on dustfall rates in this study area. It is likely that the absence of rainfall has a greater chance of enabling suitable conditions for increased dustfall; however, even under the presence of low average rainfall, below average dustfall occurs at a similar rate. As such, neither the presence of rainfall nor the absence of rainfall is conclusive in this study to determine whether dustfall rates will be increased or reduced. Additional environmental and / or operational factors may have a greater importance in determining high or low dustfall rates. Environmental factors include, wind gusts, evaporation, surface drying, surface disturbance, soil wetness and soil type (Négyesi et al. 2016; Tsai, Lee and Lin 2003; Wiggs, Baird and Atherton 2004). Operational factors could include, frequency of unpaved road usage, vehicle speeds, type of vehicle usage and additional dust suppression techniques (chemical applications and water spraying).

The low resolution of dustfall monitoring makes the interpretation and analysis of trends coarse. Higher resolution information (weekly or daily monitoring) may improve the understanding of the key factors that influence dustfall rates. Environmental conditions before and after rainfall events may provide an improved understanding of factors that contribute to higher or lower dustfall rates.

This study does suggest that dust mitigation measures should be conducted year-round and irrespective of the natural rainfall patterns. Adherence to this is likely to have a much greater impact on reducing dustfall rates compared to a reliance on natural dust suppression (rainfall).

The requirement to report rainfall in the National Dust Control Regulations (South Africa 2013) should be considered as supporting conditions other primary factors, such as wind speed, that could influence dustfall rates.

Careful consideration concerning causation effects should be taken. This study does not support that rainfall results in reduced dustfall. There is a greater likelihood that the absence of rainfall can cumulatively influence dustfall rates compared to the presence of rainfall.

The environmental factors that occur in between rainfall events are likely to have a much greater impact on dustfall results compared to the rainfall event itself. Under hot and / or windy conditions, the potential for evaporation is increased. Even under calm conditions, soil moisture content decreases sufficiently in approximately four hours becoming highly susceptible to entrainment (Tsai, Lee and Lin 2003). Furthermore, the soil types will react very differently to varying amounts of rainfall. These factors could, in a short-timespan, override the actual rainfall event by drying the soil to the point that dust entrainment can occur soon after rainfall events.

Conclusion

Rainfall has the ability to increase soil cohesion, which in turn, can reduce the potential for dust entrainment to occur. This study has attempted to improve the understanding of the potential for rainfall to reduce dustfall rates in a localised study area of Mpumalanga.

This study suggests that rainfall has little influence on reducing dustfall rates. Only 4% of the data (one month) supported the hypothesis that dustfall rates were below average when rainfall is above average. It is likely that wind speeds have a greater influence on dustfall rates, as higher wind speeds have the ability to entrain larger, greater mass particles, which will be deposited through gravitational settling.

Additional factors, such as the absence of rainfall, the intensity and frequency of rainfall, evaporation rates and local sources of pollution may influence dustfall rates more than rainfall. The environmental conditions that occur in between rainfall events may be of greater importance than the actual rainfall event.

The hypothesis that rainfall can act as a natural dust suppressant is not supported by this study in this specific location. Careful consideration should be taken on how rainfall is attributed causation effects on dustfall rates in reports.

Management of dustfall should be undertaken irrespective of rainfall events to ensure effective mitigation of dustfall.

References

ASTM. 1970, *Standard Test Method for Collection and Measurement of Dustfall (Settleable Particulate Matter)*, American Standard Test Method. West Conshohoken: ASTM International. (ASTM D1739).

Department of Water and Sanitation 2008, *Hydrological Services - Surface Water (Data, Dams, Floods and Flows)*, viewed 15 September 2016, <https://www.dwa.gov.za/hydrology/>.

Han, Q, Qu, J, Zhang, K, Zu, R, Niu, Q and Liao, K 2009, 'Wind tunnel investigation of the influence of surface moisture content on the entrainment and erosion of beach sand by wind using sands from tropical humid coastal southern China', *Geomorphology*, 104, 3–4. 230-7.

McKenna-Neuman, C and Nickling, WG 1989, 'A theoretical and wind tunnel investigation of the effect of capillary water on the entrainment of sediment by wind', *Canadian Journal of Soil Science*, 69, 1. 79-96.

Namikas, SL and Sherman, DJ 1995, 'Effects of surface moisture content on aeolian sand transport: a review', in V Tchakerian (ed.), *Desert Aeolian Processes*, New York, Chapman & Hall, pp. 269-93.

Négyesi, G, Lóki, J, Buró, B and Szabó, S 2016, 'Effect of soil parameters on the threshold wind velocity and maximum eroded mass in a dry environment', *Arabian Journal of Geosciences*, 9, 11. 588.

South Africa. 2013, *National Dust Control Regulations*. National Environmental Management: Air Quality Act 39 of 2004. (Notice 827). Government Printer, 36974, 1 Nov.

Thompson, RJ and Visser, AT 2002, 'Benchmarking and management of fugitive dust emissions from surface-mine haul roads', *Mining Technology*, 111, 1. A28.

Tsai, C-J, Lee, C-I and Lin, J-S 2003, 'Control of Particle Re-Entrainment by Wetting the Exposed Surface of Dust Samples', *Journal of the Air & Waste Management Association (Air & Waste Management Association)*, 53, 10. 1191-5.

Wiggs, GFS, Baird, AJ and Atherton, RJ 2004, 'The dynamic effects of moisture on the entrainment and transport of sand by wind', *Geomorphology*, 59, 1–4. 13-30.

Wiggs, GFS and Holmes, P 2011, 'Dynamic controls on wind erosion and dust generation on west-central Free State agricultural land, South Africa', *Earth Surface Processes and Landforms*, 36, 6. 827-38.

13

Towards the development of a GHG emissions baseline for the Agriculture, Forestry and Other Land Use (AFOLU) sector, South Africa

Luanne B. Stevens[1,2], Aidan J. Henri[*1] Martin Van Nierop[1], Elanie van Staden[1], Jared Lodder[1] and Stuart Piketh[2]

[1]Gondwana Environmental Solutions, 562 Ontdekkers Road, Florida, Roodepoort, 1716, South Africa, info@gesza.co.za
[2]Unit for Environmental Sciences and Management, North-West University, Potchefstroom, 2520, South Africa

Abstract

South Africa is a signatory to the United Nations Framework Convention on Climate Change (UNFCCC) and as such is required to report on Greenhouse gas (GHG) emissions from the Energy, Transport, Waste and the Agriculture, Forestry and Other Land Use (AFOLU) sectors every two years in national inventories. The AFOLU sector is unique in that it comprises both sources and sinks for GHGs. Emissions from the AFOLU sector are estimated to contribute a quarter of the total global greenhouse gas emissions. GHG emissions sources from agriculture include enteric fermentation; manure management; manure deposits on pastures, and soil fertilization. Emissions sources from Forestry and Other Land Use (FOLU) include anthropogenic land use activities such as: management of croplands, forests and grasslands and changes in land use cover (the conversion of one land use to another). South Africa has improved the quantification of AFOLU emissions and the understanding of the dynamic relationship between sinks and sources over the past decade through projects such as the 2010 GHG Inventory, the Mitigation Potential Analysis (MPA), and the National Terrestrial Carbon Sinks Assessment (NTCSA). These projects highlight key mitigation opportunities in South Africa and discuss their potentials. The problem remains that South Africa does not have an emissions baseline for the AFOLU sector against which the mitigation potentials can be measured. The AFOLU sector as a result is often excluded from future emission projections, giving an incomplete picture of South Africa's mitigation potential. The purpose of this project was to develop a robust GHG emissions baseline for the AFOLU sector which will enable South Africa to project emissions into the future and demonstrate its contribution towards the global goal of reducing emissions.

Keywords

AFOLU, GHG emissions, mitigation, projected baseline

Introduction

Anthropogenic activities have contributed 40% to the increase in carbon dioxide levels since 1750 and as a result have played a crucial role in shaping current and future climate. The Agriculture, Forestry and Other Land Use (AFOLU) sector plays a vital role in food security, sustainable development and climate change adaptation. The AFOLU sector is the only sector that acts as both a source (deforestation and peatland drainage) and sink (afforestation and management practices) of Greenhouse gases (GHGs), mainly CO_2. Non-CO_2 emissions (CH_4 and N_2O) primarily result from the agricultural sector through enteric fermentation, manure management, fertilizer application and biomass burning. Anthropogenic land use activities and changes in land cover can seriously affect these natural fluxes. Globally the AFOLU sector is responsible for almost a quarter (~10-12 $GtCO_2eq/yr$) of anthropogenic GHG emissions (Smith et al., 2014). South Africa's AFOLU sector is estimated to contribute around 7% of the total national GHG emissions (DEA, 2015).

Since South Africa is transitioning towards a low carbon economy mitigation options are being investigated. There are several supply and demand side options for mitigation in the AFOLU sector that have been identified. On the supply side emissions can be reduced from land use change, land and livestock management, and terrestrial carbon stocks can be enhanced by sequestration in soils and biomass. On the demand side emissions can be reduced through changing consumption patterns of natural resources (i.e. animal products and wood products). Over the past decade South Africa has improved the quantification of AFOLU emissions and the understanding of the dynamic relationship between sinks and sources through projects such as the 2010 GHG inventory (DEA, 2014a), the Mitigation Potential Analysis (MPA) (DEA, 2014b) and the NTCSA (DEA, 2015). These projects highlight the key mitigation opportunities in the country.

However, through these projects, a gap was identified in that South Africa does not have an emissions baseline (current or projected) for the AFOLU sector against which the mitigation potentials can be measured. This means that the AFOLU sector either gets underestimated or excluded from future emissions projections, which gives an incomplete picture of South Africa's mitigation potential. Baselines are routinely used for domestic policy planning. In recent years baselines have grown in importance as some countries (including South Africa) have used them to define their mitigation pledges in terms of emissions reductions.

A well-developed baseline, more specifically a projected baseline which shows future GHG emissions levels, will have the advantage of enabling Desired Emissions Reductions Outcomes (DEROs) and Carbon Budgets to be determined. In addition it will allow South Africa to demonstrate its contribution towards the global goal of reducing emissions from the AFOLU sector.

Methods

There is currently limited information and guidance available for setting national GHG baselines with significant variability in the approaches and assumptions used by countries globally. In general the methods employed are specific to countries' goals and targets (Clapp and Prag, 2012). A baseline scenario is defined as the future GHG emission levels in the absence of future, additional mitigation actions. It can also be referred to as the 'business-as-usual' scenario. Baseline scenarios can serve different purposes and therefore can be established at different levels of aggregation (e.g. project-specific, multi-project, sectoral, regional and national) so as to accommodate the various requirements of the specific applications.

It was decided for this study to use the last year of the National GHG inventory as the base year of emissions projections. Activity data and emission calculations used in the projections were extracted from the 2010 inventory (DEA, 2014a). Projections were set to decadal intervals up until 2050. The baseline was calculated in two parts; where emissions projections were made for the Agricultural sector and Land (FOLU) sector separately, these two baselines were then combined for a total AFOLU baseline.

The emissions calculations for Agriculture were based on the expanded AFOLU sector categories as they are in the national GHG inventory (2010), which incorporates the following agricultural components:
- Livestock enteric fermentation (CH_4),
- Livestock manure management (CH_4 and N_2O),
- Liming (CO_2),
- Urea application (CO_2),
- Direct N_2O emissions from managed soils,
- Indirect N_2O emissions from managed soils,
- Indirect N_2O emissions from manure management.

The emissions calculations for Land were based on emissions (CO_2, CH_4 and N_2O) from land-use conversion which takes into consideration changes in biomass carbon, dead organic matter carbon and soil carbon (gain-loss method). The land-use conversion categories used were those from the National GHG inventory 2010 which include:
- Forest land,
- Cropland,
- Grasslands,
- Wetlands,

- Settlements,
- Other land, and
- Emissions from biomass burning.

The methodology for calculating the GHG emissions is drawn from equations stipulated in IPCC 2006 guidelines (IPCC, 2006). A series of spreadsheet-based emission models were developed from these equations. Tier 1 and Tier 2 equations were used depending on the sub-category and the amount of data available. The detailed method of the emissions calculations and activity data are described in the project report (DEA, 2016). There are numerous approaches for making projections, one of them being a modelling approach which makes projections of future values of a variable based on a function that integrates the impacts of multiple drivers on that variable. A modelled approach was considered, however, this approach is very data intensive and requires the use of models which have already been tested or calibrated for the South African system. In the AFOLU sector there are numerous variables driving change and limited data to support them. There are also gaps in activity data which complicates scenario setting.

In the Agricultural sector the Bureau for Food and Agricultural Policy (BFAP) has developed a model to project changes in agricultural commodities. The model is an economic recursive, partial equilibrium model which incorporates economic, technological, environmental, political and social factors (BFAP, 2015). It was therefore decided for this study to build on the outputs of the BFAP modelling process, for the purpose of extrapolating activity data, as it is a model which has been previously used and calibrated for South African conditions making it more robust. The model outputs of the BFAP model, however, only cover commercial livestock numbers. In order to project subsistence livestock population numbers historical data was used obtained from Agricultural Abstracts (DAFF, 2012), LACTO (2015), Meissner et al. (2013) and FAO Statistics reports. A regression was fitted to the historical population data and a logarithmic transformation completed to obtain an average rate of change which was then applied to decadal intervals up to 2050 in order to project livestock numbers into the future. In addition, game farming numbers were added to the baseline. Population numbers were based on data from Wildlife ranching SA, Pers. Comm, Cloete (2015) and Du Toit et al. (2013).

The BFAP model only projects numbers up until 2024. In order to project livestock numbers up to 2050 the rate of change between 2010 and 2024 was then extrapolated and applied to decadal intervals up until 2050. The assumptions used in setting the Agricultural baseline are presented in Appendix A – Table 1. A detailed description of the assumptions and uncertainties are described in DEA (2016).

For the land sector 2013/2014 land cover change data (DEA, 2015) recently produced for the GHG inventory was the basis for the projections. The original maps had 72 classes but for the purpose of determining emissions these classes were condensed to 17 classes. Furthermore, the land change mapping between 1990 and 2013/14 was only done on the 17 land classes. Most of the remaining classes were related to land use, many of which fall within the settlement or mine categories. Emissions were determined based on these classes, but then these were further condensed to the 6 main IPCC classes for reporting purposes.

The base maps used to determine change were the 1990 and 2013/14 land cover maps recently developed by GeoTerra Image for the DEA (GTI, 2015) from Landsat 8 imagery. As a starting point, annual rates of land cover change were calculated from these maps, and then additional data provided information to restrict or validate the change rates. The rate of change in the transformed landscapes (plantations, settlements, mines, cultivated lands, as well as the smaller indigenous forest category) were investigated in more detail by obtaining data from literature and expert opinions. This information was used to restrict the rates of change for these categories. Natural vegetation classes were then added and the area normalized to provincial areas which lead to smaller rates of change than those originally projected between the 1990 and 2014 land cover change maps. This was more appropriate as baselines should be conservative. These provincial changes were then compared to overall national land change for consistency. The assumptions used in developing the Land baseline are listed in Appendix A – Table 2. The detailed description of the assumptions and uncertainties are presented in DEA (2016).

Results

Agriculture

The agricultural baseline emissions show an increase from 50 568 $GgCO_2$eq in 2010 to 69 621 $GgCO_2$eq in 2050 (Table 1). This is a 37.7% increase. The livestock populations have the largest influence over emissions in this sector (60%) as they contribute to enteric fermentation, manure management and indirect N_2O emissions from manure management. Enteric fermentation and manure management contribute 55.4% and 3% respectively to the total agriculture baseline. The relative contribution from enteric fermentation doesn't increase, whereas it increases by 0.8% from manure management. According to the baseline livestock numbers increase by 38.7% between 2010 and 2050 as feedlot cattle, pigs, poultry and game populations are predicted to increase (BFAP, 2015, SA Feedlot Association, 2013). The contribution from game to enteric fermentation increases from 48.1 $GgCO_2$eq (3.9% of total enteric fermentation emissions) in 2010 to 190 $GgCO_2$eq (11.3%) in 2050.

Emissions from aggregated and non-CO_2 emission sources are projected to increase by 36.4% between 2010 and 2050. Direct N_2O emissions from managed soils are the largest contributor to this category, contributing 30.3% in 2010. This declines to a contribution of 27.0% in 2050, as the contribution from fertilizer application increases (Table 1). Enteric fermentation is the largest contributor, accounting for roughly 56% of the agricultural emissions. Enteric fermentation shows an increase of 37% between 2010 and 2050. The largest increase (424%) comes from emissions from urea application; however the urea consumption data is highly variable and comes with high uncertainties. Indirect N_2O emissions from manure management increase by 123% between 2010 and 2050.

Table 1: Baseline emissions (GgCO_2eq) for the Agricultural Sector

	2010	2020	2030	2040	2050
Total for agriculture	50 568.01	54 282.82	60 852.48	66 201.91	69 621.35
Livestock	29 708.32	32 256.49	36 353.45	39 516.62	41 177.52
Enteric fermentation	28 139.89	30 457.99	34 187.25	37 103.13	38 550.78
Manure management	1 568.44	1 798.50	2 166.20	2 413.49	2 626.74
Aggregated and non-CO_2 emission sources	20 859.69	22 026.32	24 499.04	26 685.29	28 443.83
Liming	585.54	577.13	594.85	640.87	718.76
Urea application	478.69	724.41	1 096.26	1 658.97	2 510.54
Direct N_2O emissions from managed soils	15 097.01	15 749.39	17 218.19	18 316.03	18 813.55
Indirect N_2O emissions from managed soils	4 212.69	4 332.67	4 759.90	5 104.78	5 317.29
Indirect N_2O emissions from manure management	485.76	642.73	829.84	964.64	1 083.70

Table 2: Baseline emissions (GgCO_2eq) for the Land Sector

	Total Land	Land	Biomass burning
2014	-21104.5	-22 920.7	1818.47
2020	-25 860.4	-27 663.2	1805.02
2030	-31 390.6	-33 169.9	1781.55
2040	-32 223.2	-33 977.9	1756.86
2050	-30 683.2	-32 407.6	1726.61

Table 3: Combined Land and Agriculture Baseline Emissions

Years	Total AFOLU	Livestock	Aggregate sources and non-CO_2 emissions sources	Land
2014	30 949.4	30727.59	21326.34	-22 920.7
2020	28 442.4	32256.49	22026.32	-27 663.2
2030	29 461.9	36353.45	24499.04	-33 169.9
2040	33 978.7	39516.62	26685.29	-33 977.9
2050	38 938.2	41177.52	28443.83	-32 407.6

Land

At the national level the land projections don't show large changes in land area, but the largest changes are around the decrease in grassland, and increase in forest land and bare ground. Since forest land plays such a focal point in carbon estimations this increasing forest land leads to increased carbon sinks.

The estimated national baseline for the land sector shows an increased sink between 2014 and 2040 (21 104 GgCO$_2$eq to 32 223 GgCO$_2$eq), after which the sink slows and becomes stable (Table 2). The increasing sink is mainly due to the predicted increase in forestland, but is also combined with the decrease in wood removal from woodlands in the period until 2030. Keeping fuel wood removal constant (i.e. assuming no reduction in wood removals due to electrification) produces a more constant sink (varying less than 3 000 GgCO$_2$eq between 2014 and 2050), but it still shows a slight increase in the sink to 2030 after which it declines to 2050. If the thicket area is increased by 1% then the sink increases by 17% by 2050, which shows the importance of understanding whether the thicket area is increasing, decreasing or remaining constant. Moving towards 2050, there is also a predicted increase in bare ground and this leads to a loss of carbon (both biomass and soil) causing the overall land carbon sink to stabilize.

Combining the land and the agriculture baseline creates a baseline which shows an 8.2% decline between 2014 and 2020, after which it increases by 37% to 38 938 GgCO$_2$eq in 2050 (Table 3 and Figure 1). The increasing land sink contributes to the slight decline in the early years while the increasing agricultural emissions combined with the stabilizing carbon sink leads to the increase between 2030 and 2050. Land sequesters almost as much as the aggregated non-CO$_2$ emission emit and so the baseline is very similar in magnitude to the enteric fermentation value.

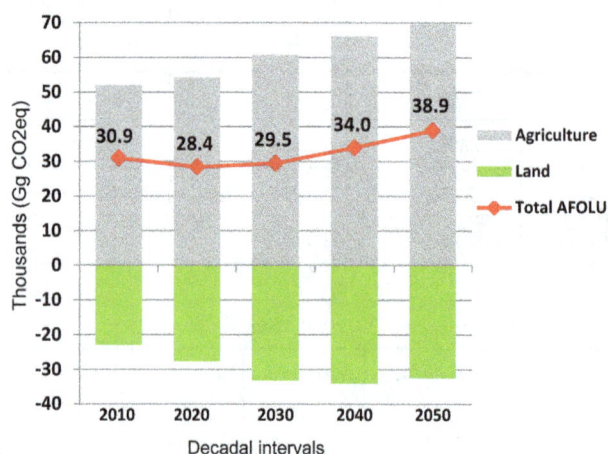

Figure 1: *Combined AFOLU Baseline emissions*

Discussion

It is not unexpected that enteric fermentation is the largest contributor of emissions, accounting for roughly 56% of the agricultural emissions. These results are consistent with the current GHG inventory trends (DEA, 2014a). Urea application had the highest increase according to the projections. The reason

being that urea consumption is determined from import and export data, and it is assumed that all urea is being applied to the field. This data thus presents a high degree of uncertainty. It is therefore, recommended that more detailed data be collected for urea consumption. One limitation of this model is that not all fertilizer types are included, and this can be another aspect which could be improved in the future. Manure management emissions increased in the projections due to future increasing feedlot cattle, piggeries and poultry.

Mitigation options for the agriculture sector are not often highlighted in terms of the AFOLU sector as they are seen to have limited potential. This does not, however, mean that these options shouldn't be considered, especially since enteric fermentation is a key category in South Africa's GHG inventory. In terms of enteric fermentation there are two options for reducing emissions; namely, increase rumen efficiency and increase livestock productivity. In terms of manure management the mitigation option of biodigesters would assist in reducing these emissions. The draft Mitigation and Adaptation Strategy of DAFF (DAFF 2015a) highlights that it is harder to reduce emissions in agriculture than it is to increase sequestration, hence the importance of land cover and land use activities.

The land baseline suggests that if forest land is increased through afforestation and thicket restoration, then the carbon sink would increase. It also indicates that if soil erosion and degradation is prevented, the future decrease in the sink would be alleviated, highlighting the importance of mitigation actions suggested in the NTCSA (DEA, 2015).

The baseline is an indication of what the expected emissions are going to be based on a business-as-usual scenario. The inventory provides information on what the actual emissions are. Both are developed based on current knowledge, and under the current reporting regime the inventory is updated every two years. However, there are still many unknown factors in the AFOLU sector, with uncertainties still being determined, therefore there are continuous improvements being made to the AFOLU sector inventory. Since the baseline is dependent on the inventory, it is suggested that the baseline be updated again in the near future, so as to incorporate any new information in this sector. The biggest unknown factor is land cover change and this is extremely difficult to predict into the future. Since land cover change drives the land sector baseline it would be necessary to update the baseline when new land change maps become available.

Conclusion

This was the first attempt at creating a baseline for the AFOLU sector in South Africa. It was a challenging task given the variability and uncertainty of the available data. Therefore the results of this study cannot be seen as the actual final baseline but rather the first step in the movement towards one in the near future. The process of developing the baseline provided many lessons. Several recommendations are listed below which would help improve the baseline in the future:

- There is a need to develop consistency in national data sets. This relates to the variability in the data sources and differences in mapping classifications, and applies both to the agricultural and land sector data.

- There are enormous challenges in predicting land cover and land use change. The method used in this study relies on historical change data and expert opinion. Land change maps can provide varied outputs depending on when in the year or in which year they were created. South Africa needs to detect change on a more regular basis, using a consistent methodology, in order to be able to have improved forward projections.

- More detailed degradation data needs to be incorporated to improve estimates. There are many different types of degradation and determining the extent of degradation is still debated. For the baseline development the rate of degradation as well as the extent is required.

- More detailed cropland data needs to be incorporated (pivot and non-pivot categories).

- Detailed livestock population estimates in the subsistence sector are still required.

- Research on nitrogen emissions is needed in order to improve the emission factors used, currently IPCC default emission factors are being used in many of the categories.

- A register of biodigesters and their fuel sources is required. The information on biodigesters is scattered, therefore, it would be useful to have a central register of this information to assist in estimating and predicting emission savings in terms of the AFOLU sector.

- A better understanding of fuelwood consumption. There is a lack of information at a national scale as to whether fuelwood removal is declining or not. It would be important to develop an understanding of the amount of fuelwood consumed at a national scale and to investigate how this is changing over time in order to improve estimates.

Acknowledgements

This project was a Department of Environmental Affairs study, funding for which was made available by the British High Commission. Thanks are given to the project steering committee and all project stakeholders for their expert input and data supply.

References

BFAP (Bureau of Food and Agricultural Policy) 2015. *BFAP Baseline Agricultural outlook 2015-2024*. Available at: www.bfap.co.za

Clapp C. and Prag A. 2012, *Projecting Emissions Baselines for National Climate Policy: Options for Guidance to Improve Transparency*. Organisation for Economic Co-operation and development: Environment Directorate International Energy Agency.

Cloete P.C., Van Der Merwe P. & Saayman M. 2015, *Game ranch profitability in South Africa*, 2nd ed, Pretoria.

DAFF, 2012: *Abstract of Agricultural Statistics*. Department of Agriculture, Forestry, and Fisheries, Pretoria, South Africa.

DEA (Department of Environmental Affairs) 2014a, *GHG Inventory for South Africa 2000-2010*. Department of Environmental Affairs, Pretoria, South Africa.

DEA (Department of Environmental Affairs) 2014b, *South Africa's Greenhouse Gas (GHG) Mitigation Potential Analysis*. Department of Environmental Affairs, Pretoria, South Africa.

DEA (Department of Environmental Affairs) 2015, *South African National Terrestrial Carbon Sinks Assessment*, Department of Environmental Affairs, Pretoria, South Africa.

DEA (Department of Environmental Affairs) 2016, *Towards the development of a GHG emissions baseline for the agriculture, forestry and other land use (AFOLU) sector in South Africa*. Department of Environmental Affairs, Pretoria, South Africa.

Du Toit C.J.L., Meissner H.H. & van Niekerk W.A. 2013, Direct greenhouse gas emissions of the game industry in South Africa, *South African Journal of Animal Science* 43(3): 376-393.

FAOSTAT, 2015a: *FAOSTAT database*. Food and Agriculture Organization of the United Nations. Available at: http://faostat.fao.org.

GTI (GeoTerra Image) 2015, Land cover maps and change between 1990 and 2013/14. *South African National Land-Cover Change*. DEA/Cardno Project number SCPF002: Implementation of Land-Use Maps for South Africa, GeoTerra Image, Pretoria.

IPCC, 2006: 2006 *IPCC Guidelines for National Greenhouse Gas Inventories*, Prepared by the National Greenhouse Gas Inventories Programme, Eggleston H.S., Buendia L., Miwa K., Ngara T. and Tanabe K. (eds). Published: IGES, Japan.

LACTO Data 2015. Statistics. Vol 18, no 1. Available at: www.milksa.co.za

Meissner H.H., Scholtz M.M. & Palmer A.R., 2013, Sustainability of the South African livestock sector towards 2050 Part 1: Worth and impact of the sector. *South African Journal of Animal Science*, 43(3):298-319.

SA Feedlot Association, 2013: Available at: http://www.safeedlot.co.za/index.asp?Content=90

Appendix A

Table 1: The assumptions used in projecting the agricultural emissions

Agricultural baseline assumptions	
Enteric fermentation	Livestock diet remains as it is
	Current Emission factor kept constant
	Herd composition kept at current ratios
	Average Emission Factor used, so no cattle species detail required
Manure management	Livestock diet remains the same
	Emission Factor kept constant
	Manure management systems continue as currently being used
	No biodigestor usage
Managed soils	Fertilizer consumption continues at current rate
	Fraction of livestock / game population in fields remains the same
	Amount of manure used for feed, fuel and construction remains constant
	Ratio of crop residues retained remain at current levels
	Emission Factors kept constant
Urea application	Rate of lime consumption continues at current level
	Rate of urea consumption continues at current level
	Emission Factor kept constant

Table 2: The assumptions used in projecting the emissions from Land

Land baseline assumptions
Thickets continue to degrade at current rate
No increase in plantations
Grasslands continue to decline at current rates
Degradation continues to increase at current rates
Tillage practices continue as they currently are
No biochar application (soil carbon reference levels remain as they are)

14

Air quality indicators from the Environmental Performance Index: potential use and limitations in South Africa

Rebecca M. Garland[1,2], Mogesh Naidoo[1], Bheki Sibiya[1], and Riëtha Oosthuizen[1]

[1]Natural Resources and the Environment Unit, Council for Scientific and Industrial Research, Pretoria, South Africa
[2]Climatology Research Group, North West University, Potchefstroom, South Africa

Abstract

In responding to deteriorating air quality, many countries, including South Africa, have implemented national programmes that aim to manage and regulate ambient air quality, and the emissions of air pollutants. One aspect within these management strategies is effective communication to stakeholders, including the general public, with regard to the state and trend of ambient air quality in South Africa. Currently, information on ambient air quality is communicated through ambient mass concentration values, as well as number of exceedances of South African National Ambient Standards. However, these do not directly communicate the potential impact on human health and the ecosystem. To this end, the use of air quality indicators is seen as a potential way to achieve communication to stakeholders in a simplified, yet scientifically defensible manner. Air quality indicators and their source data from the Environmental Performance Index (EPI) were interrogated to understand their potential use in South Africa. An assessment of four air quality indicators, together with their source data, showed improvements in air quality over the time period studied, though the input data do have uncertainties. The source data for the PM indicators, which came from a global dataset, underestimated the annual $PM_{2.5}$ concentrations in the Highveld Priority Area and Vaal Triangle Airshed Priority Area over the time period studied (2009-2014) by ~3.7 times. This highlights a key limitation of national-scale indicators and input data, that while the data used by the EPI are a well-thought out estimate of a country's air quality profile, they remain a generalised estimate. The assumptions and uncertainty inherent in such an ambitious global-wide attempt make the estimates inaccurate for countries without proper emissions tracking and accounting and few monitoring stations, such as South Africa. Thus, the inputs and resultant indicators should be used with caution until such a time that local and ground-truthed data and inputs can be utilised.

Keywords

Air quality indicators, air quality management

Introduction

Globally, air pollution is of concern and has deteriorated in many areas due to emissions from anthropogenic activities including industrial and vehicular activities, as well as from emissions from natural sources such as biomass burning. The current state of air quality has become a major threat to the health and wellbeing of people, as well as the environment in many areas around the world. In responding to deteriorating air quality, many countries, including South Africa, have implemented national programmes that aim to manage and regulate ambient air quality, and the emissions of air pollutants. In South Africa, the approach includes the declaration of priority areas for air quality management, development of national air quality standards, and an air quality monitoring programme. The aim of these programmes is to reduce air pollution-related illnesses and conditions.

One aspect within these management strategies is effective communication to stakeholders, which includes the general public, with regard to the state and trend of ambient air quality in South Africa. Currently, information on ambient air quality is communicated through ambient mass concentration values, as well as number of exceedances of South African standards. However, these do not directly communicate the potential impact on human health and the ecosystem. To this end, the use of air quality indicators is seen as a potential way to achieve communication to stakeholders in a simplified, yet scientifically defensible manner.

Environmental Performance Indicactor

The Environmental Performance Index (EPI) was developed by the Yale Centre for Environmental Law and Policy at Yale University and the Centre for International Earth Science Information Network at Columbia University (http://epi.yale.edu). The EPI aggregates over 20 indicators relating to national environmental data.

The EPI assesses two objectives, namely Environmental Health and Ecosystem Vitality. The Issue Category of "Air Quality" is within Environmental Health, though in previous EPI reports there have been Air Quality issues within the Ecosystem Vitality objective.

There have been a variety of indicators in the Air Quality issue category in the history of EPI, with recent years focussing on particulate matter (PM) and household air pollution. The EPI 2016 assessment includes an indicator on exposure to nitrogen dioxide (NO_2), and previous EPIs (e.g. 2008) have included indicators on ground-level ozone for health and for ecosystems considerations. For developing local indicators, this suite of present and historical EPI indicators should be assessed for their relevance to local air quality issues and policy priorities, and for the availability and reliability of local data.

For this study, the following indicators were selected to ground-truth air quality aspects for South Africa. This assessment is not comprehensive of all air quality indicators from the EPI, however focussed on selecting indicators that assessed different pollutants and data sources, as well as highlighting some of the pollutants and emission sources of concern in South Africa. These indicators provide information on the potential impact on human health and on ecosystems. In this analysis, only the following four indicators (that form part of the EPI) were considered. The objective that the indicator was included under in the EPI is listed in parenthesis below.

- HAP = Household Air Pollution (environmental health) - Percentage of population using solid fuel as the primary cooking fuel (%)
- SO2CAP = Air pollution (ecosystem vitality) - Sulphur dioxide emissions per capita
- SO2GDP = Air pollution (ecosystem vitality) - Sulphur dioxide emissions per Gross Domestic Product (GDP)
- PM25 = Air Pollution (environmental health) - Population weighted exposure to $PM_{2.5}$ ($\mu g/m^3$)

The SO2CAP and SO2GDP are not included in the current (2014 or 2016) estimations of EPI; they are nonetheless important indicators for South Africa and thus included here.

This study interrogated the input data into the EPI and compared this input data to publically available local data. This comparison will help to gain a better understanding of the robustness of the input data, and in turn, the indicators. These findings assisted in understanding the potential uses and limitations of the indicators, as well as provided insight into the state of air quality in the country.

Methods

EPI input data

EPI output values for indicators are reported on a national scale. The website does give links to the underlying data sources, and those sources are described here (http://epi.yale.edu/).

EPI: Solid fuel use for cooking data

The HAP data are derived from Bonjour et al. (2013), which were based upon data from the WHO Household Energy Database (2012). The EPI indicator is defined as the percentage of population using solid fuel for cooking. The percentage of population that are exposed to household air pollution was assumed to be the same as the percentage of households using solid fuels; thus the percentage of households using solid fuels is assessed and compared. These data are based on national surveys, which do report percentage of households using solid fuels for cooking.

EPI: SO_2 emissions

The SO2GDP and SO2CAP were last reported in EPI 2012, and those are the input data reported here. The SO_2 aspect is represented by total anthropogenic emissions for a country. The input SO_2 emissions were based on the research detailed by Smith et al. (2011), in which global bottom-up inventories (primarily through mass balance for combustion and metal smelting) were created for each country and constrained by any available locally derived emissions measurements or estimates. The original inventory covers years 1850-2005 and is reported in 10 year increments in Smith et al. (2011). The source sectors considered were coal combustion, petroleum combustion, natural gas processing and combustion, petroleum processing, biomass combustion, shipping bunker fuels, metal smelting, pulp and paper processing, other industrial processes, and agricultural waste burning.

EPI: Population and GDP data

Indicator SO2GDP requires that GDP be converted to international dollars using purchasing power parity rates; for the 2012 EPI, 2005 international dollars were used. These were sourced from the World Development Indicators (indicator NY.GDP.MKTP.PP.KD; World Bank, 2011) and covered the period 1980-2011. SO2CAP requires country population data and this was also sourced from the World Development Indicators (indicator SP.POP.TOTL; World Bank, 2011) and covered the period 1960-2010.

EPI: Ambient $PM_{2.5}$ simulated concentrations

The PM25 EPI indictor quantifies the population weighted exposure to $PM_{2.5}$ for the country. This indicator uses $PM_{2.5}$ ambient concentrations that were originally estimated by Van Donkelaar et al. (2015) and are available online (ACAG, 2016). The datasets used here were the "All composition" satellite-derived $PM_{2.5}$ at a relative humidity of 35% for a three-year running median.

The methodology used to estimate surface $PM_{2.5}$ concentrations was included in the method used in estimating the Global Burden of Disease that is attributable to PM (Burnett et al., 2014; Brauer et al., 2012). The methods are not the same, however, as Brauer et al. (2012) did use multiple data sources, including ground-based data.

The methodology followed to estimate $PM_{2.5}$ ambient concentrations is detailed in Van Donkelaar et al. (2010), Van

Donkelaar et al. (2015) and Boys et al. (2014). Briefly, aerosol optical depth (AOD) from the combination of Moderate Resolution Imaging Spectroradiometer (MODIS), Multiangle Imaging Spectroradiometer (MISR) and Sea-viewing wide field-of-view sensor (SeaWIFS) satellite instruments were used together with global chemistry transport model simulations using the Goddard Earth Observing System model with Chemistry (GEOS-Chem). Ground-level concentrations of $PM_{2.5}$ were estimated by developing an AOD conversion factor (accounting for aerosol size, aerosol type, diurnal variation, relative humidity and the vertical structure of aerosol extinction) based on GEOS-Chem simulations. The results were daily values coinciding with satellite overpass time; these were aggregated into three year moving median values. The median values were used to reduce the noise in the data from the satellite retrievals (Van Donkelaar et al., 2015).

Local data

This section details the local data sources that were compared to the "international" data from the EPI. As discussed in the section below, the national SO_2 and solid fuel use data for the EPI and the "local" data have similar sources.

Local: Solid fuel use for cooking data

The local data were provided by the South African Department of Environmental Affairs (DEA) and included information on the distribution (in percentage) of households that use domestic fuels (paraffin, wood, and coal) for household activities such as cooking, heating and lighting. These data were compiled from the 2014 General Household Survey data from Statistics South Africa (Stats SA) (Statistics South Africa, 2015).

Figure 1 displays the percentage of households using paraffin, wood or coal for cooking in South Africa for 2002-2014. For comparison to the EPI, the percentage of households using solid fuels for cooking was defined as those using coal and wood. It should be noted that EPI includes the burning of crop residues, dung and charcoal, which were not included here due to lack of local data.

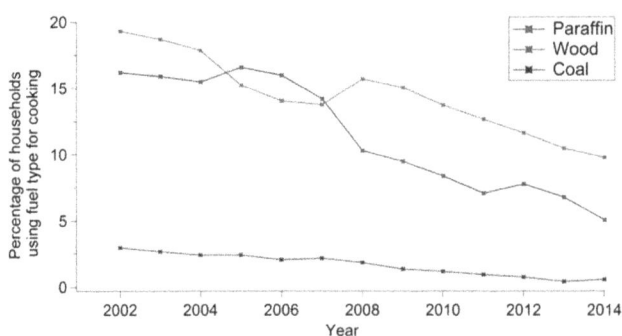

Figure 1: Stats SA data for percentage of households in South Africa using coal (grey line), wood (brown line) or paraffin (blue line) for cooking.

Local: SO_2 emissions

There are no locally derived data for a complete national SO_2 emission inventory. As detailed in section 2.1.2, the EPI used data from Smith et al. (2011). However, there is much room for improvement regarding SO_2 emissions as there are high levels of uncertainty in Smith et al. (2011) estimates for South Africa. This is due to lack of local emissions reporting, and in uncertainty due to assumptions in bottom-up calculations such as fuel sulphur content and activity data (i.e. the actual amount of fuel used). Smith et al. (2011) specify uncertainties of up to 54% for South Africa (included in the "Other Countries" grouping for uncertainty analysis) for the sources included.

Klimont et al. (2013) built on the methodology and work by Smith et al. (2011) and estimated global SO_2 emissions through the Greenhouse Gas–Air Pollution Interactions and Synergies (GAINS; Amann et al. 2011) model for the period 2000-2011. This period is relevant to assessing a South African emissions profile due to rapid development. These published outputs are used in this comparison.

Klimont et al. (2013) also report a still significant amount of uncertainty may exist in this newer inventory; however, a quantitative estimate was not produced. The assumption around this large uncertainty was based on the inclusion of regional activity and fuel data from developing countries and within the international shipping sector. Ideally, a locally derived estimate, which includes local fuel specifications and activity data, must be provided to the general public such that researchers can include these data into their studies and indices.

Local: Population and GDP data

Economic and demographic data are readily available for most countries through either the World Bank or United Nations Populations Division. While it is possible to refine the World Bank Development Indicator population estimates using local census data, the difference is marginal for years 1996 (1.5% underestimate), 2001 (0.2% overestimate) and 2011 (0.4% underestimate) when compared to Stats SA census releases (Statistics South Africa, 2012). In order to have a full time series of population data, the EPI data source for population were used (World Bank, 2011).

In evaluating and comparing the SO2GDP indicator it is necessary to use the same units specified within the EPI methodology. The 2012 EPI methodology specifies GDP in 2005 international dollars. The only readily available data representing this are the World Development Indicators (same as used in EPI); these data could not be easily found from a local source directly. It is assumed here that these represent accurate local estimates of GDP and thus were used in calculating the local indicator (indicator SP.POP.TOTL; World Bank, 2011).

Ambient $PM_{2.5}$ monitored data

The South African Air Quality Information System (SAAQIS) data were used to ground truth the EPI $PM_{2.5}$ data to verify its level of accuracy. The SAAQIS data used were for the Vaal Triangle Airshed Priority Area (VTAPA) and DEA Highveld Priority Area (HPA) air quality monitoring networks. The VTAPA network has been running since 2007, and HPA network since 2008. The networks have a relatively continuous monitoring record, and the measured data are quality controlled and managed by the South African Weather Service (SAWS) through the SAAQIS. The networks are in priority areas, and thus are in the more polluted areas of South Africa.

The $PM_{2.5}$ data were provided by SAAQIS as one hour averages for 1 Jan 2008 – 1 October 2015. These data were quality checked (QC) by CSIR, and then averaged to monthly values, and processed to a corresponding three year moving median as reported in EPI data. The quality control included removing negative values and repeating values. Table 1 displays the number of data points (N) before the QC procedure was applied and after the QC procedure was applied. Further analyses were only performed on the valid hourly values (i.e. after the QC procedure was applied).

It is best practice when averaging monitored data to use a "data completeness threshold." This threshold indicates the percentage of data that must be present in order to derive a representative average. For example, for a 70% data completeness rule, if fewer than 70% of 1-hour $PM_{2.5}$ data were recorded in one day, then the daily average could not be calculated and would be left blank. In this analysis, thresholds from 75% to 50% were applied and tested to calculate a three year moving median. However the loss of data was large and analysis presented for this criterion would not have been possible at all sites. Thus, no threshold was applied when averaging the 1-hour values from SAAQIS for this analysis.

Table 1: *Number of data points (N) per station before QC was applied and after QC was applied.*

Site		Lat	Lon	N (before QC)	N (after QC)
HPA	Ermelo	-26.4934°	29.9681°	55 949	55 949
	Hendrina	-26.1509°	29.7168°	39 611	39 592
	Middelburg	-25.7961°	29.4636°	53 438	53 419
	Secunda	-26.5486°	29.0801°	52 063	52 028
	Witbank	-25.8778°	29.1887°	48 789	48 789
VTAPA	Diepkloof	-26.2507°	27.9564°	42 939	42 737
	Kliprivier	-26.4203°	28.0849°	48 762	48 755
	Sebokeng	-26.5878°	27.8402°	41 739	41 725
	Sharpeville	-26.6898°	27.8678°	51 685	51 474
	Three Rivers	-26.6583°	27.9982°	45 852	45 803
	Zamdela	-26.8449°	27.8551°	57 731	57 580

Figure 2 shows the geographical location of the VTAPA and HPA monitoring stations (as points with names of stations indicated) on the 10 km x 10 km grid from the $PM_{2.5}$ data used by the EPI. The $PM_{2.5}$ 3-year running median from each site was compared

to the corresponding EPI value from the grid cell where the site is located.

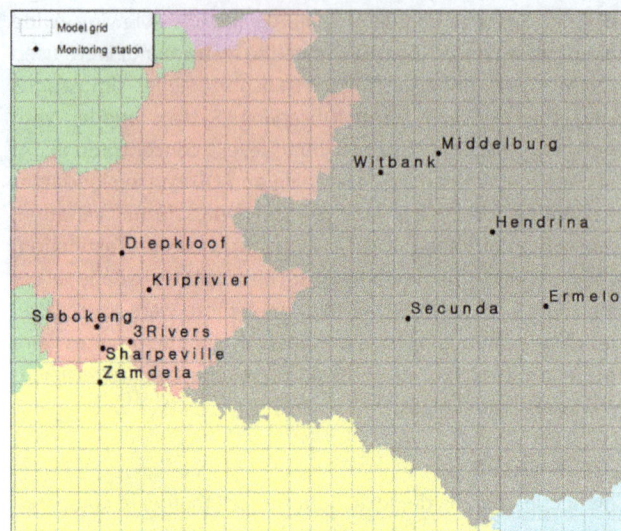

Figure 2: *Map of SAAQIS monitoring stations as points with their names indicated on the 10km resolution model grid of the EPI $PM_{2.5}$ input data. The colours indicate the provinces. The VTAPA monitoring stations are on the west and the HPA are on the east.*

Results and Discussion

Overview of South Africa's standing in the Air Quality Issue Category

The scoring of countries' performance within the EPI is relative to the top-performing country, which receives a score of 100. The other countries' indicator scores are normalised to this top-performer. Thus in the score, a larger number indicates better performance. The top-performing country also receives a rank of 1, thus in rank a lower score indicates better performance. Scoring and ranking occur at each level within the EPI (i.e. indicators, issue categories, objectives and total score).

As the EPI Issue Category Scores are normalised to the top-performing country, over time, a country can have a worsening score even if the air quality is improving, if other countries are improving at a faster rate.

In the 2016 EPI Air Quality issue category, South Africa's score was reported as 88.84 (out of 100), which led to a ranking of 49 out of 180 countries.

South Africa was calculated to have improved its Air Quality issue category score over the past ten years from 74.47 (resulting in a rank of 61) in 2006. The rank of South Africa for PM25 was 60 in 2006, and fell to 69 in 2016. For the HAP indicators, the rank improved from 114 in 2006 to 94 in 2016.

Compared to sub-Saharan Africa's performance as a region, South Africa's 2016 Air Quality score was 18.98 points higher than the region's average score. Thus according to EPI, in general, South Africa for this issue category is performing well for the region, and has on average been improving.

HAP

Figure 3 displays the comparison of the EPI input data and local Stats SA data for the percentage of households using solid fuels for cooking in South Africa. The two datasets compare well, which was expected as the EPI input data source does rely on national surveys. Figure 3 also does highlight the large decrease in households who report using a solid fuel as their primary source for cooking, which can have positive implications for indoor and ambient air quality.

A limitation in these data is that the Stats SA data are limited to primary fuel only. In South Africa, low-income households rely on multiple fuels (e.g. Madubansi and Shackelton, 2007; Llyod et al., 2004; Thom, 2000; Davis 1998). A national survey found that 48% of South African households rely on multiple fuels for cooking (DoE, 2012). Thus, these data won't capture those who do use solid fuels but do not consider it their primary source, nor those who use multiple fuels.

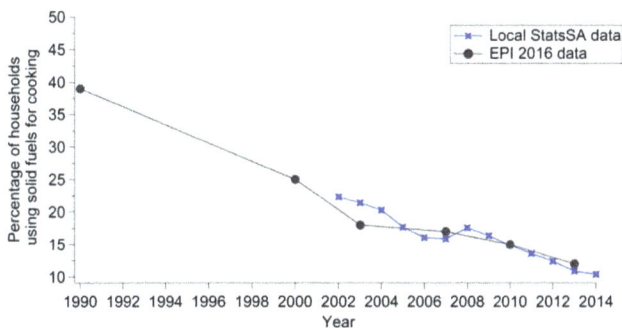

Figure 3: *Percentage of households using solid fuel for cooking from the EPI 2016 input data (grey line) and local Stats SA data (blue line).*

SO2CAP and SO2GDP

There are no local data to compare to the EPI 2012 SO_2 indicators. However, Klimont et al. (2013) has updated the emission estimates from the methodology used in the EPI 2012 indicator; those two datasets are explored here. The economic and population data used to calculate these indicators were the same for EPI 2012 and "Klimont" reported results (Figure 4).

As seen in Figure 4, anthropogenic SO_2 emissions from the datasets agree well, though the Smith et al. (2011) data had a larger dip in emissions in 2001-2002. The anthropogenic SO_2 emissions have been estimated to have increased from 1860 Gg in 1980 to 2795 Gg in 2011. However, during this time, SO2CAP and SO2GDP have both had an overall decreasing trend, though recently the SO2CAP has appeared to hover ~55 kg/person.

The SO2GDP indicator in particular highlights that South Africa's SO_2 economic intensity has decreased, i.e. the GDP growth has been decoupled from the growth emissions in SO_2. This is a positive trend; however as SO_2 emissions are still increasing, there is a continuing negative impact on air quality. However, without a comprehensive national emissions inventory, it is not possible to validate this trend using bottom-up local data.

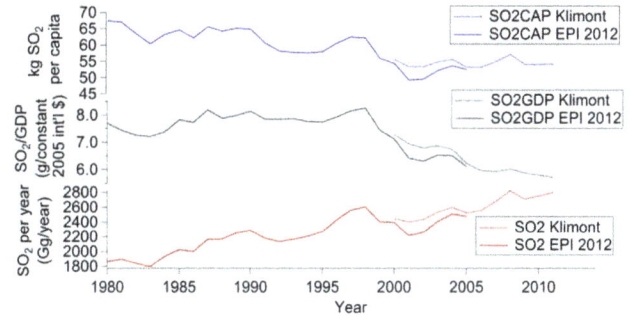

Figure 4: *SO_2 emissions per year (red lines), SO_2 emissions per GDP (black lines) and SO_2 emissions per capita (blue lines)*

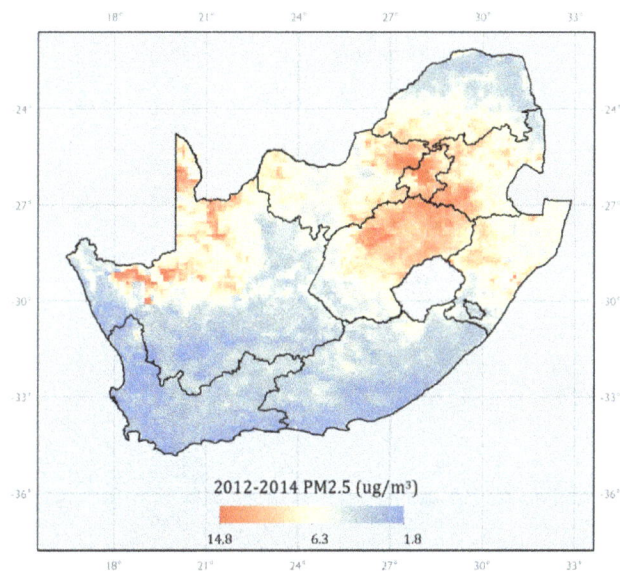

Figure 5: *2013 three year running median of surface $PM_{2.5}$ mass concentrations that were used as input into the EPI.*

PM25

Figure 5 displays the 2013 three-year running median of ground-level $PM_{2.5}$ ambient mass concentrations that was used as input in the PM25 indicator. The running median will be reported by the midpoint year (i.e. 2012-2014 in Figure 5). In this study, the medians for 2010-2013 were compared for all sites. While the magnitude of the medians does change in these averaging periods, the general spatial distribution is similar to Figure 5 across years. This spatial distribution is what would be expected, with higher PM loadings in Gauteng, HPA, and VTAPA, where anthropogenic emissions of air pollution are high. In addition, peaks are seen in the Northern Cape, which in the input dataset are attributed to dust.

Figure 6 displays the average three-year running mean $PM_{2.5}$ concentration of the HPA and the VTAPA stations' for both observations (blue) and EPI database (red). Table 2 displays the $PM_{2.5}$ mass concentrations for the EPI input dataset and from the monitored data from SAAQIS for each station within the HPA and VTAPA (labelled as "Monitored" data). The EPI uses a three-year running median of annual averages. The monitored data from SAAQIS did not have consistent data completeness at all sites across years. This inconsistent completeness may bias the median, as well as inter-annual comparisons. This could particularly have impact in areas that have a strong seasonal

Table 2: Ground-level $PM_{2.5}$ concentrations (µg/m³) for EPI input data and monitored data from SAAQIS per site of the HPA and the VTAPA

Network	Site	Year (midpoint three-year running median)	EPI Input annual $PM_{2.5}$ concentrations (µg/m³) (Three-year running median)	Monitored annual $PM_{2.5}$ concentrations (µg/m³) (Three-year running median)	Monitored annual $PM_{2.5}$ concentrations (µg/m³) (Annual average)	Number of monthly values used in annual average
HPA	Ermelo	2010	7.6	30.0	30.0	12
		2011	7.1	29.2	29.2	12
		2012	7.1	28.5	28.5	11
		2013	7.4	24.6	24.6	12
	Hendrina	2010	7.0	20.5	23.3	12
		2011	7.0	18.8	18.8	11
		2012	6.8	18.4	18.4	11
		2013	6.6	18.4	12.5	6
	Middelburg	2010	7.9	22.5	22.2	12
		2011	7.5	23.8	26.0	12
		2012	7.4	23.8	23.8	11
		2013	7.4	19.2	19.2	12
	Secunda	2010	10.2	41.0	40.2	12
		2011	9.3	40.2	46.9	12
		2012	8.9	34.7	34.7	11
		2013	9.3	29.5	29.5	12
	Witbank	2010	8.4	30.2	29.5	12
		2011	7.8	29.5	30.2	12
		2012	7.6	24.5	24.4	11
		2013	7.3	24.3	24.5	12
VTAPA	Diepkloof	2010	10.4	46.8	64.6	7
		2011	10.3	29.7	29.7	9
		2012	9.7	27.0	27.0	11
		2013	9.7	23.0	23.0	12
	Kliprivier	2010	8.6	52.4	52.4	8
		2011	8.0	44.4	44.4	5
		2012	7.4	37.8	37.8	11
		2013	7.5	35.2	30.0	12
	Sebokeng	2010	8.2	51.5	51.5	12
		2011	8.2	51.5	65.0	6
		2012	7.6	33.9	33.9	11
		2013	7.4	30.1	29.3	12
	Sharpeville	2010	9.8	43.8	48.0	9
		2011	9.8	43.0	43.0	5
		2012	9.0	40.7	40.7	11
		2013	8.9	38.7	34.8	12
	Three Rivers	2010	9.6	27.5	32.2	12
		2011	8.8	26.3	21.8	3
		2012	8.7	24.3	26.3	11
		2013	8.7	25.0	24.3	12
	Zamdela	2010	10.4	31.9	36.4	12
		2011	9.8	31.9	31.9	10
		2012	9.2	29.8	29.8	11
		2013	9.2	29.8	29.0	11

cycle, and thus a missing month(s) would strongly impact the annual average and thus the three-year running median. Thus, the annual average per year from the monitored data and the number of monthly values used to calculate each average are also shown in Table 2. While there are differences between the median and the mean values (e.g. 2010 for Diepkloof), it is clear from Table 2 that both values at all sites at all years are much

higher than the annual values in the EPI dataset.

Both Table 2 and Figure 6 highlight that the EPI input data underestimates the ground-level $PM_{2.5}$ at all of these sites. On average, the monitored PM is ~3.7 times that used by the EPI, with a range of 2.4 to 6.3 Due to varying data completeness, comparisons between sites and between years were not made.

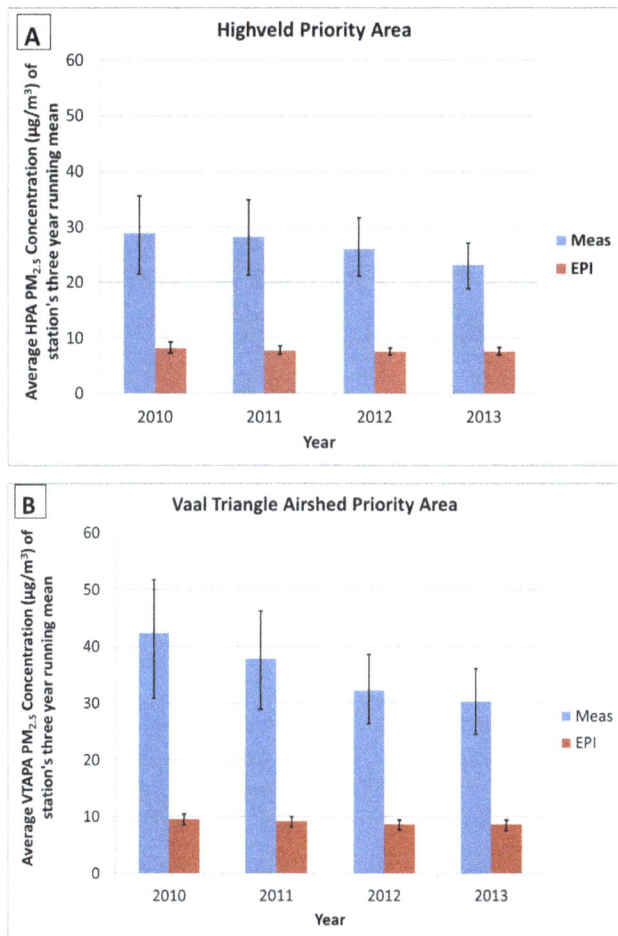

Figure 6: *Average PM$_{2.5}$ concentration of (A) the Highveld Priority Area and (B) the Vaal Triangle Airshed Priority Area stations' three-year running means from observations (blue) and EPI database (red). The bars extend from 25th to 75th percentile values of the stations' averages.*

At all sites, the EPI and the monitored data do show a decrease over the time period studied; however, due to the short length of the data set and the poor data recovery across years, the significance of this trend was not tested.

If it is assumed that this ~3 to 4 times underestimation of the EPI is valid for all of South Africa, then the resultant national PM25 values become similar to Laos, which has a rank of 172 for PM25 (out of 180). This assumption is a simplification; however, it provides a point of comparison of how such underestimation in PM25 could impact South Africa's score and assessment of the quality of the air.

The method to derive the EPI input PM2.5 data was also incorporated into the method used in the Global Burden of Disease (Burnett et al., 2014; Brauer et al., 2012). The methods are not the same, however, as Brauer et al. (2012) did use multiple data sources, including ground-based data. The underestimation of ground-level PM2.5 does still occur in the Brauer et al. (2012) dataset (Supplementary Material Table S1). Thus, it is likely that the health impacts attributable to PM in such studies are underestimated; and further research is required to refine those estimates.

Conclusion

The EPI uses indicators at a national level, as it is a comparison of 180 countries across the globe. For air quality management in South Africa, such national indicators could also be useful to understand and communicate the national state of air quality. In order to identify hotspots, however, it would be necessary to spatially resolve indicators. In addition, as can be seen here, trend analyses of indicators over time are particularly useful to understand progress. Thus, for domestic purposes, indicators that can be spatially resolved and calculated over multiple years are ideal.

Potential uses and limitations of indicators in South Africa

For the indicators assessed here, the potential use and limitations are indicator-dependent.

HAP – This index has the strongest local data sources, and in addition, spatially and temporally resolved local data are available. As domestic burning impacts indoor and ambient air pollution, an indicator of this type can be useful in South Africa to track high-level progress of solid fuel use as a proxy for air pollution exposure. This analysis could be tailored to fuels and uses in South Africa (e.g. cooking and heating assessed separately). Local Stats SA data on this are available and could be used; however, in order to understand trends the same questions must be used across surveys or else Stats SA must "backcast" usage when the question changes. However, it must be noted that Stats SA data are limited to primary fuel only, and thus do not capture the fact that a large proportion of households, in particular low-income households, rely on multiple fuels. This use of multiple fuels should be included in a local indicator.

SO2GDP and SO2CAP – These provide a helpful perspective on the intensity of SO$_2$ emissions. However, there are no locally derived data for comprehensive national SO$_2$ emissions; thus there is a strong need for local, bottom-up estimates to understand how robust findings using international data are. The trend in SO2GDP looks promising; these numbers should be ground-truthed with known emission sources.

PM25 – There are no locally derived and validated products of PM$_{2.5}$ concentrations for South Africa with comprehensive spatial coverage that could be used to estimate this indicator. Thus, in order to quantify the national PM25 indicator, global products for air quality would be needed together with local gridded data of population (such as in the EPI). However, from this study, it is clear that these underestimate PM$_{2.5}$ concentrations in the two priority areas.

It is not clear why there is this underestimation, though there may be many potential reasons for error. This comparison is comparing one sampling point to a grid cell, which assumes that the sampling point is representative of the full grid cell. The sampling stations have been sited to avoid strong local sources; however there would be spatial differences in the PM concentrations across the grid cell. As emissions are not well-quantified on a national level for South Africa, there would be uncertainties in the emissions information used in GEOS-Chem simulations. In addition, the AOD–ground-level relationship is not well quantified for South Africa, and that may lead to uncertainties in deriving ground-based concentrations from satellite information (Hersey et al., 2015). Ford and Heald (2016) estimated an uncertainty of ~20% in deriving $PM_{2.5}$ burden of mortality from satellite retrieved data due to uncertainties in the AOD–ground-level relationship alone. In addition, there are a lack of freely available and continuous ambient $PM_{2.5}$ measurements in South Africa, that can be used for ground-truthing. Even this comparison is constrained to a few sites in heavily polluted areas in South Africa.

Since PM is a pollutant of concern in South Africa, indicators based on PM exposure are key to understanding and tracking air quality. Thus, there is a critical research need to develop input data for a national assessment, as well as at disaggregated spatial scales to identify hotspots and trends in such areas. This would need more continuous measurements of $PM_{2.5}$ and modelling.

Data needs and recommendations

Basing indicators on locally measured and derived data is important. However, collecting and compiling local data for national indictors is not trivial. Bottom-up emission estimates need data and input from a variety of sources at a national scale. In addition, as can be seen here, an important analysis of such data and indicators are the assessment of their trends; this requires regular data collection and analysis for emissions estimates, and continuous monitoring for ambient concentration analysis. This can be resource intensive. However, without such data and analyses, it will not be possible to fully understand the state and trends of air quality and its impacts in South Africa. A starting point may be to focus on a small number of locally important indicators where local data are missing, and work to collect the necessary information for a first bottom-up estimate. Such estimates can then be compared to international estimates and data, which can help to identify missing sources (McLinden et al., 2016) and to decrease the uncertainties in both the local and international estimates.

It is recommended to focus on developing local information for a small number of indicators that are considered key for South Africa (e.g. SO_2 and $PM_{2.5}$). These indicators would be useful to South Africa and air quality management as they do present additional information than just ambient concentrations and exceedances. However, the strength of the indicator, and its trends, are in the underlying data.

Acknowledgements

This study was funded by the Department of Environmental Affairs (DEA) under the Rapid Response Research Component of the DEA-CSIR MOU. The authors would like to acknowledge Peter Lukey and Judy Beaumont from DEA for their assistance in formulating the research question and approach. The authors would like to thank SAAQIS and SAWS for the air quality data.

References

Amann, M., Bertok, I., Borken-Kleefeld, J., Cofala, J., Heyes, C., Höglund-Isaksson, L., Klimont, Z., Nguyen, B., Posch, M., Rafaj, P., Sandler, R., Schöpp, W., Wagner, F. and Winiwarter, W. 2011, 'Cost-effective control of air quality and greenhouse gases in Europe: Modeling and policy applications', *Environmental Modelling & Software*, 26:1489-1501

Atmospheric Composition Analysis Group (ACAG), 2016, Dalhoise University, Surface PM2.5, http://fizz.phys.dal.ca/~atmos/martin/?page_id=140. Accessed February 2016.

Bonjour, S., Adair-Rohani, H., Wolf, J., Bruce, N.G., Mehta, S., Prüss-Ustün, A., Lahiff, M., Rehfuess, E.A., Mishra, V. and Smith, K.R. 2013, 'Solid fuel use for household cooking: Country and regional estimates for 1980-2010', *Environmental Health Perspectives*, 121:784–790.

Burnett, R.T., Pope III, C.A., Ezzati, M. Olives, C., Lim, S.S. et al. 2014, 'An integrated risk function for estimating the global burden of disease attributable to ambient fine particulate matter exposure' *Environmental Health Perspectives*, 112:391-403.

Boys B.L., Martin, R.V., van Donkelaar, A., MacDonnell, R.J., Hsu, N.C., Cooper, M.J., Yantosca, R.M., Lu, Z., Streets, D.G., Zhang, Q. and Wang, S.W. 2014. 'Fifteen year global time series of satellite-derived fine particulate matter'. *Environmental Science and Technology*, 48:11109–11118.

Brauer M., Amann, M., Burnett, R.T., Cohen, A., Dentener, F., Ezzati, M., Henderson, S.B., Krzyzanowski, M., Martin, R.V., Van Dingenen, R., van Donkelaar, A., and Thurston, G.D. 2012, 'Exposure assessment for estimation of the global burden of disease attributable to outdoor air pollution', *Environmental Science and Technology*, 46: 652–660.

Center for International Earth Science Information Network - CIESIN - Columbia University. EPI 2014: Environmental Health Objective- Air Quality Palisades, NY: NASA Socioeconomic Data and Applications Center (SEDAC). http://sedac.ciesin.columbia.edu/downloads/maps/epi/epi-environmental-performance-index-2014/epi2014-eh-air-quality.png. Accessed February 2016.

Emerson, J.W., Hsu, A., Levy, M.A., de Sherbinin, A., Mara, V., Esty, D.C. and Jaiteh, M. 2012. The 2012 Environmental Performance Index and Pilot Trend Environmental Performance Index. New Haven: Yale Center for Environmental Law and Policy.

Hsu, A. et al. 2016, '2016 Environmental Performance Index'. New Haven, CT: Yale University. Available: www.epi.yale

Klimont, Z., Smith, S.J. and Cofala, J. 2013. The last decade of global anthropogenic sulfur dioxide: 2000–2011 emissions. *Environmental Research Letters*, 8-014003. 6pp

McLinden C.A., Fioletov, V., Shephard, M.W., Krotkov, N., Li, C., Martin, R.V., Moran, M.D., and Joiner, J. 2016, 'Space-based detection of missing sulfur dioxide source of global air pollution', *Nature Geoscience*, DOI: 10.1038/NGEO2724

Smith, S.J., van Aardenne, J., Klimont, Z., Andres, R.J., Volke, A. and Delgado Arias, S. 2011. 'Anthropogenic sulfur dioxide emissions: 1850–2005', *Atmospheric Chemistry and Physics*, 11:1101–1116.

Statistics South Africa. 2012. 'Census 2011 – Census in brief', *Public report number 03-01-41*. http://www.statssa.gov.za/census/census_2011/census_products/Census_2011_Census_in_brief.pdf (accessed 09/02/2016)

Statistics South Africa. 2015. 'General household survey 2014' report P0318.

Van Donkelaar, A., Matin, R.V., Brauer, M., Kahn, R., Levy, R., Verduzco, C., and Villeneuve, P.J. 2010, 'Global estimates of ambient fine particulate matter concentrations from satellite-based aerosol optical depth: Development and application', *Environmental Health Perspectives*, 118:847–855.

Van Donkelaar, A., R. V. Martin, M. Brauer and B. L. Boys. 2015, Use of Satellite Observations for Long-Term Exposure Assessment of

Global Concentrations of Fine Particulate Matter', *Environmental Health Perspectives*, doi: 10.1289/ehp.1408646

World Bank. 2011. 'World Development Indicators', International Bank for Reconstruction and Development/ The World Bank: Washington DC, USA. Data source: http://users.cla.umn.edu/~erm/data/sr486/data/rawdata/wdi/csv/

World Health Organization (WHO) 2012, 'Household Energy Database http://www.who.int/indoorair/health_impacts/he_database/en/

Supplementary Material

Global gridded pollution and population estimates used in the Global Burden of Disease 2013 are available as Supporting Information from Brauer et al. (2012). These data were downloaded (Supporting Information 005) and the $PM_{2.5}$ concentrations from the VTAPA and HPA sites were extracted for 2005, 2010, 2012 and 2013.

Table S1 below highlights the annual average $PM_{2.5}$ concentrations from this database (significant figures as reported in database). The 2005 concentrations were used in Burnett et al. (2014).

Table S1: : *Ground-level annual averaged $PM_{2.5}$ concentrations (µg/m³) from Brauer et al. (2012). Significant figures are as reported in database.*

	Site	2005	2010	2012	2013
HPA	Ermelo	12.91724	12.81185	12.39297092	12.18869538
	Hendrina	11.9542	12.44675	12.4082598	12.38905936
	Middelburg	12.80226	13.46134	13.17713612	13.03729226
	Secunda	14.44574	14.57748	14.65277698	14.6905712
	Witbank	14.13451	15.9503	15.78367744	15.70102002
VTAPA	Diepkloof	24.72513	26.35033	26.69545272	26.86970548
	Kliprivier	15.9039	16.79124	16.38826719	16.19042202
	Sebokeng	16.0523	17.43382	17.58054743	17.65437359
	Sharpeville	19.50312	21.43347	22.18445547	22.56975871
	Three Rivers	19.45811	21.39796	21.41572369	21.42461106
	Zamdela	14.60568	15.44814	14.81980258	14.51528373

15

Air quality management in Botswana

Modupe O. Akinola[1,2,*], M. Lekonpane[2,3], and Ebenezer O. Dada[1]

[1]Environmental Biology Unit, Department of Cell Biology and Genetics, Faculty of Science,
University of Lagos, Nigeria.
[2]Department of Environmental Science, University of Botswana, Botswana.
[3]Aqualogic (Pty) Ltd., Plot 182, Unit 1, Commerce Park, Gaborone, Botswana.
*Corresponding author (Email address: moakinola@unilag.edu.ng; mayomi12@yahoo.com

Abstract

This paper examines air pollution situation and the history of air quality management in Botswana. The current air quality management in Botswana is still largely underpinned by the Atmospheric Pollution Prevention Act of 1971, supplemented by the more recently enacted legislations such as the Environmental Impact Assessment (EIA) Act of 2010 and the Ambient Air Quality - Limits for Common Pollutants of 2012 published by the Botswana Bureau of Standards. Though commendable efforts have been made toward legislating against air and other forms of pollution, these have not yielded expected results in view of the prevailing levels of air pollutants like sulphur dioxide and fine particulate matters in the country's atmospheric environment. Legislation as a sole measure may not be effective in tackling this challenge. Rather, government should also address some root-causes of the problem by making policies and programmes that will reduce unemployment and increase the earning capacity of citizenry. This will, among other things, effectively check poverty-induced biomass burning in the country. The paper looks at some other challenges of air pollution management and suggestions are made to tackle the identified problems.

Keywords

Air pollutants, atmosphere, environment, carbon monoxide, mining, sulphur dioxide, nitrogen dioxide

Introduction

The environment refers to the totality of all the conditions that affect living organisms, including humans, in their habitat (Santra 2013; Girard 2014). Air is a component part of the environment that plays important life-sustaining roles, but in spite of this, it is still constantly subjected to pollution abuse. A report released by the World Health Organization (WHO) indicated that in 2012, about one in eight of global deaths was attributable to exposure to air pollution; this constituted a total of around 7 million deaths (WHO 2014). The importance of adequate air quality management cannot therefore, be over-emphasized. The increasing rate of urbanisation without the proportional infrastructural development has worsened air pollution situation in Africa (Baumbach et al. 1995; Eliasson et al. 2009). Sub-Sahara African countries, of which Botswana is one, is especially prone to soil-derived particulate matter pollution because it is largely a dry continent.

Air pollution in Botswana

Botswana is a landlocked country with an approximate area of 582,000 km². It is located in the centre of southern Africa with geographic coordinates of 22 00 S, 24 00 E. Census figures showed that the population of Botswana in 2011 was 2,024,904;

this figure was projected to rise to 2,230,905 in 2016 (Botswana Central Statistics Office 2016). Botswana climate is semi-arid with warm winters and hot summers. The land terrain is predominantly flat to gently rolling tableland with Kalahari Desert in the southwest. The country experiences periodic droughts and August winds blow from the west, carrying sand and dust across the country. The country is faced with natural and anthropogenic environmental issues including drought, overgrazing, desertification and limited freshwater resources. Mean monthly temperatures as high as 32°C to 35°C are recorded in the summer months of October to January (Shaikh et al. 2006; Shaikh et al. 2006; Eliasson et al. 2009). These invariably contribute to air pollution problems in the country. Outdoor and indoor air pollution problems in the country are brought about by biomass burning, vehicular emissions, smelting activities, and population growth (Jayaratne and Verma 2001; Shaikh et al. 2006; Eliasson et al. 2009; Verma et al. 2010).

Botswana has continued to witness steady technological and population growth in recent years. Citing figures released by the Central Statistics Office in 2004, Shaikh et al. (2006) indicated that the country witnessed an increase in maintained roads from 8,000 km in 1988 to 11,000 km in 2002, while the number of vehicles increased from 65,00 to 163,000 in the same period. Meanwhile, latest figures released by the same Central Statistics Office and accessed in 2016, showed that government-maintained road network increased from the 11,000 km of 2002, to 18,507 km in 2014; which corresponds to a 7,507 km (68 %) increase. Moreover, the number of visitors that entered the country increased from 1, 200,000 in 1998 to 1,800,000 in 2002. Latest figures put number of visitors at 2,082,521 in 2014. Such human and vehicular increases must have resulted in increased fuel usage and its attendant emissions.

Jayaratne and Verma (2001) carried out a study to investigate the impact of biomass burning on the environment of Gaborone, the capital city of Botswana, using two automatic laser scattering particle counters. The size range of the particles monitored was between 0.1 μm and 5.0 μm. The mean daily particle concentrations were found to vary from about 200 particles cm^{-3} on clear visibility days during the summer to a high of over 9000 particles cm-3 on cold winter evenings. The results also showed that the size and concentrations of aerosols were consistently higher in the highly populated areas relative to low density locations. They stated that due to the absence of proper legal restrictions, majority of the inhabitants use firewood for cooking and heating purposes. In many homes, logs of wood are used for indoor heating purposes during winter, leading to a pall of smoke hanging over the city in the evenings, with a marked influence on atmospheric visibility. Follow-up studies including those of Verma and Thomas (2007); and Verma et al. (2010) have reported increasing atmospheric concentrations of aerosol particles, CO, and particulate matter of size range of 0.3-5.0 μm especially in the low income residential areas.

In addition to biomass burning and vehicular emission, a major threatening source of particulate matter air pollution in Botswana is the mining industry. Ekosse et al. (2004) noted that the growth in mining activities in Botswana may have generated corresponding increase in particulate matter. One identified major site where mining activities pose environmental contamination challenge is the Selebi Phikwe mine area (Ekosse et al. 2004). The town of Selebi Phikwe was established in the early 1960s following the onset of copper-nickel mining activities. The first official activities related to the discovery of the Selebi Phikwe mineral deposits was in 1959, when surface prospecting started over an area of about 67,000 square kilometers. Geochemical exploration techniques were used to discover the Selebi deposit in 1963 and this gave an impetus to the commencement of robust mining activities in 1970. A large proportion (26 %) of the country's labour force, dominated mainly by males, is engaged in mining of Ni-Cu (Ekosse et al. 2006). However, over the past years, there has been growing concern over environmental pollution threat posed by mining activities in the town. Ekosse et al. 2004 carried out a chemical and mineralogical characterization of particulate matter (PM) at the Selebi Phikwe Ni-Cu area in Botswana. They reported that the air in the mining area was polluted by heavy metals including cadmium (Cd), cobalt (Co), chromium (Cr), copper (Cu), nickel (Ni), selenium (Se), and zinc (Zn). The results of Ekosse et al. 2004 study were also indicative of traces of very fine

quartz (SiO_2), pyrrhotite ($Fe1_{-x}S$), chalcopyrite ($CuFeS_2$), albite ($NaAlSi_3O_8$) and djurleite ($Cu_{31}S_{16}$). A consortium of the Institute of Environmental Management and Assessment (IEMA) carried out an Environmental Impact Assessment (EIA) of Selebi Phikwe mining area. They reported atmospheric sulphur dioxide (SO_2) concentrations of up to 100 μg/m^3 in the mining complex as against the critical level of 20 μg/m^3 set by the European Union. The IEMA consortium found a convincing correlation between the concentration of SO_2 in the air with the apparent plant growth reduction and foliar damage in the area (IEMA 2012).

Earlier, Ekose (2005) investigated the general health status of residents of Selebi Phikwe. Data generated revealed that the most frequent common health complaints in the area included frequent coughing, headaches, influenza/common colds and chest pains. These are respiratory tract related problems, suspected to be linked to the effects of air pollution caused by the emission of SO_2 from mining and smelting activities.

Criteria pollutants of concern in Botswana

Criteria pollutants are a set of air pollutants that are considered harmful to health and the environment, and that can cause property damage. They are typically emitted from many sources including industry, mining processes, transportation, agriculture, and electricity generation. The six criteria air contaminants that were the first set of pollutants recognized by the United States (US) Environmental Protection Agency (EPA) are sulphur dioxide (SO_2), particulate matter (PM), ozone (O_3), carbon monoxide (CO), lead (Pb), and nitrogen dioxide (NO_2) (Office of Air Quality Planning and Standard, OAQPS 2013). In Botswana, some of these criteria pollutants and their environmental and health effects have been reported. These include sulphur dioxide (SO_2), particulate matter (PM), carbon monoxide (CO), and nitrogen dioxide (Jayaratne and Verma, 2001; Ekose et al. 2004; Ekosse 2005; Ekosse et al. 2006; Shaikh et al., 2006; Eliasson et al., 2009; Verma et al. 2010). For instance, Ekosse 2005; Ekosse et al. 2006 associated some respiratory health problems identified among the residents of Selebi Phikwe to particulate emissions from mining activities in the region.

Though, many authors have, in the past, identified the sources of CO pollution in Botswana viz: vehicular emissions and biomass burning, especially among the low income group households (Jayaratne and Verma 2001; Verma et al. 2010), one potential source of CO pollution in Botswana that might not have attracted attention is emission from cattle and sheep. Methane (CH_4) is produced in the stomachs of ruminants and intestine of termites. As cattle digest food, methane is produced in their intestines. The methane enters the cow's bloodstream; and when the blood gets to the lungs, the methane is released and exhaled in the normal breathing process. Oxygen in the atmosphere subsequently oxidises the exhaled methane to CO via the equation: $2 CH_4 + 3 O_2 \rightarrow 2 CO + 4 H_2O$ (Girard 2014).

History of air quality management in Botswana

There has been very little development in air quality legislation in Botswana. The Atmospheric Pollution Prevention Act which came into effect from 14th May, 1971 was promulgated to prevent the pollution of the atmosphere through emissions from industrial processes. The act empowers the Air Pollution Control Officer, appointed by the minister, to sanction any unauthorised person who carries out an industrial process capable of causing or involving the emission, into the atmosphere, of objectionable matter within a controlled area. The penalty for any convicted person under the Act ranges from a fine of P500-1000 or a prison term of 6-12 months or both, depending on whether it is a first or second conviction.

The Department of Sanitation and Waste Management was established in April, 1999 under the provisions of the Waste Management Act, 1998. Prior to this, the government developed the Botswana's Strategy on Waste Management. The Department of Sanitation and Waste Management was merged with the Air Control Division of Department of Mines in 2005, with a responsibility of implementing the Atmospheric Pollution Prevention Act, 1971. The merger gave rise to the Department of Waste Management and Pollution Control. The Department of Waste Management and Pollution Control formulates and provides policy direction and leadership to all issues pertaining to waste management, while the implementation of the policies is done by the local authorities.

The Botswana Environmental Impact Assessment Act was passed in 2005, thus making Environmental Impact Assessment (EIA) mandatory for specified projects. The EIA Act, 2005 defines the Department of Environment and Conservation as the competent authority that is responsible for administering and controlling EIA activities in Botswana. This department was renamed as the Department of Environmental Affairs (DEA) and the functions previously assigned to the National Conservation Strategy Agency (NCSA) relating to EIA in the country were transferred to DEA.

The Environmental Assessment Act, 2010 passed by the National Assembly on the 12th April, 2011 and gazetted in June, 2011, is a consolidation of that of 2005. The Act provides for EIA to be used to assess the potential effects of planned developmental activities with a view to determining and providing mitigation measures for any significant adverse effects on the environment. The Act also provides for monitoring and evaluation of the environmental impacts of implemented activities. The Act stipulates sanctions ranging from a fine of P100,000 or a prison term not exceeding 5 years or both for anyone who undertakes a developmental activity without undergoing necessary processes as put in place by the Act. Moreover, such a convicted person shall rehabilitate the area affected by the adverse environmental impact of the implemented activities. Failure to comply will attract a further fine of P1,000,000 or a term of imprisonment not exceeding 15 years, or both.

The Botswana Bureau of Standards published Ambient Air Quality – Limits for Common Pollutants (BOS 498:2012). This Standard specifies limit values (Table 1) for common air pollutants to ensure that the negative effects of such pollutants on human health and the environment are prevented or reduced.

Limit values for common air pollutants in Botswana

Pollutant	Limit value ($\mu g/m^3$)	Average period
Sulphur dioxide (SO_2)	350 120	1 hour 24 hours
Nitrogen dioxide (NO_2)	200 40	1 hour 1 year
Carbon monoxide (CO)	30,000 10,000	1 hour 8 hours
Particulate matter (PM_{10})	200 100	Monthly 1 year
Ozone (O_3)	120	8 hours
Lead (Pb)	0.5	1 year
Benzene (C_6H_6)	5.0	1 year

Source: Botswana Bureau of Standards (2012)

The Botswana limit values for common air pollutants as presented in the table above compares favourably with the limits in South Africa, a neighbouring country, except for PM_{10} one year limit value which is 100 % higher the South African value of 50 $\mu g/m^3$ (South African National Standard, 2011). In spite of this, the Botswana government may still have to consider a downward review of some these pollutants, especially SO_2. This opinion is hinged on the earlier mentioned IEMA Environmental Impact Assessment of Selebi Phikwe mining area which found a correlation between the 100 $\mu g/m^3$ SO_2 atmospheric concentration with the prevalent plant growth reduction and foliar damage in the area (IEMA 2012).

Challenges facing air pollution management in Botswana

Like in many other developing countries, air pollution management in Botswana is faced with several challenges, some of which have already been mentioned earlier in this paper. These challenges include increasing vehicle numbers and road networks (Shaikh et al. 2006), biomass burning (Jayaratne and Verma 2001; Verma et al. 2010), growing mining and smelting activities (Ekosse et al. 2004; Ekosse et al. 2006), inadequate data to monitor and appraise pollution levels, lack of cohesive air quality policies, and weak or no legal restrictions (Jayaratne and Verma 2001). The measures put in place to manage air quality; including the Atmospheric Pollution Prevention Act of 1971, the related but more recent Environmental Impact Assessment Acts of 2005 and 2010, and the Ambient Air Quality - Limits for Common Pollutants of 2012 published by the Botswana Bureau of Standards might not have yielded expected results in view of the prevailing levels of air pollutants like sulphur dioxide and fine particulate matters in the country's atmospheric environment. This conclusion is partly based on published air quality monitoring studies (Jayaratne and Verma 2001; Verma and Thomas 2005, 2007), some of which are outdated by a couple of recent legislations in the country. There is the need for regular empirical data to allow for proper evaluation of the policies and legislations put in place to combat air pollution in the country.

The future of air quality management in Botswana and ways forward

Air quality management in Botswana requires a holistic evaluation and review to bring about significant reduction in the level of atmospheric pollution in the country. Air pollution is already taking its toll on the health and, by extension, the economy of residents, especially around Selebi Phikwe Ni-Cu plant and in the highly populated areas where the level of pollutants is high (Ekosse, 2005; Ekosse et al. 2006). Apart from strengthening existing laws on pollution prevention, effective enforcement measures should be put in place. Some of the challenges facing air pollution management in Botswana and measures that government can adopt or put in place to mitigate them are enumerated hereunder.

Increasing vehicle numbers and road network

To reduce vehicular emissions to the atmosphere, government should be more concerned in investing in quality public transport system. This will discourage the use of personal vehicles and subsequently bring about a reduction in the average number of vehicles on the roads. Importation and use of vehicles should be limited to new and fuel efficient ones. The legislation that has been put in place to set emission limits for automobile, industrial and other engines should be strengthened to ensure compliance. The use of bio-fuels which are environment friendly should be encouraged.

Biomass burning

The prevalent use of firewood, cow dung, plastic bags, and Chibuku cartons for cooking and heating purposes is fuelled by poverty and inadequate legislation (Jayaratne and Verma 2001; Verma et al. 2010). Legislation as a sole measure cannot be effective in tackling this challenge. Rather, government should address the root cause of the problem by making policies and programmes that will reduce unemployment and increase the earning capacity of citizenry. This can then be complemented by mass enlightenment and legislation. In the interim, government should find ways of making liquefied petroleum gas (LPG) and electricity affordable to the masses, probably by subsidizing the costs of these products. Electricity and LPG are more efficient and environment friendly alternatives to biomass fuels. Uncontrolled burning of wastes should be tackled by the local authorities.

Growing mining and smelting activities

The Selebi Phikwe Ni-Cu mine area is one of the most reported sources of air pollution in Botswana (Ekosse 2004; Ekosse et al. 2004, 2006). In view of the reported and potential health and environmental consequences posed by the mining plant, government and other stakeholders should work in synergy to curtail atmospheric pollutants emanating from the plant. It has been noted that a whopping 26% of the country's labour force, dominated mainly by males, is engaged in mining of Ni-Cu. Steps should therefore be taken by government and relevant bodies to reduce the over-reliance on mining, by diversifying the economy of the country and of Selebi Phikwe. Government should set sustainable emission standards for mining activities, and more importantly, ensure compliance. The huge SO_2 generated in the mining process can be captured and channelled to other productive uses such as manufacture of fertilizers and reagents like H_2SO_4. It may not be out of place for government to consider the idea of resettling the residents of Selebi Phikwe. By so doing, Selebi Phikwe will become a dedicated area for mining.

Inadequate data to monitor and appraise pollution levels

It has been observed that urban cities in developing countries are likely to have fine particulate matter concentrations up to 10-fold higher than the US National Ambient Air Quality Standards. Meanwhile, proper assessment of air pollution in these countries is difficult because of lack of cohesive air quality policies, poor environmental monitoring, and paucity of disease surveillance data (Shah et al. 2013). These observations capture the situation in Botswana, where environmental monitoring is weak and data are not available for regular evaluation of air quality. Government, stakeholders, and policy makers should set reference threshold limits values for ambient air quality for pollutants in the country. In addition, a body should be designated to monitor air quality in the country and its report should be published yearly to allow for evaluation.

The adverse effects of pollution, especially atmospheric, are now known not to be limited to the immediate locality where pollutant level is high (Dada et al. 2016). Air pollution is therefore better tackled not in isolation, but in collaboration with other neighbouring countries. Air quality monitoring data and policies in Botswana should be compared and shared with neighbouring countries – South Africa, Namibia and others.

Conclusion

Efforts to monitor air quality and control atmospheric pollutants in Botswana have not yielded expected results in view of the prevailing high levels of aerosol particles and associated compounds in the country's environment. The air pollution problem is fuelled by vehicular emissions, poverty-induced biomass burning, and mining activities among others. Since the potential adverse effects of these air pollutants on the citizens and environment may be too grievous to neglect, government and stakeholders should, as a matter of urgency, take proactive and holistic steps to tackle the identified causative problems and ensure a cleaner atmosphere. Legislation as a sole measure cannot be effective in tackling this challenge. Rather, government should address the root cause of the problem by making policies and programmes that will reduce unemployment and increase the earning capacity of citizenry. This will effectively check poverty-induced biomass burning in the country. Failure to do these may amount to a time bomb, going by the pollution-associated health challenges already identified among the citizens.

References

Atmospheric Pollution Prevention Act of Botswana 1971, *Atmospheric pollution prevention*.

Baumbach G., Vogt U., Hein K.R.G., Oluwole A. F., Ogunsola O. J. and Olaniyi H. B. 1995, 'Air pollution in a large tropical city with high density – results of measurements in Lagos, Nigeria', *The Science of the Total Environment*, 169:25-31.

Botswana Central Statistics Office 2016, Statistics Botswana, Botswana Central Statistics Office, Gaborone, *www.gov.bw*

Botswana Environmental Assessment Act 2005.

Botswana Environmental Assessment Act 2010.

Botswana Bureau of Standards 2012, Ambient air quality – Limits for common pollutants, BOS 498: 2012.

Dada E. O., Njoku K. L., Osuntoki A. A. and Akinola M. O. (2016). Heavy metal remediation potential of a tropical wetland earthworm, *Libyodrilus violacues* (Beddard), *Iranica Journal of Energy and Environment*, 7(3): 247-254.

Ekosse G. 2005, 'General health status of residents of the Phikwe Ni-Cu mine area, Botswana', *International Journal of Environmental Health Research*, 15(5):373-381.

Ekosse G. E., de Jager L., Heever D. and Vermaak E. 2006. Pulmonary health status of residents of a Ni-Cu mining and smelting environment based on spirometry', *Journal of Environmental Health Research*, Chartered Institute of Environmental Health, Vol. 5, no. 1.

Ekosse G., van den Heever D. J., de Jager L. and Totolo O. 2004, 'Environmental chemistry and mineralogy of particulate matter around Selebi Phikwe nickel-copper plant, Botswana', *Minerals Enginneering*, 17:349-353.

Eliasson I., Jonson P. and Holmer B. 2009,. 'Diurnal and intra-urban particle concentrations in relation to windspeed and stability during the dry season in three African cities', *Environmental Monitoring Assessment*, 154:309-324. DOI 10.1007/s10661-008-0399-y.

Girard J. E. 2014, *Principles of Environmental Chemistry* 3rd edition, Jones and Barlett Learning, USA, 711 pp.

Institute of Environmental Management and Assessment 2012, Combating air pollution in Botswana. *EIA Quality Mark in Botswana*, IEMA article.

Jayarantne E R. and Verma T. S. 2001, The impact of biomass burning on the environmental aerosol concentration in Gaborone, Botswana. *Atmospheric Environment*, 35:1821-1828.

Office of Air Quality Planning and Standard (OAQPS) of the United States Environmental Protection agency, 2013, Air quality standard. *OAQPS Air Pollution Monitoring*. <www.epa.gov>

Santra S. C. 2013, *Environmental Science* 3rd Edition, New Central Book Agency (P) Ltd, London, 1529 pp.

Shah A. S. V., Langrish J. P., Nair H., McAllister D. A., Hunter A. L., Donaldson K., Newby D. E. and Mills N. (2013), Global association of air pollution and heart failure: a systematic review and meta-analysis, *The Lancet*, 382(9897): 1039-1048.

Shaikh M., Moleele N., Ekosse G. E., Totolo O. and Atlhopheng J. (2006). Soil heavy metal concentration patterns at two speed zones along the Gaborone-Tlokweng border post highway, Southeast Botswana. *Journal of Applied Sciences and Environmental Management*, 10(2):135-143.

South African National Standard (2011). South African National Standard: Ambient air quality-limits for common pollutants, SANS 1929: 2011.

Verma T S. and Thomas T. A. (2007). Atmospheric aerosol concentration due to biomass burning, *International Journal of Meterology*, 32: 226-230.

Verma T. S., Chimidza, S. and Molefhi, T. 2010, 'Study of indoor pollution from household fuels in Gaborone, Botswana', *Journal of African Earth Sciences*, 58:648-651.

World Health Organization, 2008, 7 million premature deaths annually linked to air pollution, News Release, http://who.int/mediacentre/releases/2014/air-pollution/en

16

Uncertainty of dustfall monitoring results

Martin A. van Nierop[1], Elanie van Staden[*1], Jared Lodder[1], and Stuart J. Piketh[2]

[1]Gondwana Environmental Solutions, 562 Ontdekkers Road, Florida, Roodepoort, 1716, South Africa, info@gesza.co.za
[2]Unit for Environmental Sciences and Management, North-West University, Potchefstroom, 2520, South Africa

Abstract

Fugitive dust has the ability to cause a nuisance and pollute the ambient environment, particularly from human activities including construction and industrial sites and mining operations. As such, dustfall monitoring has occurred for many decades in South Africa; little has been published on the repeatability, uncertainty, accuracy and precision of dustfall monitoring. Repeatability assesses the consistency associated with the results of a particular measurement under the same conditions; the consistency of the laboratory is assessed to determine the uncertainty associated with dustfall monitoring conducted by the laboratory. The aim of this study was to improve the understanding of the uncertainty in dustfall monitoring; thereby improving the confidence in dustfall monitoring. Uncertainty of dustfall monitoring was assessed through a 12-month study of 12 sites that were located on the boundary of the study area. Each site contained a directional dustfall sampler, which was modified by removing the rotating lid, with four buckets (A, B, C and D) installed. Having four buckets on one stand allows for each bucket to be exposed to the same conditions, for the same period of time; therefore, should have equal amounts of dust deposited in these buckets. The difference in the weight (mg) of the dust recorded from each bucket at each respective site was determined using the American Society for Testing and Materials method D1739 (ASTM D1739). The variability of the dust would provide the confidence level of dustfall monitoring when reporting to clients.

Keywords

Buckets, confidence, dust, dustfall, monitoring, precise, uncertainty

Introduction

Fugitive dust is a nuisance and a source of air pollution (Datson, Hall and Birch 2012). Anthropogenic sources of fugitive dust include, but are not limited to, construction, industrial and mining activities. These sources are regulated under the National Dust Control Regulations (NDCR) of 2013 (NEMA: AQA 2013). The purpose of the NDCR is to prescribe general measures for the management and monitoring of dustfall using the American Society for Testing and Materials method D1739:1970 (ASTM D1739: 1970) or equivalent internationally approved method.

Little has been published on the repeatability, uncertainty, accuracy and precision of dustfall monitoring. The aim of this study was to improve the understanding of the uncertainty and the confidence level of dustfall monitoring using the ASTM D1739: 1970 method.

Methods

A dustfall monitoring network was established along the perimeter of a lime processing facility in Gauteng and monitored for 12 months. The network consisted of 12 directional dustfall samplers that were modified by removing the rotating lid. Each sampler contained Four buckets (A, B, C and D) with the dimensions 238 mm (height) and 175 mm (diameter). (Figure 1).

The basic premise with the four buckets per stand was to ensure that each bucket would be exposed to the same conditions and for the same period; therefore, should have equal amount of dust deposition. This assumes that dustfall rates for each of the four buckets are not impacted by the close proximity of the four buckets to each other on the stand. This is an untested limitation of this study. The difference in the weight (mg) of the dust recorded from each bucket at each respective site is observed.

Figure 1: Converted Directional Dust Bucket Stands.

Statistical analysis

The variability of each bucket at each site was calculated to determine the difference in the dust collected for each bucket by calculating the standard deviation for each sampler. This gave an indication of precision. Box plots for all of the sites for every month show the distribution of the data.

A margin of error for each site was calculated using the following

$$E = (t_c)\frac{\sigma}{\sqrt{n}} \qquad\qquad (1)$$

Where; E = margin of error
t = critical value for confidence level c (at 90%)
σ = standard deviation
n = amount of samples

To calculate the uncertainty of the results, the mean of each site was determined. The upper and lower limits (plus/minus 10% from the mean) was used to determine what percentage of samples were outside this band.

Thereafter, the relative standard deviation (%RSD) was calculated to compare the precision of the absolute deposition values between sites.

Results

Some of the results are presented in this section, the balance can be found in appendix A to C.

The standard deviation of 144 samples (12 sites monitored for 12 months) was calculated (Figure 2). 91% of the data points had a standard deviation below 400 mg/m²/day, 81 % of the data points had a standard deviation below 300 mg/m²/day, and 38% had standard deviations below 100 mg/m²/day, this gives an indication of the range of deviation for the entire data set.

The analysis of variance for the results is presented using box plots (Figures 3 and 4). These plots (representing two of the 12 months sampled) are a visual representation of the spread of the data collected for eash site. The smaller the box plot, the lower the variance, and in this case the uncertainty.

Outliers are those data points that are statistically uncertain.

A second method of measuring the uncertainty was to plot the 90% confidence interval (Figures 5 and 6) and to determine the percentage of data points that fell outside of this interval. The majority of data points (51%) at all site fell outside of the 90% confidence level.

The third method of measuring the uncertainty was to provide a band of plus/minus 10% from the mean of the four data points and determine the number of samples lying outside of the band. This is represented graphically for sites 4 and 11 (Figures 7 and 8). 28% of the 288 results were outside the band.

Finally, the relative standard deviation is calculated to compare the precision of the absolute deposition values between sites. A high RSD value indicates a high uncertainty. The average RSD for all sites and for all months was calculated at 11.69%. Most of the sites have a low percentage RSD indicating a small spread between the points (Table 1). There are some points within the dataset that have a higher variability. The cell shading in Table 1 represent the following:

- No colour: RSD below 15%
- Light red: RSD between 15 and 20%
- Red: RSD above 20%
- Dark Red: RSD above 40%

Discussion

Standard deviation is used to show how far the data spreads from the mean. The higher the standard deviation the more spread out the data is. A low uncertainty would be represented by a standard deviation of less than ±5% of the mean. The buckets at each site were exposed to the same environments; therefore, it is expected that they should collect the same amount of dust.

The box plots are a visual way of representing the data from the sample. It shows the minimum, maximum, median, interquartile ranges and outliers. They are only able to show the outlier with the greatest or the smallest value. This is due to the small data groups (populations of 4). Therefore, when the area of the box is minimal, it indicated a closely spaced dataset, which in turn means precise data, i.e. lower uncertainty. Whereas a large area within the box represents spread data with large ranges between the results, i.e. greater uncertainty. It should be considered that the amount of dust per site would vary; therefore, only the size of the box should be taken into consideration and not its position on the y-axis of the graph.

The area in which the test was conducted has a dust standard of 1,200 mg/m²/day (NEMA: AQA, 2013). The margin of error was calculated to see if it is possible for the value of the reading to shift around this standard. That is, if the weight was just below or above the standard, would it be possible for the actual dust deposition to be above or below the standard, respectively. This confidence interval (Figures 5 and 6) indicates that for some of the samples with readings close to the standard it is possible for the result to provide a false exceedence or false conformance to the Standard.

The ASTM D1739–98 reported a standard deviation of 18% in the recovery measurements of water insoluble dustfall from Project Threshold (ASTM D1739–98. 1998), and that there was no link found between dustfall rate and reproducibility or repeatability. Repeatability and reproducibility was not conducted in this current study; however, it is aligned with the Project Threshold study. No link between the dustfall rate and repeatability (standard deviation) was found. The RSD was used to obtain an uncertainty for the entire process whereas Project Threshold

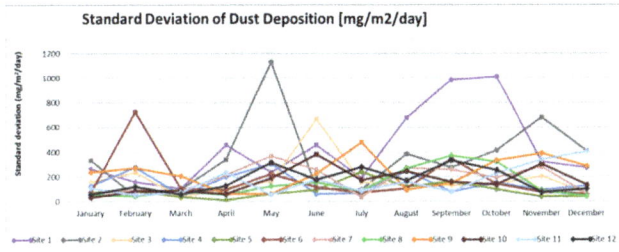

Figure 2: *Standard deviation of all the sites over 12 months.*

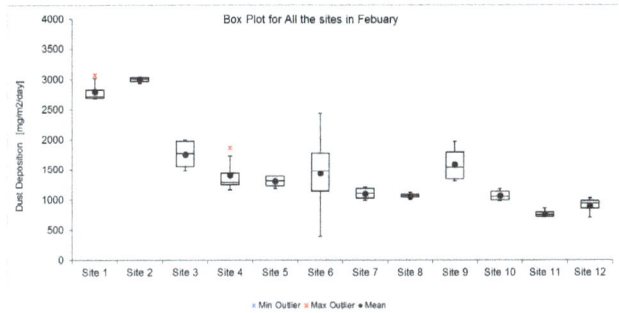

Figure 3: *Box Plot indicating the data distribution for all the sites in February.*

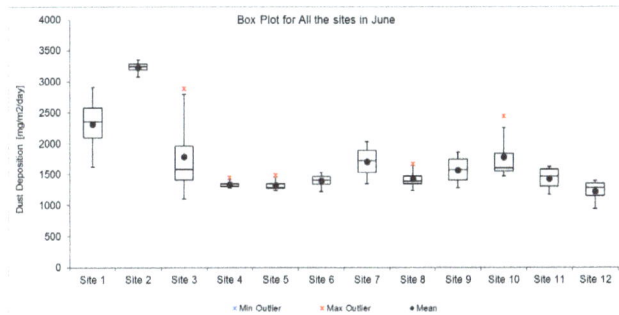

Figure 4: *Box Plot indicating the data distribution for all the sites in June.*

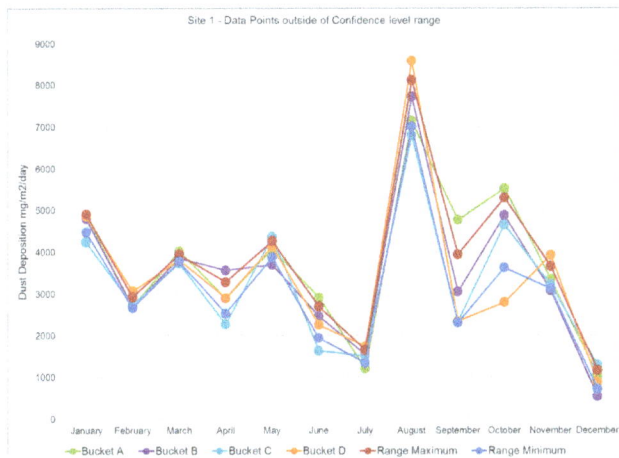

Figure 5: *Indication of data with respect to a 90% confidence interval.*

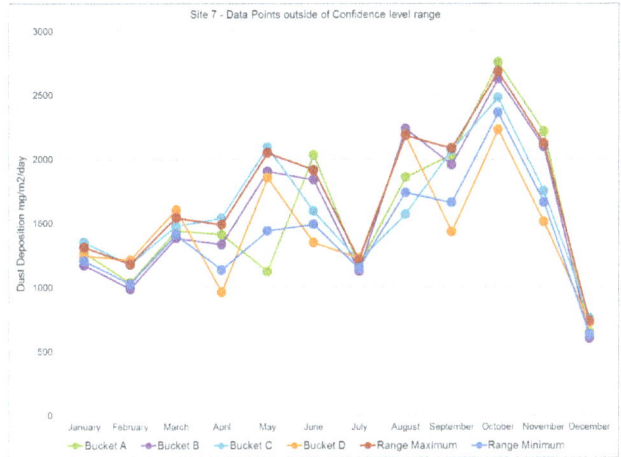

Figure 6: *Indication of data with respect to a 90% confidence interval.*

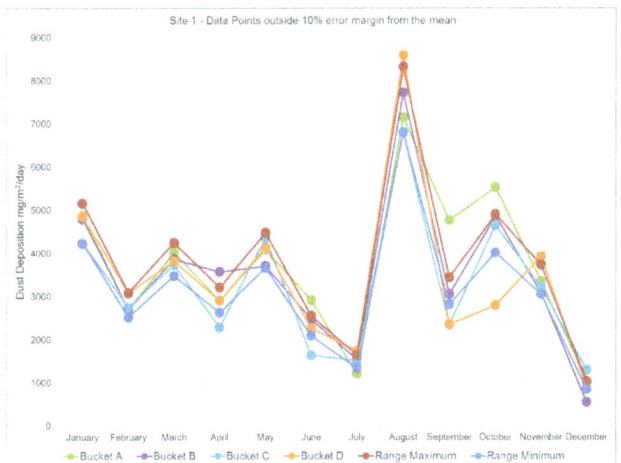

Figure 7: *Indication of data with respect to a 10% margin from the mean.*

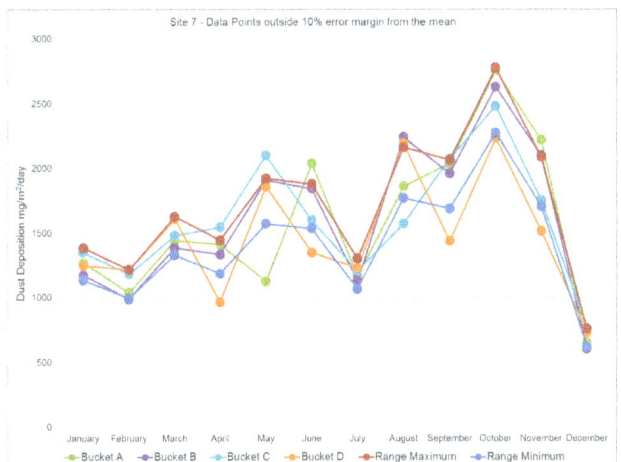

Figure 8: *Indication of data with respect to a 10% margin from the mean.*

Table 1: *The relative standard deviation*

	January	February	March	April	May	June	July	August	September	October	November	December
Site 1	5.65%	5.61%	2.75%	15.69%	5.82%	19.84%	12.77%	8.98%	31.69%	22.76%	9.53%	29.13%
Site 2	10.60%	1.53%	3.03%	13.76%	33.08%	3.11%	7.22%	14.44%	11.25%	10.28%	26.15%	22.47%
Site 3	8.10%	13.14%	1.99%	9.10%	12.22%	37.30%	10.38%	8.61%	6.53%	6.36%	12.80%	5.06%
Site 4	7.88%	19.60%	3.76%	19.86%	25.38%	4.68%	5.94%	12.64%	4.20%	6.85%	7.65%	19.05%
Site 5	5.41%	6.67%	2.89%	1.25%	4.88%	7.05%	17.68%	7.03%	10.12%	5.81%	3.09%	5.81%
Site 6	4.94%	50.16%	7.13%	5.39%	9.78%	8.15%	4.57%	7.56%	21.47%	9.11%	7.44%	9.68%
Site 7	5.09%	8.56%	5.52%	16.45%	21.19%	15.12%	3.29%	13.75%	13.67%	7.73%	14.64%	9.04%
Site 8	4.80%	4.22%	6.29%	4.66%	8.94%	10.95%	7.20%	18.79%	20.88%	16.96%	7.38%	9.66%
Site 9	14.36%	17.18%	11.56%	4.80%	4.29%	14.35%	19.44%	6.64%	7.99%	15.89%	14.39%	20.46%
Site 10	2.76%	7.84%	6.55%	4.50%	8.80%	21.61%	9.84%	16.80%	9.43%	7.78%	12.94%	20.25%
Site 11	10.49%	7.64%	5.93%	11.14%	7.03%	12.72%	4.35%	13.70%	5.17%	17.57%	24.69%	50.76%
Site 12	6.98%	13.28%	5.07%	3.97%	11.04%	14.28%	19.14%	8.91%	41.00%	9.07%	9.04%	15.69%

reported on the laboratory component of dustfall monitoring only. The current study identifies environmental conditions that have a greater contribution to the calculated uncertainty of the method.

Conclusion

The dustfall rate for each group of four samplers per site was expected to have a low variability given that they were exposed to the same conditions. However, variation in the dustfall rate indicates some level of uncertainty. The results of this study show that there is uncertainty in the results from the dustfall samplers. Although some uncertainty could be attributed to sample handling, the majority is considered to be from environmental factors.

The proximity of the four buckets on each stand could affect the flow pattern around these buckets and potentially affect the deposition into the bucket. For this study it was assumed that the effect each bucket has on the others is equal. Future work for this study will correlate the highest mass of the four buckets with the dominant wind direction.

References

ASTM D1739-70 1970, Standard Test Method for Collection and Measurement of Dustfall (Settleable Particulate Matter), ASTM International, (Reapproved 2004), West Conshohoken.

ASTM D1739-98 1998, Standard Test Method for Collection and Measurement of Dustfall (Settleable Particulate Matter), ASTM International, (Reapproved 2010), West Conshohoken.

Datson, H, Hall, D and Birch, B 2012, 'Validation of a new method for directional dust monitoring', *Atmospheric Environment*, 50, 1-8.

NEMA: AQA 2013. National Environmental Management: Air Quality Act (39/2004): National Dust Control Regulations. No 827 of 2013, Government Gazette. 827(36974). 1 November, Government Notice 827. Cape Town: Government Printer.

Vertex42 LLC 2014, Box and Whisker Plot Template; *Create a Box and Whisker Plot using Microsoft® Excel ®*, accessed 6 August 2014, <http://www.vertex42.com/ExcelTemplates/box- whisker-plot.html>

Appendix A: Monthly Box Plots

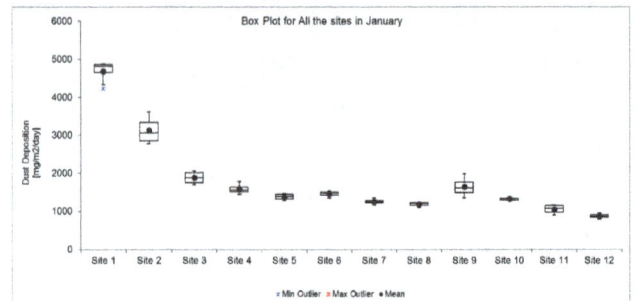

Box Plot for All the sites in January

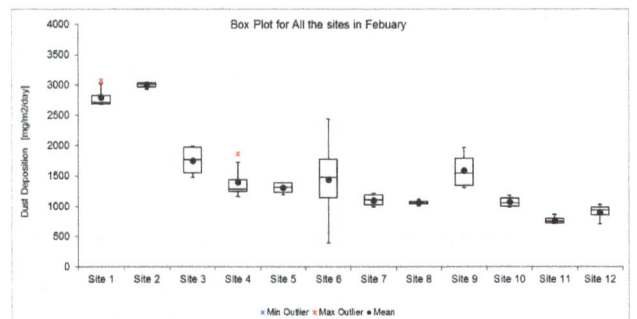

Box Plot for All the sites in Febuary

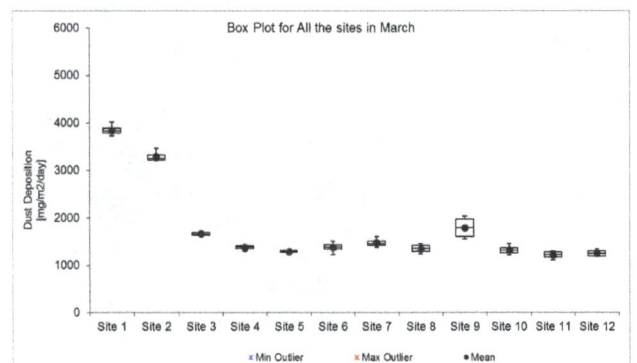

Box Plot for All the sites in March

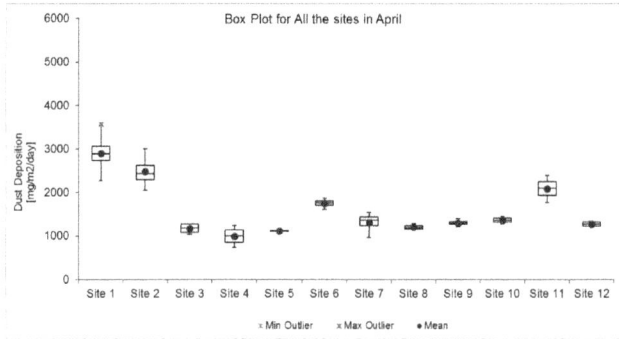

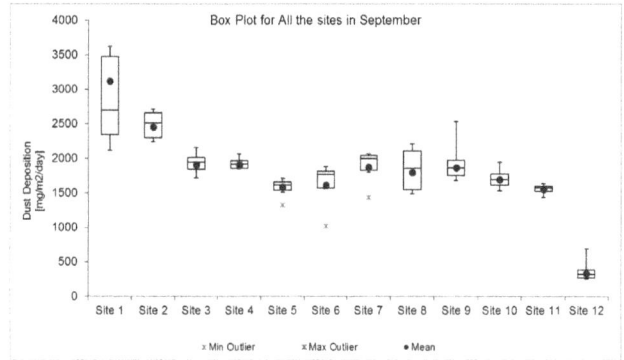

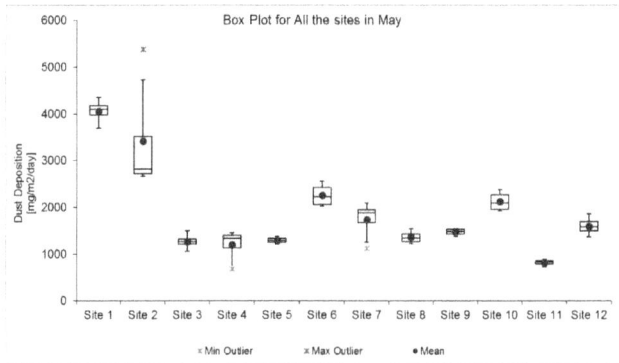

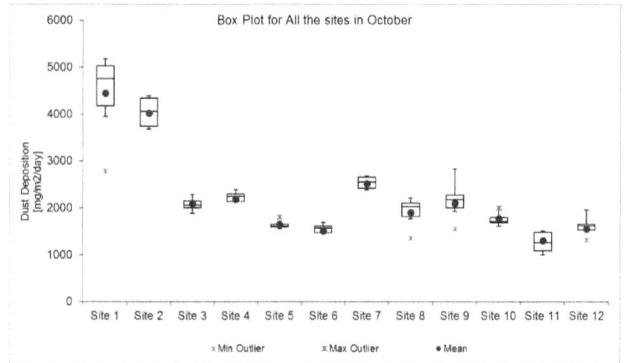

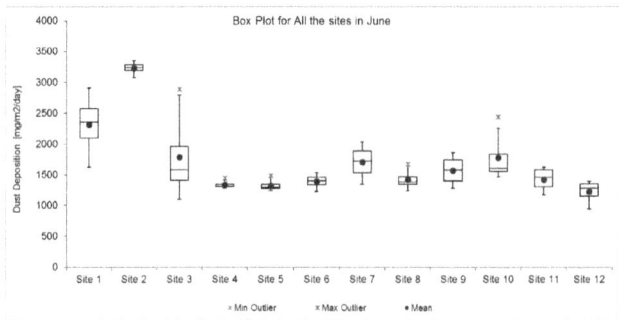

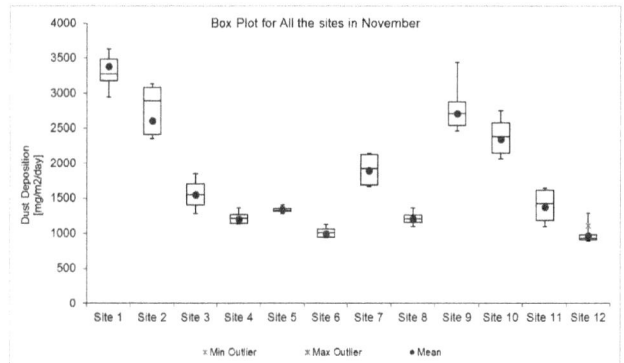

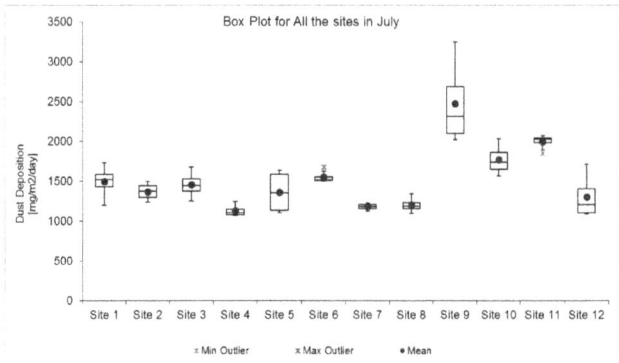

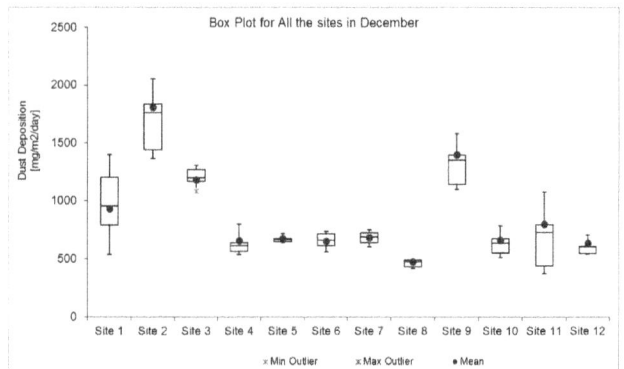

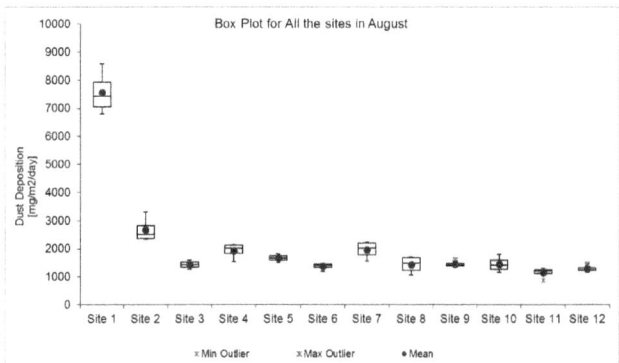

Appendix B: 90% Confidence Interval Graphs

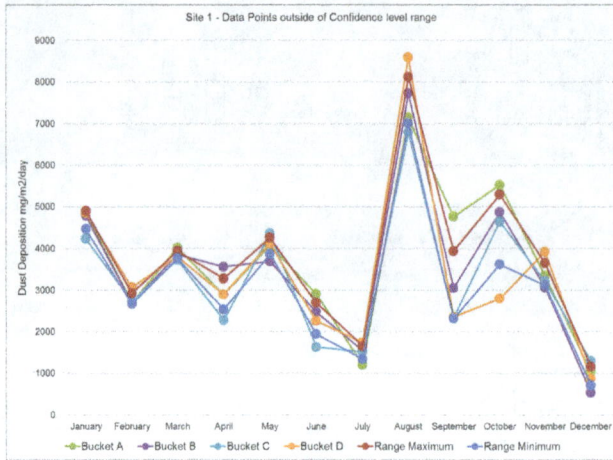

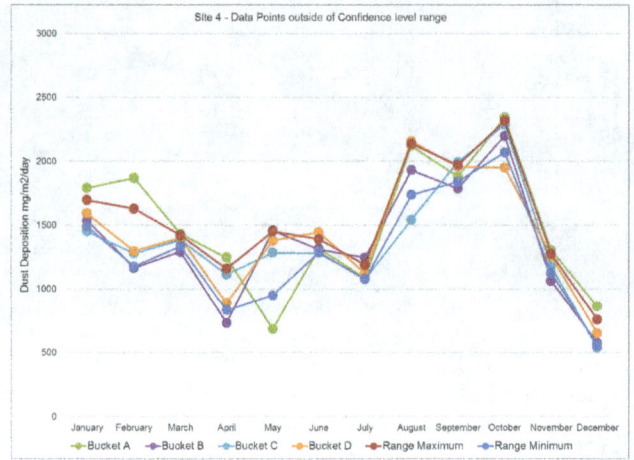

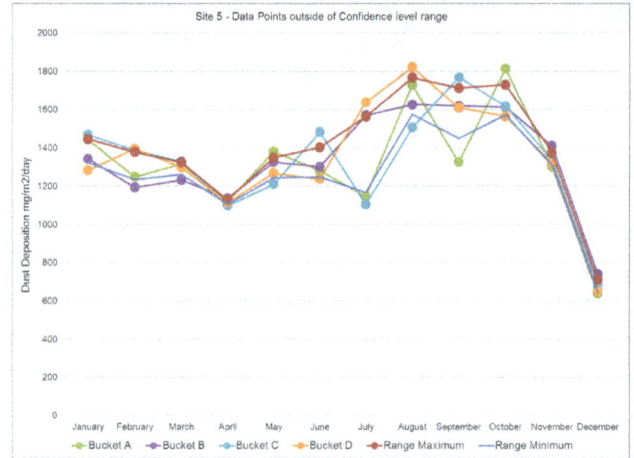

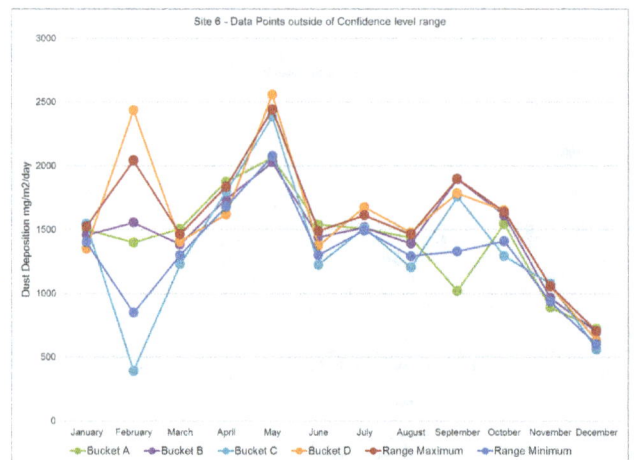

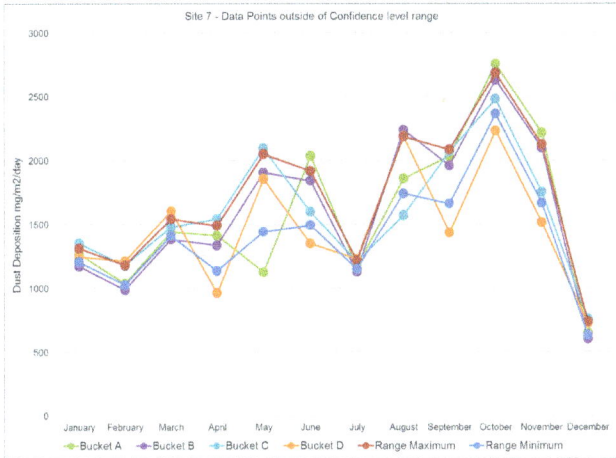
Site 7 - Data Points outside of Confidence level range

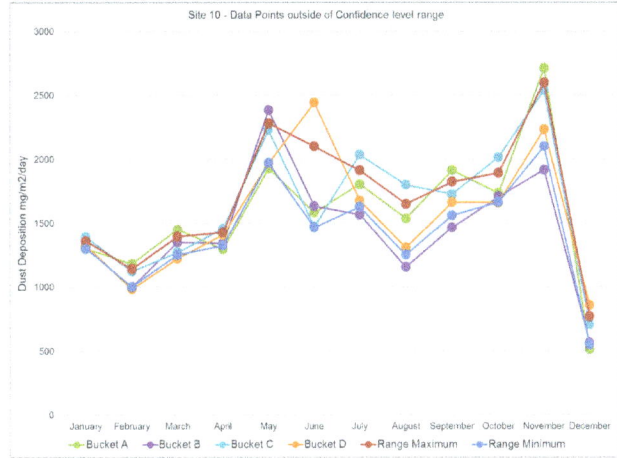
Site 10 - Data Points outside of Confidence level range

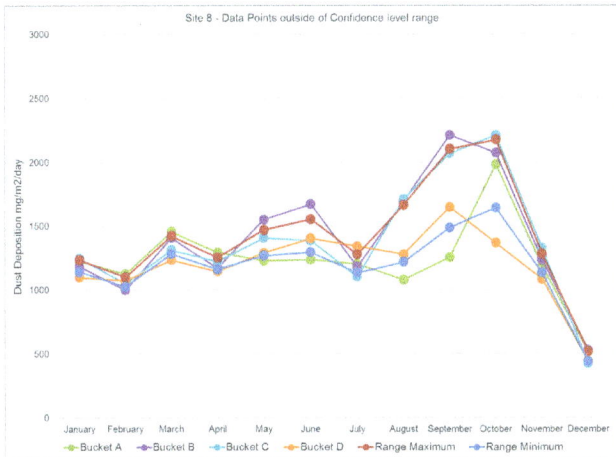
Site 8 - Data Points outside of Confidence level range

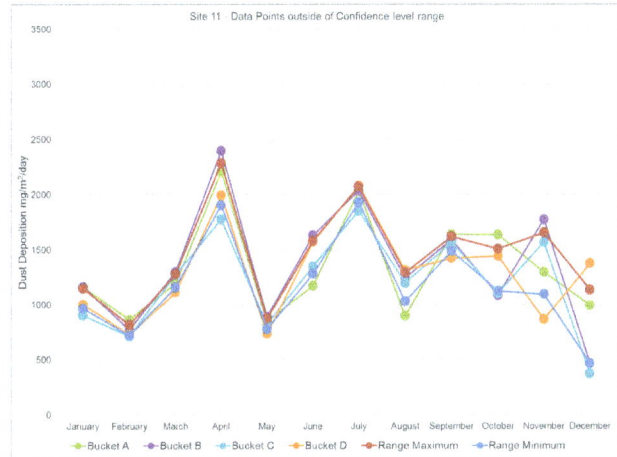
Site 11 - Data Points outside of Confidence level range

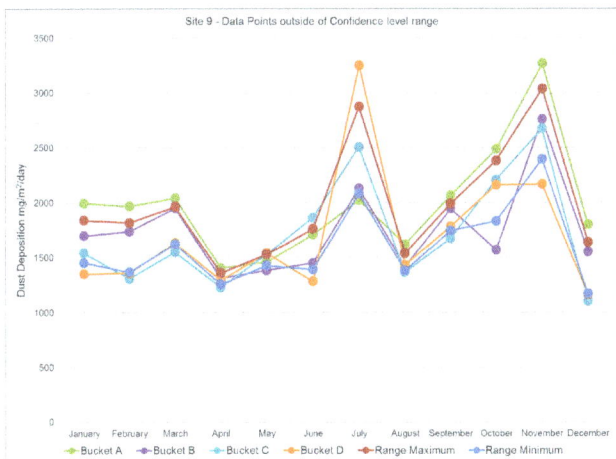
Site 9 - Data Points outside of Confidence level range

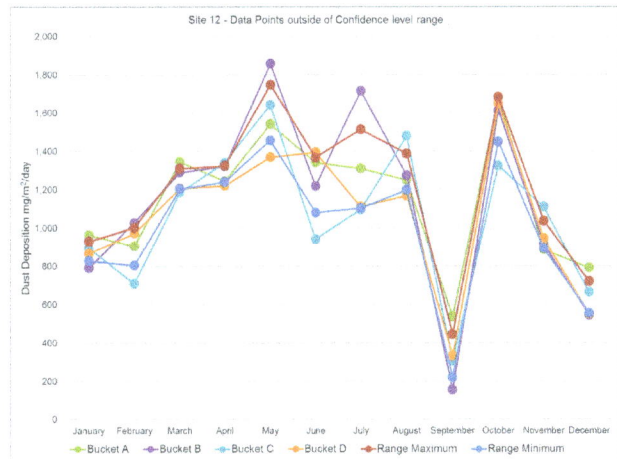
Site 12 - Data Points outside of Confidence level range

Appendix C: 10% Error Margin Graphs

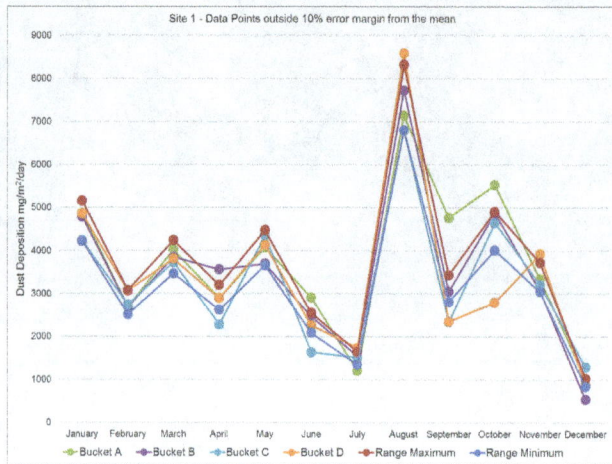

Site 1 - Data Points outside 10% error margin from the mean

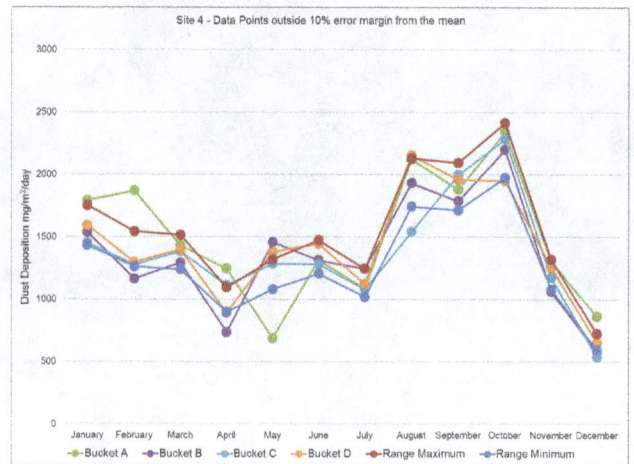

Site 4 - Data Points outside 10% error margin from the mean

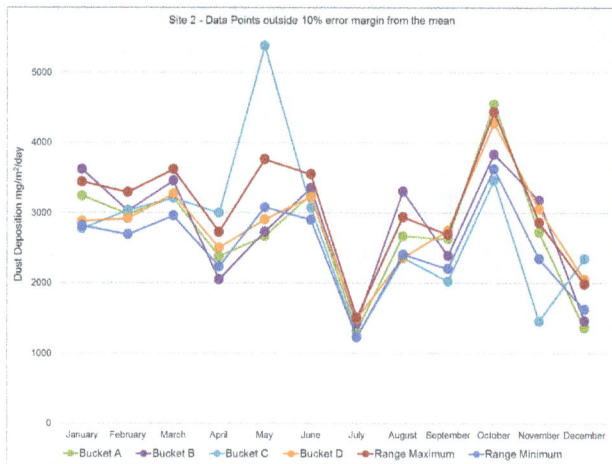

Site 2 - Data Points outside 10% error margin from the mean

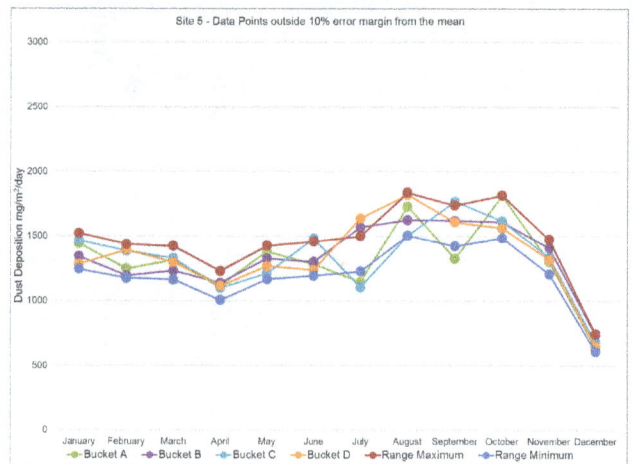

Site 5 - Data Points outside 10% error margin from the mean

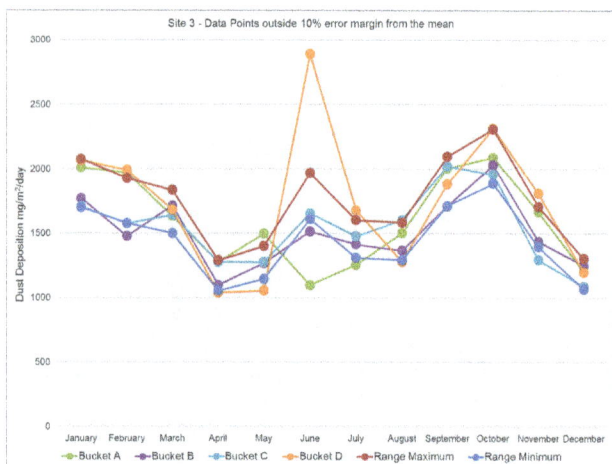

Site 3 - Data Points outside 10% error margin from the mean

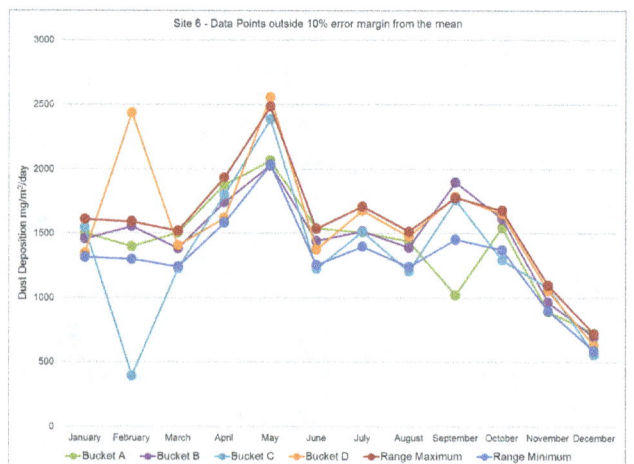

Site 6 - Data Points outside 10% error margin from the mean

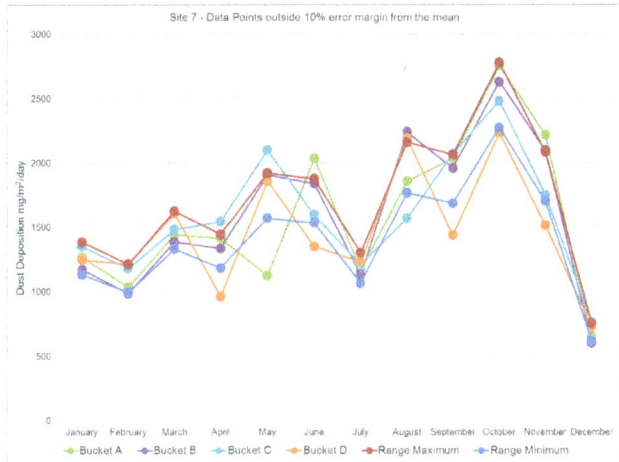

Site 7 - Data Points outside 10% error margin from the mean

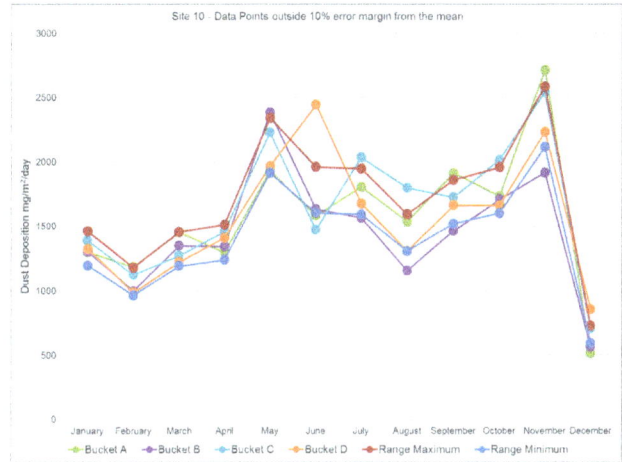

Site 10 - Data Points outside 10% error margin from the mean

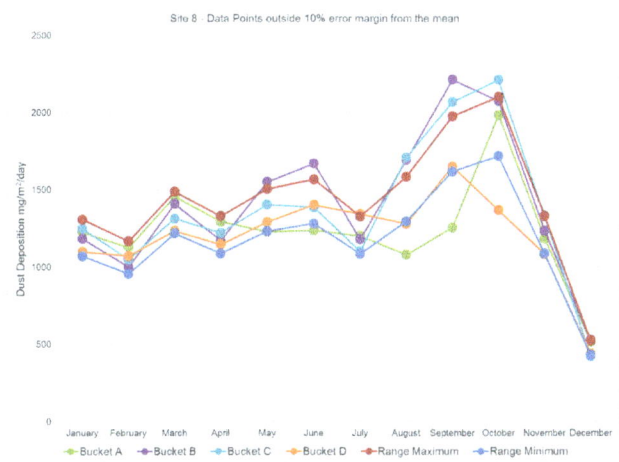

Site 8 - Data Points outside 10% error margin from the mean

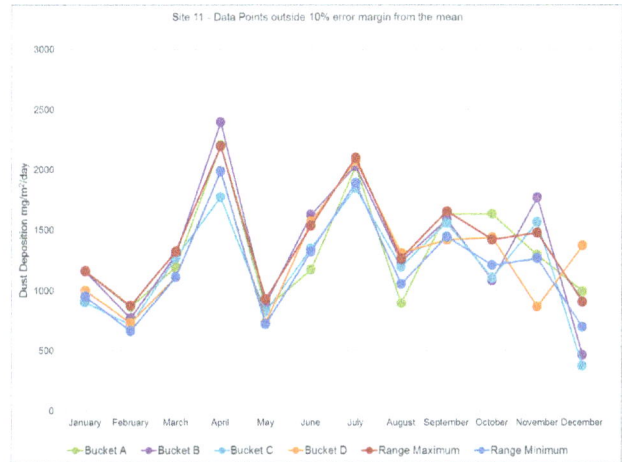

Site 11 - Data Points outside 10% error margin from the mean

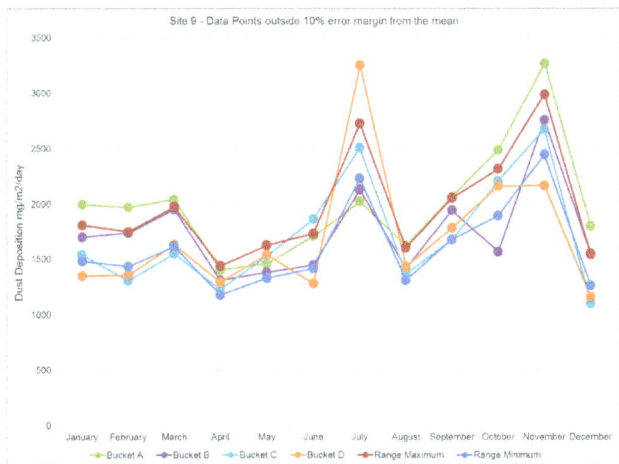

Site 9 - Data Points outside 10% error margin from the mean

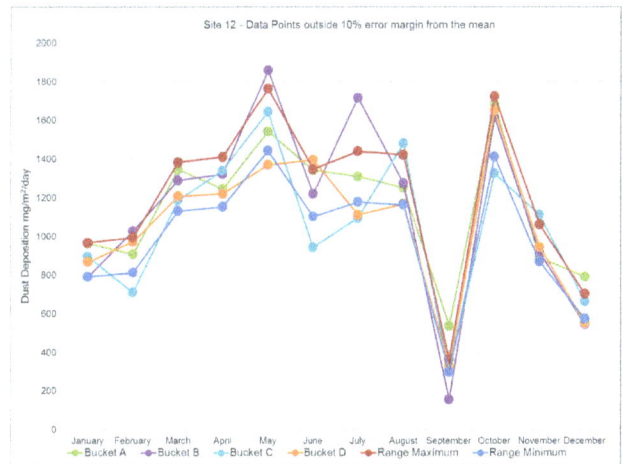

Site 12 - Data Points outside 10% error margin from the mean

17

A Nairobi experiment in using low cost air quality monitors

Priyanka deSouza[*1], Victor Nthusi[2], Jacqueline M. Klopp[3], Bruce E. Shaw[4], Wah On Ho[5], John Saffell[6], Roderic Jones[7], Carlo Ratti[8]

[1]Research Fellow, Senseable City Lab MIT, Cambridge MA 02139 desouzap@mit.edu
[2]Science Division, United Nations Environment Program(UNEP), UN Avenue, Nairobi, Kenya
[3]Center for Sustainable Urban Development, Earth Institute, Columbia University, 475 Riverside Dr. Suite 520 New York NY 10115
[4]Lamont Doherty Earth Observatory, Columbia University, Palisades, New York, USA
[5]Senior Research Scientist, Alphasense Ltd, Sensor Technology House, 300 Avenue West, Skyline 120, Great Notley, Essex CM77 7AA, UK
[6]Technical Director, Alphasense Ltd, Sensor Technology House, 300 Avenue West, Skyline 120, Great Notley, Essex CM77 7AA, UK
[7]University of Cambridge, UK
[8]Director, Senseable City Lab MIT, Cambridge MA 02139

Abstract

Many African cities have growing air quality problems, but few have air quality monitoring systems in place. Low cost air quality sensors have the potential to bridge this data gap. This study describes the experimental deployment of six low cost air quality monitors consisting of an optical particle counter Alphasense OPC-N2 for measuring PM_1, $PM_{2.5}$ and PM_{10}, and Alphasense A-series electrochemical (amperometric) gas sensors: NO2-A43F, SO2-A4, NO-A4 for measuring NO_2, NO and SO_2 in four schools, the United Nations Environment Program (UNEP) headquarters and a community center in Nairobi. The monitors were deployed on May 1 2016 and are still logging data. This paper analyses the data from May 1 2016 to Jan 11 2017. By examining the data produced by these sensors, we illustrate the strengths, as well as the technical limitations of using low cost sensors for monitoring air quality. We show that despite technical limitations, sensors can provide indicative measurements of air quality that are valuable to local communities. It was also found that such a sensor network can play an important role in engaging citizens by raising awareness about the importance of addressing poor air quality. We conclude that these sensors are clearly a potentially important complement but not a substitute for high quality and reliable air quality monitoring systems as problems of calibration, certification, quality control and reporting remain to be solved

Keywords

outdoor air quality, low cost sensors, Nairobi, citizen science, African cities

Introduction

Poor air quality is the world's single largest environmental health risk. Exposure to air pollution in 2012 was responsible for an estimated seven million premature deaths and this problem is growing (World Health Organisation 2014). Given the large public health costs of air pollution, many countries are putting in place more measures to improve air quality, including laws, regulations, monitoring systems and public awareness campaigns (http://web.unep.org/airquality/). As further impetus for these efforts, the new Sustainable Development Goals includes as global targets, reducing annual mean levels of urban fine particulate matter (PM_{10} and $PM_{2.5}$) and the mortality rate attributed to household and ambient air pollution.

These efforts at monitoring and research are uneven across the globe. In sub-Saharan Africa, air quality data often do not exist, and regulations and laws are often not in place to curb air pollution; or if in place, are not implemented, even though existing research shows that the annual mean fine particulate matter in these cities often exceeds World Health Organisation standards (Njee et al., 2016; Petkova et al., 2013). Few African cities operate air monitoring systems, and most cities lack any air quality monitoring capabilities (Schwela, 2012a, Njee et. al 2016). Currently, only Ghana and South Africa operate comprehensive and well organized air quality monitoring programs (Amegah and Agyei-Mensah, 2016). In addition, the air quality data that does exist is not always made public and/or communicated effectively, which limits public awareness and effective policy (Petkova et al., 2013).

Although systematic, long term monitoring is missing in most African cities, existing studies show a serious and growing problem in urban air quality due to rapid urbanization coupled with industrialization, increasing motorization and the continued use of biomass fuel as the household energy source

(UNEP, 2016; Lindén et al., 2012; Fiore et al., 2012; Schwela, 2012b). The worst urban air pollution may actually be in sub-Saharan African countries (Schwela, 2012a). There is thus an urgent need to monitor urban air quality in this region so that the health effects of pollutants can be better understood and quantified, leading to cost-effective abatement strategies and greater public awareness and pressure.

For many African cities, cost is one barrier to investing in air quality monitoring (Amegah and Agyei-Mensah, 2016 ; Schwela, 2012a). The cost of reference air quality monitoring systems (AQMS) is high (costing between US $5000 and US $200,000 for each AQM), and training and AQMS maintenance, as well as managing and analysing the data can also be expensive (Kumar et al., 2015; Mead et al., 2013). This means that even in those countries that have air quality standards and laws, there are often no monitoring systems to measure compliance.

Within this context, low-cost sensors (costing between US $100 and US $3000 for each node) appear to have the potential to help us move from a paradigm of high cost, highly accurate, sparsely located reference air quality monitors, to a dense, low cost, reasonably accurate air quality monitoring network that can also involve citizen science. However, currently no standards or certification criteria exist for such sensors, and there are concerns about the quality of the data (Lewis and Edwards, 2016). Further, the flood of low cost sensors onto the market makes it difficult to determine the reliability of each model. Complicating the challenge of certification, cheap sensors from the same manufacturer often have variable performance.

A US Environmental Protection Agency study of low cost sensors on the market found that either 'no lower cost sensors currently meet [the EPA's] strict requirements or have not been formally submitted to the EPA' (Williams et al., 2014). The US EPA in their study tested these sensors in a clean environment in North Carolina, but how these sensors will perform in the polluted, hot, humid environments frequently found in the developing world is unknown. This is because temperature and humidity can affect the sensitivity of some of these sensors-especially low cost electrochemical gas sensors. Therefore, more work is needed to quantify the accuracy of these sensors under different conditions. Overall, more research is needed on the performance of low cost air quality networks in the field to address the need for monitoring in many of the world's cities (Kumar et al., 2015, Lewis and Edwards 2016).

This paper presents the results and lessons learned from an experiment in using a low cost air quality monitoring network in Nairobi, Kenya. The main aim of this work is to contribute to the growing and important conversation about the role of low cost sensors in air quality monitoring efforts in cities (Kumar et al., 2015; Lewis and Edwards 2016, Kotsev et al. 2016, Piedrahita et al., 2014; Popoola 2012). We were interested in exploring the feasibility of deploying such networks in African cities as a means of gathering some basic data in a quick and efficient way that also involves citizens.

A collaboration between UNEP, the company Alphasense, the University of Cambridge, NASA-GLOBE, the Wajukuu Arts Collective and the Kibera Girls Soccer Academy, resulted in the deployment of a pilot, six node air quality network in four schools, UNEP and one community center in the city of Nairobi, Kenya. The collaboration also aimed to share the experience of air quality monitoring with interested citizens. The sensors include an optical particle counter (Alphasense OPC-N2) that measures PM_1, $PM_{2.5}$ and PM_{10}, Alphasense A-series electrochemical gas sensors (NO2-A43F, NO-A4, SO2-A4) for measuring NO_2, NO and SO_2, temperature and humidity sensors, and a SIM card to transmit data in near real time via the GSM network.

These pollutants were chosen to be measured because particles with aerodynamic diameters less than 10 μm, when inhaled, become embedded in soft tissue and have major health effects. Particulate matter in the environment can have hundreds of different sources. NO and NO_2 are the two oxides of nitrogen that majorly affect human health. NO typically rapidly oxidizes to NO_2. However, the direct emission of NO from vehicles can result in high levels of NO close to roads. SO_2 also negatively affects health. It also reacts with other compounds in the atmosphere to form fine particulates. SO_2 is typically emitted from power plants, industrial facilities, and from the burning of diesel with high sulphur content.

This network started running on May 1, 2016 and is still in operation at the time of writing this paper. After a brief review of air pollution in Nairobi, where no continuous monitoring system yet exists, we present our methods and analyse data collected from this network. Drawing on this experimental deployment, we discuss lessons learned for the potential of low cost air quality networks to support air quality monitoring in African cities.

Background to the Nairobi Case Study

The capital of Kenya, Nairobi is a rapidly growing metropolitan area with an estimated 4 million people living or working within its city boundaries. By 2030, this population may grow to as much as 6 million (World Bank 2016). Air pollution has accompanied this urban growth. Sources include vehicles, open air burning of solid waste, industrial activity and domestic cooking using biomass (Gatari 2009, Kinney et al. 2011, Muindi et al. 2016). Despite growing air quality regulations, such as in the Environmental Management and Coordination Act (Air Quality) Regulations 2014, Nairobi, like most Africa cities, does not have an institutionalized air quality monitoring system.

Scientists at the University of Nairobi, African Population and Health Research Center and their international collaborators (Gaita et al., 2014; Gatari et al., 2009; Gatari and Boman, 2003; Kinney et al., 2011; Muindi et al., 2014; Ngo et al., 2015; Vliet and Kinney, 2007) have taken a number of measurements in Nairobi. These are, however, short-term observations at limited points around the city (background, industrial, roadways, and households in informal settlements) and limited numbers of pollutants, mostly $PM_{2.5}$. In many cases, levels of $PM_{2.5}$ appeared well above the World Health Organization (WHO) 24-h average guideline of 25 mg/m³ and an annual average guideline of 10 μg/m³. However, some measurements were not always comparable with these guidelines, as continuous monitoring was not taking place (Kinney et al., 2011; Ngo et al., 2015).

Methods

Air quality monitors were bought from the company: Atmospheric Sensors Ltd. in the UK (The product catalogue is found here: http://atmosphericsensors.com/products/product-brochures/remote-air-quality-monitor/view). The monitors comprised of an optical particle counter (Alphasense OPC-N2) and Alphasense A-series electrochemical gas sensors, temperature and humidity sensors, and a SIM card to transmit data in near real time via the GSM network. The OPC-N2 (costing USD 450 each) measures particle counts in 16 bins ranging from 0.38 μm to 17.5 μm. It does this by illuminating one particle at a time using focused light from a laser, and measuring the intensity of light scattered from aerosol particles. The amount of scattering from a particle is a function of particle size and composition, which can be calibrated using mono-disperse particles (Sousan et al., 2016). The number of particles per volume for each of these bins can be obtained by dividing the particulate counts of each bin by flow rate and sample time. Alphasense provides a partially proprietary algorithm that makes assumptions about the particle density to calculate PM_1, $PM_{2.5}$ and PM_{10} data from the particle count data. The OPCs in this deployment turn on and run for 20s every 60s; there is 15s of warm up then 5s of actual measurement. The sampling flow rate (SFR) is typically 3.7 mL/s, but varies with temperature. The accuracy of these monitors depends on the size distribution of particulates present, environmental factors such as humidity and the hygroscopicity of the particulates present (Sousan et al., 2016). Without this detailed information, the uncertainty in measurements of the OPC-N2 cannot be quantified.

The Alphasense A-series electrochemical (amperometric) gas sensors (NO2-A43F, SO2-A4, NO-A4): 4-electrode, 20 mm diameter aperture sensors (USD 50-75 each), measure NO_2, NO, SO_2. The electronics of the node used to convert the current of the electrochemical gas sensors to volts and the analogue-to-digital conversion of this voltage signal is proprietary to Atmospheric Sensors. The gas sensors log data every minute at the same time as the OPC. The monitors were coupled to a UPS in order to maintain instrument sensitivity during short power failures. Electrochemical gas sensors exhibit cross-interferences with other pollutants. For example, the NO sensor is sensitive to NO2 (Data sheet: http://www.alphasense.com/WEB1213/wp-content/uploads/2016/03/NO-A4.pdf). The NO_2 sensor is extremely sensitive to O_3 (Data sheet: http://www.alphasense.com/WEB1213/wp-content/uploads/2016/04/NO2-A43F.pdf). The SO_2 sensor is most sensitive to H_2S and NO_2 (Data sheet: http://www.alphasense.com/WEB1213/wp-content/uploads/2013/12/SO2A4.pdf).

Changes in ambient temperature and humidity also affect the sensitivity and sensor gain. The sensors were pre-calibrated at the Alphasense laboratory in the UK, and calibration curves were provided for the gas sensors in order to convert the signals into gas concentrations, expressed as parts per billion by volume (ppb). Alphasense also provided the temperature correction factors for the gas sensors. Research has shown though, that although the sensor manufacturer's correction is effective for sensitivity-dependent temperature correction, it is not effective for temperature-dependent baseline change. Research has also shown that this baseline effect is more pronounced for the NO sensor than for the NO_2 sensor (Popoola et al., 2016). This shall be discussed further when the results are presented.

Data was pulled from the Alphasense server via a file transfer protocol.

One of the biggest drawbacks of our network is that co-location of the low cost monitors with a reference air quality monitor was not conducted. Thus, we have no way to test the accuracy of data from our monitors. We also did not calibrate the electrochemical gas sensors in the ambient environment. We tried to conduct a qualitative appraisal of the data we gathered by going to each site and talking to the people there about what they observed. However, we acknowledge that co-locating at least one of our monitors with a reference air quality monitor would have significantly enhanced our results. We present our analysis and data in this context.

This is the first time air quality was monitored in schools in Nairobi. We specifically engaged with three schools that were part of the NASA GLOBE community (https://www.globe.gov/), which is an international program that allows students the opportunity to participate in data collection. We did this to leverage the existing citizen science program in the schools. We also hoped that we could use our collaboration with NASA GLOBE to expand our deployment in other GLOBE schools in the future.

We selected our sites in a variety of locations. We deployed air quality monitors in low-income schools in the informal settlements: a) Kibera Girls Soccer Academy situated in the informal settlement: Kibera near the railway tracks and b) Viwandani community center in an informal settlement in the industrial area of Nairobi. We also deployed monitors at c) St Scholastica, situated 20 meters away from the notoriously congested Thika Highway in order to capture pollutants from vehicular emissions, d) at UNEP located in Gigiri, which is a relatively green, low density residential, and wealthy part of the city. At e) All Saints Cathedral School which is close to a major road, Mbagathi road, as well as several small shops and industries. Finally, we deployed a monitor at the elite national school, f) Alliance Girls School, located in Kikuyu, a small town to the North of Nairobi as an urban background site. By deploying our monitors in this range of sites, we hoped to capture the signature of different sources in the city as well as get a sense of the differing conditions between very poor and wealthier neighbourhoods. Figure 1 shows the geographic locations of the sites in the city.

The monitors were deployed on walls 1.5-2 meters above the ground so that they would be at close to adult breathing height, but out of reach of the casual passer-by. Note that as the monitors were mounted on walls instead of poles, the sensors only measure pollutants from air masses for a swath of 180 degrees. A plastic shield provided by Alphasense was used to protect the monitor from rain as seen on the upper right-hand side of Figure 1. As of January 11, 2017, the OPCs at all the sites, except for that at Viwandani, which experienced power outages for a few days in May and June and an extended power outage past July 2016, are logging data. The monitor at Alliance Girls School experienced a power outage for most of the month of September 2016, and the one at Kibera Girls School Academy experienced a few hours of power shortage on 19 August 2016, but otherwise have been logging data.

Figure 1: This shows a map of the six deployment sites with a photograph of what each site looks like. The upper right figure is a photograph of the monitor deployed at each site.

We hoped that by engaging with schools and community centers, we would be able to involve the public in air quality monitoring. Participation by residents in the monitoring is an important way to communicate the science of air pollution to citizens (Ngo et al. 2015a). Studies show that Nairobi residents from poor neighbourhoods appear to have a wide variety of often-inaccurate perceptions about air pollution, in part because they have very little information about it (Egondi et al., 2013; Muindi et al., 2014; Ngo et al., 2017, 2015). Nevertheless, a 2015 telephone survey of a representative sample of Nairobi residents, revealed that a majority of Nairobi's adult citizens perceive the air in the city as bad or very bad (69%) and among those who consider the air bad, 93% believed it had an impact on their health. This makes the idea of involving people in monitoring, especially through learning institutions, a viable and potentially important approach that we wanted to test

Finally, we presented preliminary data analysed using the 'OpenAir' package in R version 3.3.2 (Carslaw and Ropkins, 2012) to school children at each deployment site in order to raise awareness about air pollution as well as to brainstorm potential pollution management strategies for the community. We sourced large-scale wind data (that is not local, canyon-influenced wind data) averaged over a period of two minutes, half hourly for this analysis from the Wyoming Weather Website (http://weather.uwyo.edu/surface/meteorogram/), for the Jomo Kenyatta airport site to the south of the city at an elevation of 1624 meters.

Results

Figure 2 shows the hourly averaged PM data obtained from monitors at each of the six sites from May 5 2016 to Jan 11 2017. The monitors started running at different times on May 1 so for consistency, we ignore the data for the first 5 days of measurement. Raw minute wise PM data has been plotted in Fig 1A in the Appendix. The raw PM data showed peaks that were as high as a few 1000 $\mu g/m^3$. It is extremely unlikely for PM readings to reach such high values in a natural environment. However, without co-location with a reference instrument, it is impossible to distinguish the signal from the noise.

From Figure 2, we see that the PM readings at the informal settlements Kibera and Viwandani are routinely very high. A summary of the average minute wise PM readings for each site are provided in Table 1. We note that the difference in $PM_{2.5}$

in up-scale schools such as St Scholastica and Alliance Girls School, and the $PM_{2.5}$ recorded by the monitors at the sites in the informal settlements: Kibera and Viwandani are not very high. We also observe that particulate matter pollution recorded by our monitors at UNEP and All Saints are lower than at the other sites.

The hourly averaged periodic spikes in PM_{10} at the Alliance site are observed to reach a few 1000 $\mu g/m^3$. As mentioned previously, it is unlikely that PM_{10} reaches such high values in the natural environment. These peaks in pollution could indicate a source of pollution very close to the sensor. On going to the site, we found that the school did indeed burn wood very close to the site. This 'ground-truthing' shall be discussed further. It is interesting to note that peaks of the same magnitude as seen in the PM_{10} data were not seen in the finer particulate observations. More information on the kind of burning is required to speculate why this is the case.

We analysed the temperature corrected gaseous pollutant data using the Alphasense calibration and temperature correction at each site. Figure 3 shows temperature-corrected hourly-averaged NO_2, SO_2 and NO data at each site. Raw minute wise gaseous pollutant data for each site can be found in Figure 2A in the Appendix.

We see from Figure 3 that a significant number of gaseous pollutant observations were < 0. Table 1 shows how much of the gaseous data recorded was < 0. This appears to be an issue of instrument calibration and also perhaps consistency between instruments, which we will discuss in more detail in later sections.

The raw data in Figure 2A in the Appendix also shows that for each site some NO observations go to a few -100 ppb. This seems to correspond to the value of NA for the NO sensor. We have applied a filter and eliminated NO values less than -100 ppb from our analysis from here onwards.

As mentioned before, no ambient calibration was carried out for the electrochemical gas sensors and therefore the gaseous pollutant values have to be viewed with skepticism. We present them here to see if any useful signal can be gleaned from the pollutant data.

Table 1 provides a summary of the pollutant data at each site. A multi-pollutant approach of analysing air quality in Nairobi can be useful in identifying common-sources across the city, as well as in identifying possible health effects that could arise from exposure to multiple pollutants, and not just a single pollutant. (Dominici et al., 2010).

From Figure 3 we note that SO_2 is measured to be highest at Viwandani. This seems reasonable. Our site is located in the industrial area of Nairobi. Community members informed us that several factories existed in the vicinity of the site ranging from a paint factory, a factory that manufactured electrical connections and a factory that produced the raw materials for tear gas. Given this background, the high SO_2 values that we saw were not unexpected. We were, however, surprised to see the peak in SO_2 levels at St Scholastica. More work is required to verify this peak, and to identify a potential source. We posit that

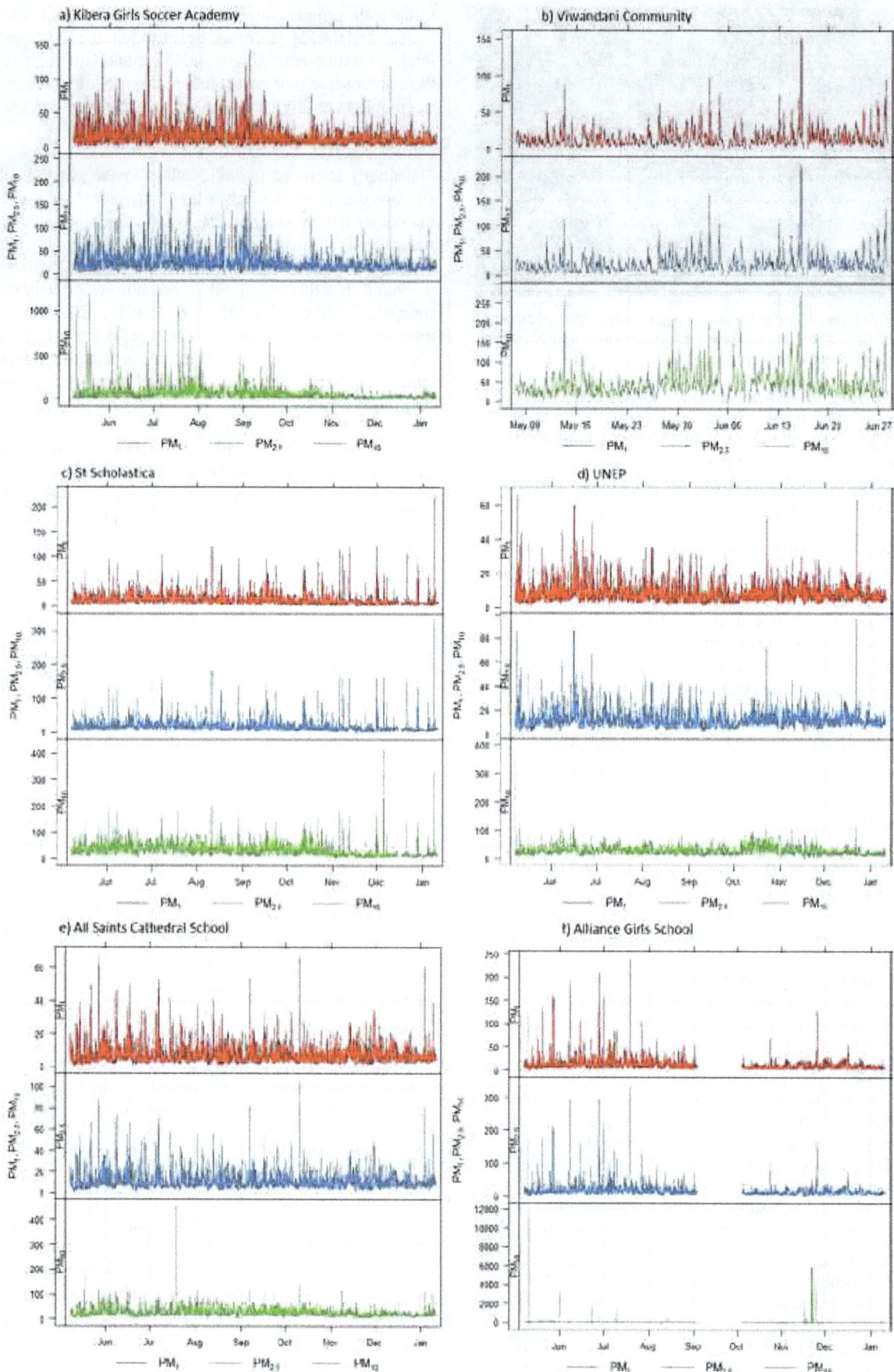

Figure 2: *Hourly averaged PM₁ (red), PM₂.₅ (blue) and PM₁₀ (green) time series plots for each site in units of µg/m³   a) Kibera Girls Soccer Academy, b) Viwandani Community Center, c) St Scholastics, d) UNEP, e) All Saints Cathedral School, f) Alliance Girls School from May 5 2016 to January 11 2017.*

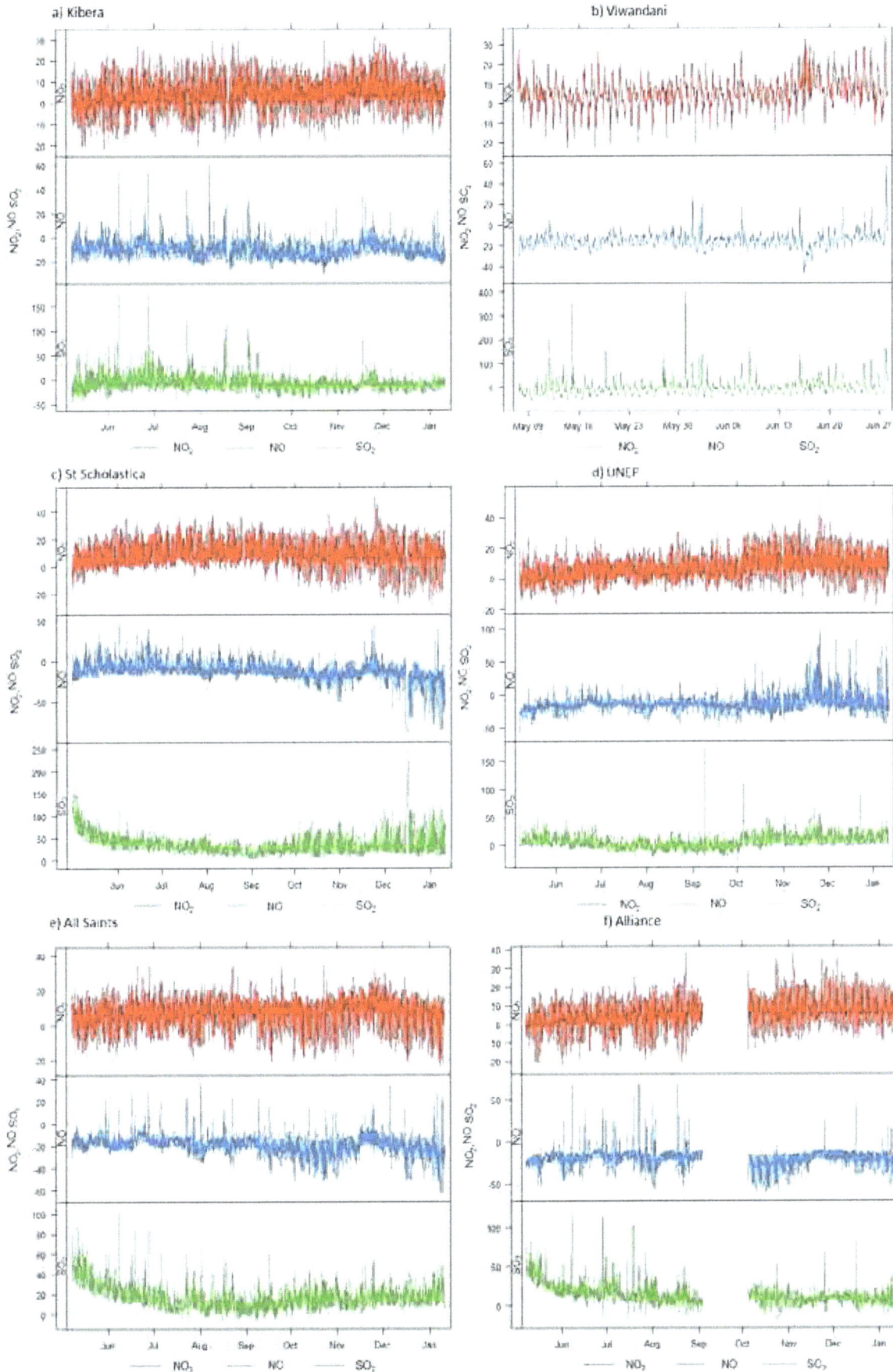

Figure 3: Hourly averaged NO_2 (red), NO (blue) SO_2 (green) time series plots for each site in units of ppb for the sites a) Kibera Girls Soccer Academy, b) Viwandani Community Center, c) St Scholastics, d) UNEP, e) All Saints Cathedral School, f) Alliance Girls School from May 5 2016 to January 11 2017.

Table 1: *Summary of air quality statistics at each of the six sites. We have rounded pollutant values to the nearest whole number to avoid reporting insignificant figures. Note that for calculating correlations (R) involving the gaseous pollutants we used raw values. We only applied a filter to remove NO values that were < -100 ppb at all sites.*

	Kibera	Viwandani	St Scholastica	UNEP	All Saints	Alliance
Total #	355274	72950	352926	355662	357168	312844
Mean PM_1 µg/m³	15	14	11	8	7	12
Mean $PM_{2.5}$ µg/m³	23	21	17	12	11	17
Mean PM_{10} µg/m³	59	44	30	28	26	43
# NO_2>0	255692	54958	308370	288084	276668	248772
# SO_2>0	66192	21687	352638	220297	339850	283234
#NO>0	20782	1885	35812	27324	6202	5826
Mean NO_2 in ppb for values >0	8	9	12	10	10	8
Mean SO_2 in ppb for values >0	19	40	35	13	18	16
Mean NO in ppb for values >0	11	10	10	19	13	21
Correlation of temperature with humidity	-0.83	-0.86	-0.86	-0.89	-0.86	-0.86
Correlation of PM_1 with temperature	-0.11	-0.2	-0.13	-0.21	-0.19	-0.18
Correlation of $PM_{2.5}$ with temperature	-0.08	-0.18	-0.12	-0.18	-0.17	-0.17
Correlation of PM_{10} with temperature	0.039	-0.017	-0.054	0.12	0.05	-0.025
Correlation of NO_2 with temperature	-0.2	-0.33	-0.11	0.059	-0.51	0.015
Correlation of SO_2 with temperature	-0.38	-0.33	0.66	0.28	0.28	-0.0021
Correlation of NO with temperature	-0.39	-0.51	-0.69	-0.43	-0.74	-0.82
Correlation of PM_1 with humidity	0.09	0.16	0.09	0.25	0.2	0.17
Correlation of $PM_{2.5}$ with humidity	0.047	0.13	0.07	0.2	0.17	0.15
Correlation of PM_{10} with humidity	-0.06	-0.05	-0.045	-0.15	-0.12	0.05
Correlation of NO_2 with humidity	0.09	0.26	-0.05	-0.09	0.45	-0.079
Correlation of SO_2 with humidity	0.28	0.25	-0.5	-0.19	-0.12	0.069
Correlation of NO with humidity	0.40	0.44	0.44	0.41	0.74	0.72
Mean $PM_{2.5}/PM_1$	1.6	1.6	1.53	1.53	1.63	1.45
Standard Deviation $PM_{2.5}/PM_1$	0.29	0.2	0.15	0.15	0.18	0.26
Correlation between $PM_{2.5}$ and PM_{10}	0.43	0.71	0.87	0.64	0.62	0.12
Correlation between PM_1 and $PM_{2.5}$	0.96	0.99	0.99	0.99	0.99	0.99

as St Scholastica is situated next to the notoriously congested Thika Highway, fumes from the burning of diesel with high sulphur contents could have resulted in such high values of SO_2 being observed.

NO_2 values don't vary greatly across sites. The lowest NO_2 values were observed at Viwandani and Alliance. This could be because our monitors were located far away from the main roads at both sites. The highest NO_2 values recorded were at St Scholastica, and we again posit that this could be because of the proximity of this site to Thika Highway.

We were also surprised to see the high NO values at UNEP and Alliance.

More work is required to identify how many of the values we see were signal as opposed to noise. More work is also required to identify contributing sources.

Dependence of measurements on temperature / humidity

PM

Pearson correlation coefficients (R) of the pollutants with temperature and humidity were calculated at each site. These values are also summarized in Table 1. Note that we applied a filter to the NO data and eliminated records that were < -100 ppb. Otherwise we used the raw data to calculate correlations involving gaseous pollutants- including negative observations.

Table 1 shows that except for Kibera and St Scholastica, there is a small correlation between temperature/humidity and $PM_{2.5}$ and PM_1. When we plotted $PM_{2.5}$ versus temperature at each site in Figure 3A, it was not clear that peaks in $PM_{2.5}$ corresponded to low temperatures. More research is required to identify how this temperature dependence affects the measurements. One possible reason for this correlation is the lower the temperature, the higher the humidity (temperature and humidity are correlated strongly). If the particles at the site are hygroscopic, the particle size increases, and the OPC detects bigger particles and thus overestimates $PM_{2.5}$. We see that the correlation between temperature/humidity and PM_{10} is negligible.

Gaseous pollutants

As stated previously, temperature and humidity greatly impact the electrochemical gas sensors performance. We thus analyse the data from our sensors in relation to these parameters in order to identify temperature and humidity ranges in which the data is more likely to be less dependent on effects of these environmental factors.

We note that NO_2 and SO_2 and NO are strongly correlated with temperature at each site as can be seen from Table 1. The correlation of each pollutant with temperature varies widely across sites. In addition we note that the sign of the correlation also is not constant across sites for NO_2 and SO_2. We plotted the time series of NO_2, SO_2 and NO at each site with the temperature at each site determining the colour scale in Figures 4A to 7A in order to examine this correlation in more detail.

From Figure 4A and 6A, we clearly see that high temperatures (roughly > 20° Celsius) correspond to negative values of NO_2 and NO being recorded. As mentioned previously, we know that the Alphasense temperature correction does not adequately account for the baseline temperature correction of electrochemical sensors, especially for NO. We also know from the chemistry of electrochemical sensors that the effect of temperature is higher at higher temperatures (Popoola et al., 2016). We thus posit that this is the reason we observe negative gaseous values. Co-location with a reference instrument is required to test this hypothesis.

We note that the sign of the correlation between NO_2 and temperature is positive for the sites UNEP and Alliance in Table 1, because although all negative values of NO_2 recorded are at high temperatures, some high temperatures also correspond to positive NO_2 values, and there are fewer negative NO_2 values for these sites. (Note that in Figure 4A, we have applied a filter and removed NO values< -100 ppb. Figure 5A in the Appendix includes these values).

From Figure 7A, we see that negative values of SO_2 correspond to high temperature readings for the sites Kibera, Viwandani and Alliance. However, for St Scholastica, UNEP and All Saints we find that very few of the temperature corrected values are < 0 (refer to Table 1) and thus we do not see the same negative correlation. We are not sure why this is the case. It could be possible that the temperature correction factor for the SO_2 sensors for the latter three sites are better than for the former, which raises the question of potential consistency across these sensors; or it could mean that cross-interference with other

pollutants are affecting the data at the sites at which they occur in significant quantities. To address questions of potential consistency between sensors, it would be helpful to test a number of these sensors at the same site.

Note, that in Table 1A, when only gaseous pollutant values > 0 were used, we see that the correlation obtained between the gaseous pollutants and temperature/humidity change dramatically, and this time, are the same sign across all sites. In addition, we find that the magnitude of correlation between the measurements that are > 0 and temperature/humidity is low indicating that the signal the sensors are picking up is more likely due to pollutants. We will thus work with these gaseous pollutant values for the rest of this analysis.

Table 1 and 1A also shows correlations between all observations of pollutants. Table 1 shows that PM1 and $PM_{2.5}$ are strongly correlated at each site. From Table 1A, we see a significant correlation between SO_2 and $PM_{2.5}$, NO and $PM_{2.5}$ (except at St Scholastica and UNEP), and NO_2 and $PM_{2.5}$ (except at Alliance), NO_2 and SO_2 (except at Alliance).

Intra Urban Variation of Pollution

We next examined the intra-urban variability in each pollutant across out sites. We note from the correlation between $PM_{2.5}$ for each site-pair in Table 2, that most site-pairs correlate with one other to a not-insignificant manner.

The correlation of $PM_{2.5}$ at each site-pair is not based on distance between sites. The sites at UNEP, St Scholastica, All Saints and Kibera are < 10 km away from each other. The site at Alliance Girls School is ~15 km away from all the sites. However, we note that correlations between Alliance, and UNEP and All Saints are relatively high, in spite of Alliance being far from these sites.

In order to understand if this correlation is due to wind, we produce continuous bivariate plots of normalized $PM_{2.5}$ for each site as a function of wind speed and wind direction using the package OpenAir as seen in Figure 3. Note by using smoothing techniques (via the polarPlot function in the openair R package) to produce the bivariate plots, we are able to identify and group similar features to help identify sources. Wind speeds are zero at the origin and increase radially in each plot. The black arrow in each plot corresponds to the direction in which the monitor at each site is facing.

We note here that for Viwandani and Alliance, for example, there appears to be a source of pollution existing in the west, so that winds from that direction result in the OPC logging high values of PM. This could partially explain the correlation in $PM_{2.5}$ we see across sites. However, we do not see any correlation between $PM_{2.5}$ at Kibera and St Scholastica even though there appear to be a source in the south-east for both sites. This could be because we are not using site-specific wind data. Local canyon effects could profoundly affect our results. In the future, we recommend using site-specific wind data for this analysis.

It must also be noted here that as our monitors were wall-mounted, their swath as mentioned before is 180 degrees instead of 360 degrees. By indicating the direction which each monitor is pointing, we can also examine if there is a directionality bias for

Table 2: Correlation (R) between PM_1, $PM_{2.5}$, PM_{10}, NO_2, SO_2, NO for each pair of sites. Gas values > 0 are considered only.

PM_1	Kibera	Viwandani	Scholastica	UNEP	All Saints	Alliance
Kibera	1	0.13	0.09	0.18	0.19	0.15
Viwandani		1	0.11	0.25	0.29	0.14
Scholastica			1	0.18	0.09	0.11
UNEP				1	0.30	0.26
All Saints					1	0.22
Alliance						1
$PM_{2.5}$	Kibera	Viwandani	Scholastica	UNEP	All Saints	Alliance
Kibera	1	0.13	0.08	0.16	0.18	0.13
Viwandani		1	0.10	0.24	0.28	0.14
Scholastica			1	0.16	0.09	0.10
UNEP				1	0.28	0.23
All Saints					1	0.21
Alliance						1
PM_{10}	Kibera	Viwandani	Scholastica	UNEP	All Saints	Alliance
Kibera	1	0.04	0.07	0.06	0.08	0
Viwandani		1	0.15	0.15	0.28	0.01
Scholastica			1	0.24	0.2	0
UNEP				1	0.25	0.02
All Saints					1	0.01
Alliance						1
NO_2	Kibera	Viwandani	Scholastica	UNEP	All Saints	Alliance
Kibera	1	0.53	0.49	0.53	0.67	0.51
Viwandani		1	0.32	0.46	0.62	0.38
Scholastica			1	0.53	0.32	0.45
UNEP				1	0.41	0.58
All Saints					1	0.35
Alliance						1
SO_2	Kibera	Viwandani	Scholastica	UNEP	All Saints	Alliance
Kibera	1	0.29	0.11	0.12	0.18	0.22
Viwandani		1	0.044	0.09	0.17	0.12
Scholastica			1	0.17	0.56	0.14
UNEP				1	0.15	0.018
All Saints					1	0.21
Alliance						1
NO	Kibera	Viwandani	Scholastica	UNEP	All Saints	Alliance
Kibera	1	0.23	0.03	0	0.17	0.32
Viwandani		1	0.18	0.18	0.11	0.33
Scholastica			1	0.099	0	-0.06
UNEP				1	0.09	-0.13
All Saints					1	0.10
Alliance						1

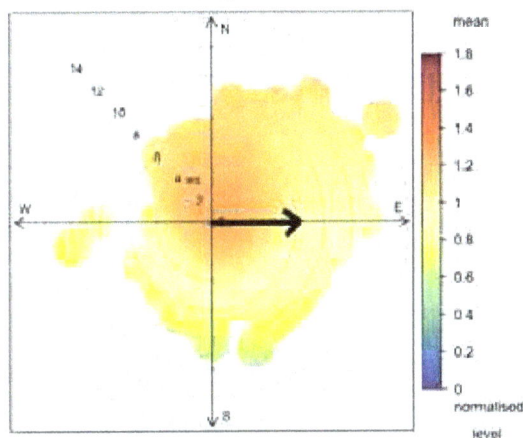

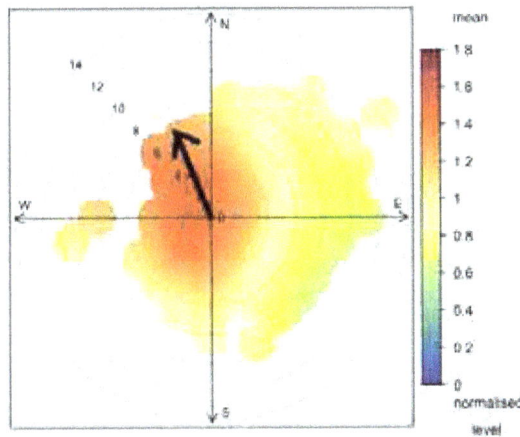

Figure 4: *Bivariate plot of PM_{2.5} normalised by dividing by their mean value from 5 May 2016 to 11 January 2017 plotted against wind speed and wind direction for the sites a) Kibera Girls Soccer Academy, b) Viwandani Community Center, c) St Scholastica, d) UNEP, e) All Saints Cathedral School, f) Alliance Girls School. Wind speed is zero at the origin and increases radially. The color scale indicates the PM_{2.5} concentration. The black arrow in each plot points in the direction each monitor is facing.*

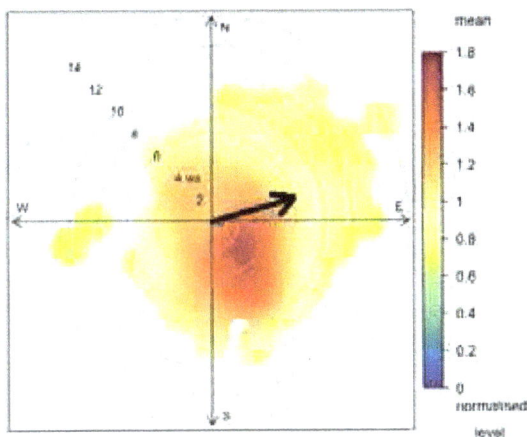

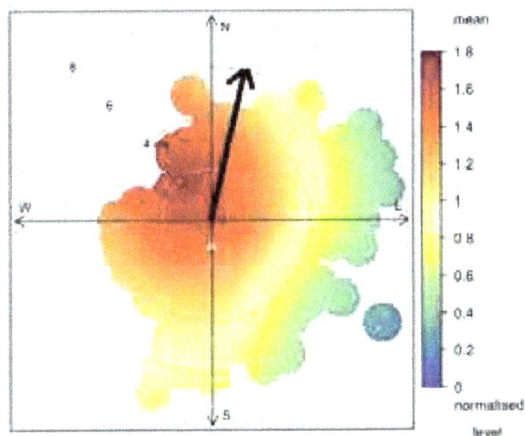

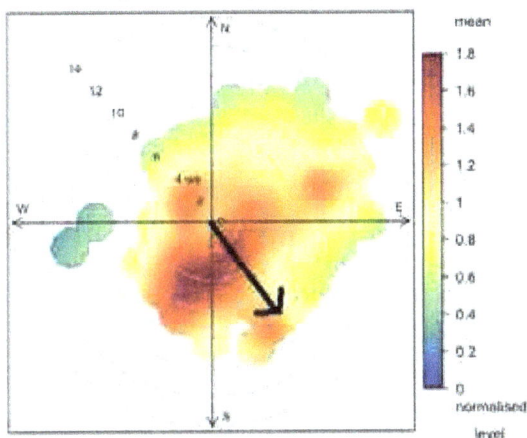

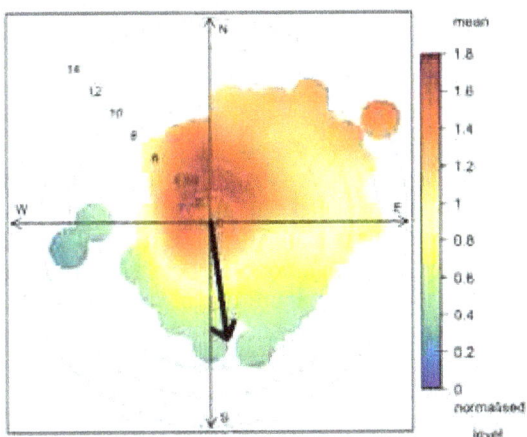

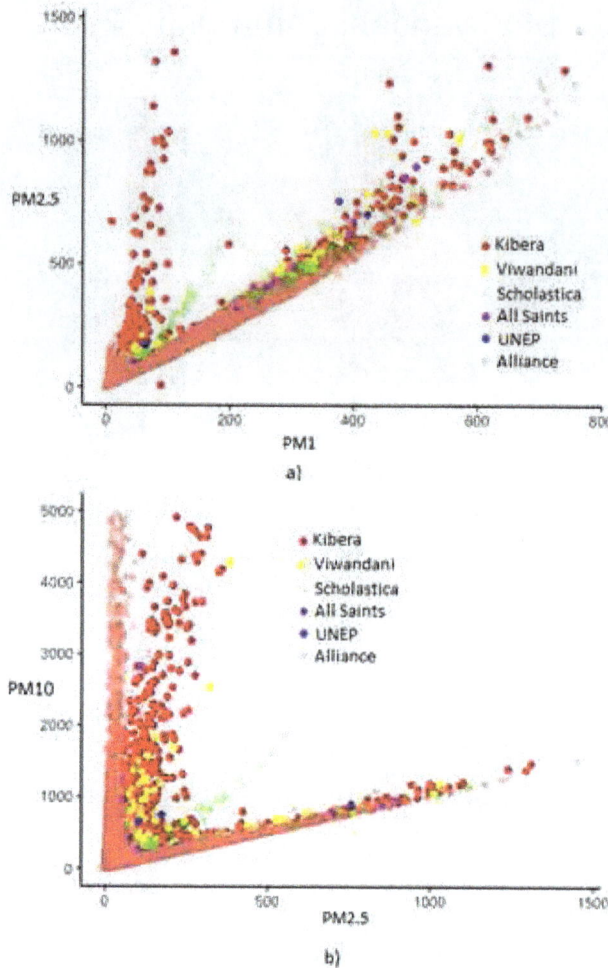

Figure 5: *Scatter plots of PM$_{2.5}$ versus PM$_1$ and PM$_{10}$ for each site. Units are in μg/m³*

each monitor. It appears that the monitors record pollution in the direction in which they are facing, indicating that the limited swath of our monitors could also affect our results.

Analyzing the chemical composition of PM$_{2.5}$ at each site to conduct a source apportionment could also provide further insights into the correlation of PM$_{2.5}$ between sites.

Figure 4 provides us with further insights in itself. We see that for our sites in Kibera and Viwandani, fine particulates impinge on the monitor from many directions, even at low wind speeds, with the greatest pollution coming from the south-east for the former and north-west for the latter.

From Figure 4 we also note that for St Scholastica, there is a major source of pollution to the south of the site for fairly high wind speeds. Thika Super Highway is to the south-east of the monitor, and it is possible that most of the fine particulates that the monitor has registered are from vehicular emissions coming from the highway.

We note from Table 2 that the correlation for NO$_2$ across all sites is high. The correlation for NO across all sites is low on the other hand. NO is a chemical that persists in the atmosphere for a very short time before being oxidized to NO$_2$. This could explain the low correlation between NO across all sites. However, NO$_2$

persists longer and is mainly emitted from vehicles. Traffic patterns in Nairobi are roughly the same across the city at all sites and this could explain the high correlation in NO$_2$ across all sites.

We next look at the minute-wise PM$_{2.5}$/PM$_1$ and PM$_{10}$/PM$_{2.5}$ at each site as shown in Figure 5. These values are summarized in Table 1. We see that PM$_1$ and PM$_{2.5}$ correlate strongly. Figure 5 shows that observations from all sites can be viewed in 2 clusters. The bulk of the observations have a PM$_{2.5}$/PM$_1$ ratio between 1.4 and 1.7. A small cluster of observations have a much higher PM$_{2.5}$/PM$_1$ ratio ~ 5. St Scholastica, is unique in that the monitor at this site records some observations that have a PM$_{2.5}$/PM$_1$ of ~2.5.

This could indicate a unique source at this site. A visit to each site is required to test this hypothesis. Table 1 is a summary table that provides the mean PM$_{2.5}$/PM$_1$ at each site and the standard deviation of each ratio. We also note the clusters of data seen in the plot of PM$_{10}$/PM$_{2.5}$.

In order to understand the latter more closely we look at the variation in the ratio of PM$_{10}$/PM$_{2.5}$ with respect to wind speed and wind direction. This will allow us to look at the signature of different sources of pollution, located in different directions and different distances from each site.

We thus plotted PM$_{10}$ versus P$_{M2.5}$ for each site versus wind speed and wind direction as seen in Figure 6. We see that the ratio of PM$_{10}$/PM$_{2.5}$ is somewhat dependent on wind speed and wind direction.

We repeat the same analysis for the gaseous pollutants and have plotted observations of NO$_2$, NO and SO$_2$ that are > 0 versus PM$_{2.5}$ for each site as seen in Figure 7. We see that the SO2-PM$_{2.5}$ ratio is correlated more strongly than any of the other pollutant combinations in Figure 7.

We will now examine the pollutants at each site in detail.

Site Analysis

Kibera Girls Soccer Academy

Figure 8 shows the raw PM concentration variations averaged over a week and over a single day, for the measurement timeframe: May 5, 2016 to January 11, 2017 at the Kibera Girls Soccer Academy site in the informal settlement of Kibera. We see pollution peaks in the morning shortly before 6 am and in the afternoon on weekdays. However, on Saturday, we see another sharp peak at noon. Pollution appears to reduce on Sundays.

The pollution at this school is far worse than at the other sites. PM$_{10}$ goes up to 100 μg/m³ frequently and exceeds this value during peak hours. Kibera was the only site where on certain days, the 24-hour limit value for PM$_{10}$ (100 μg/m³) set out in Kenya's EMCA (Air Quality) Regulations (2014) was exceeded. This limit was exceeded for 17 days for the time-period of measurement.

The high concentration of particulates could indicate the presence of a significant local source of pollutants. The practise of burning waste due to inadequate waste collection is common here and could be the cause of the high values of pollutants recorded. $PM_{2.5}$ is above the WHO standard, an average of 20 µg/m³ over the course of a day. These results are similar to the high PM2.5 levels measured in the poor neighbourhood of Mathare (Ngo et. al 2015a).

PM counts decrease at night, indicating that most of the PM pollution is due to daytime human activity.

In order to examine the various sources for this site in more detail, we again used bivariate plots using the OpenAir package (Carslaw and Beevers, 2013), and mapped all the pollutants with respect to wind speed and wind direction to identify common sources. Figure 9 shows continuous bivariate plots produced of each pollutant recorded with respect to wind speed and wind direction at the site. Although we are aware that the gas pollutant data in particular is suspect, we believe that by plotting bivariate plots of each pollutants and common sources are identified, that could give us some indication if we are observing any signal in the gaseous pollutant data and thus allow us to vet this data crudely. It is with this perspective that we examined the data. As mentioned earlier, smoothing techniques is used in producing these plots to identify similar groups of pollutants. We used a smoothing parameter of 100 (low smoothing) for producing plots for particulate matter and NO_2, while we used a smoothing parameter of < 50 (high smoothing) for NO and SO_2 as we did not have enough data to produce smooth continuous surfaces for higher cluster sizes.

It can be seen from Figure 9 that there appears to be a major source of particulate matter, some NO and some NO_2 in the south-east. When we visited the site, we learnt that a significant amount of burning was happening to the south of our site near the railway track and this could be a potential source of the particulates. There is a source of SO_2 and NO_2 pollution from the north-west for high wind speeds. The main road is to the west of the site, and this could be a source of these pollutants.

It is not clear if, given that the monitor faces the north-east, there is a bias in the directionality of particulates the monitor registers. Further studies will be needed to determine this.

Viwandani Community Center

Unfortunately, the OPC at this site stopped recording values in early July. However, for the months of May and June, we repeated the above analysis and obtained Figure 10.

We see that here, as at Kibera, PM_{10} levels are higher than at other sites. The monitor here is situated in an informal settlement in the industrial area of Nairobi that is highly polluted, which explains the high values of pollutants recorded. We see that particulate pollution peaks in the morning before 6 am, and in the evening around 6 pm. Pollution reduces on average on Sundays but peaks on Saturdays. Given that the pollution here too reduces in the night, we can conclude that pollution is driven by human activities. Thus, the time variation of the pollution provides us an idea of the time at which activities (cars on the street, burning of waste) take place at this site.

We also see that here, unlike in Kibera, $PM_{2.5}$ and PM_1 track PM_{10} more closely (the correlation between $PM_{2.5}$ and PM_{10} as shown

in Table 2 is 0.71 as opposed to 0.43 for Kibera). An examination of the different sources in this area needs to be conducted to determine why this is the case.

Here too we plot bivariate plots for each pollutant at the site to identify common sources in Figure 11. We see that there appears to be a common source of fine and coarse particles as well as SO_2, NO_2 from the north of the site. We note that there appear to be multiple clusters of pollutants indicating the presence of multiple sources of pollution close to the site. Here, as in Kibera, we used a smoothing parameter of < 50 for SO_2 (note we did not have enough data to plot NO), while for the other pollutants we used a smoothing parameter of 100.

St Scholastica School

The time variation of PM_1, $PM_{2.5}$ and PM_{10} at St Scholastica School as shown in Figure 12. Here we see that PM peaks in the morning and in the evening. These peaks corresponds to the flow of traffic of people coming to work in the morning, and leaving in the evening, implying that emissions from vehicles is a major source of pollution at this site. The peaks in particulates are far more pronounced than at UNEP (Figure14) which is also next to a road. Indeed, $PM_{2.5}$ during the day is as much as 15 µg/m³ higher than during the night. This could be because Thika Highway, a major highway closer to the school, accommodates far more traffic than UN Avenue where UNEP is located. Pollutant concentrations decrease on average on both Saturday and Sunday, and not just on Sunday as seen at UNEP. PM peaks in June as at UNEP.

Figure 13 shows bivariate plots of all pollutants. It can be seen that there is a source of particulate matter in the south. We see there appear to be multiple sources of SO_2 from different directions. There also appears to be a source of NO_2 and NO in the west for high wind speeds. Thika Highway running from the south-west to the north-east of the monitor could be a major source of pollutants.

UNEP

Figure 14 shows the variation of minute-wise particulate matter concentrations over a typical week and a typical day for our other urban background site at UNEP.

Note that typical $PM_{2.5}$ concentrations vary between 10-15 µg/m³ for this site in keeping with the studies done earlier by (Gaita et al., 2014). Here, as at St Scholastica, particulate matter concentrations tend to peak in the morning and evening from Monday to Saturday, corresponding to the flow of traffic of people coming to work in the morning, and leaving in the evening, implying that emissions from vehicles is a major source of pollution at this site. The peak registered on Saturday evenings is surprisingly high, and we still have to account for its cause. We speculate that it is due to people visiting the nearby mall Village Market. Note that on Sunday, pollutant levels are low. We can also see that PM levels are also highest in June.

Figure 15 shows the bivariate plots of each pollutant with respect to wind speed and wind direction. Note to produce the plots we used a smoothing parameter of 100 for all pollutants except for NO, where we used a smoothing parameter of < 50. We see that there is a source of fine particulates close to the monitor. We see there is a major source of coarse particulates

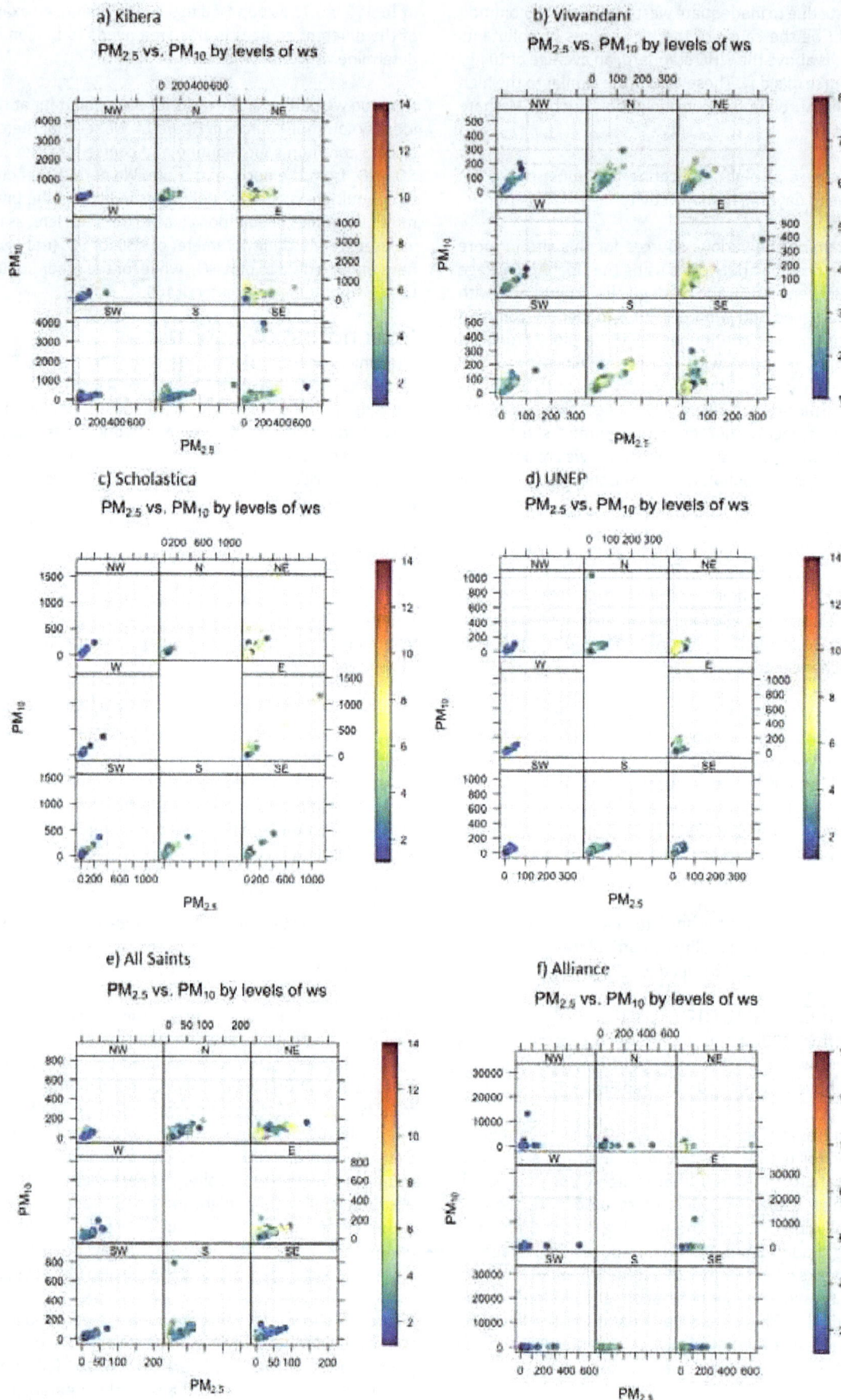

Figure 6: *Scatter plots of PM$_{2.5}$ versus PM$_1$ and PM$_{10}$ for each site with the color scale indicating wind speed, broken up by wind direction. Units are in μg/m³.*

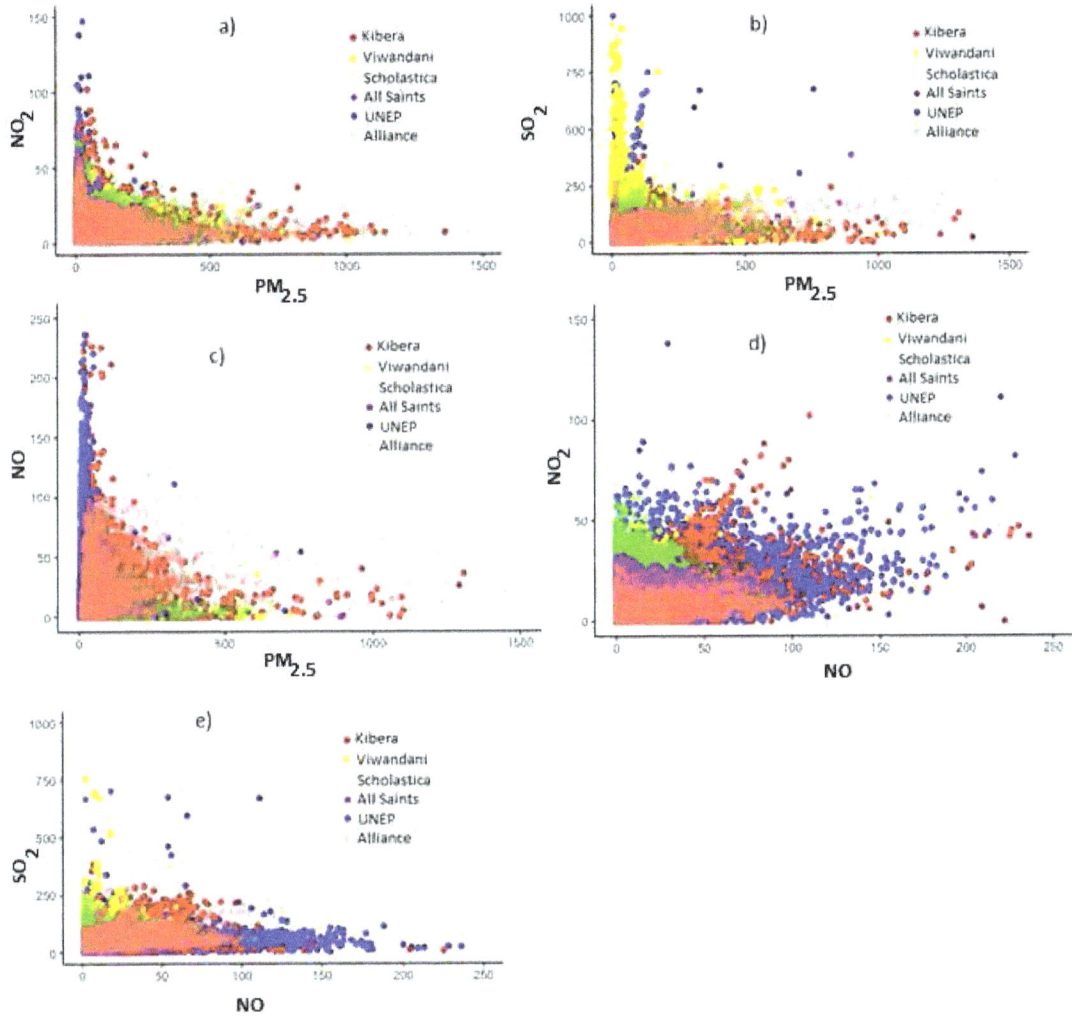

Figure 7: Scatter plots of a) NO_2 b) SO_2 and c) NO (units in ppb) versus $PM_{2.5}$ (units are in µg/m³) for each site, d) NO_2 versus NO at each site, e) SO_2 versus NO at each site (all gases are reported in ppb).

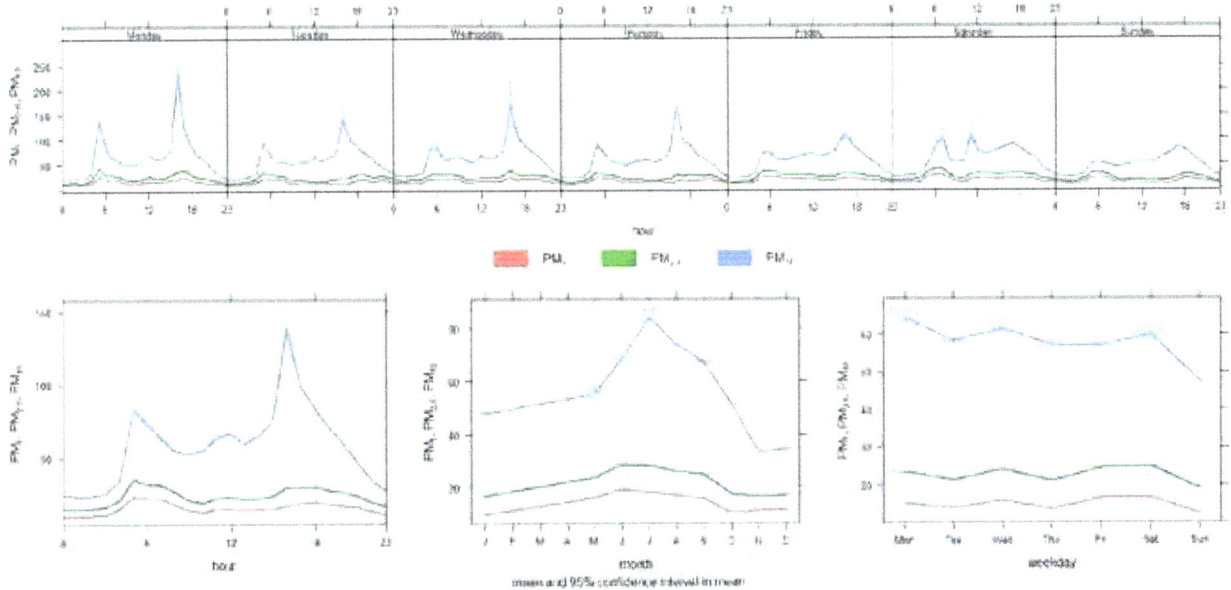

Figure 8: Kibera Girls Soccer Academy. The top panel shows the variation of PM_1, $PM_{2.5}$ and PM_{10} over the course of an average week in units of µg/m³. The panel on the bottom left shows these concentrations varying over the course of an average day. The bottom middle figure shows the variation of PM over 8 months (May 5, 2016 to Jan 11, 2017). The bottom right figure shows concentrations during an average week. The shadings in the plot indicate 95% confidence intervals.

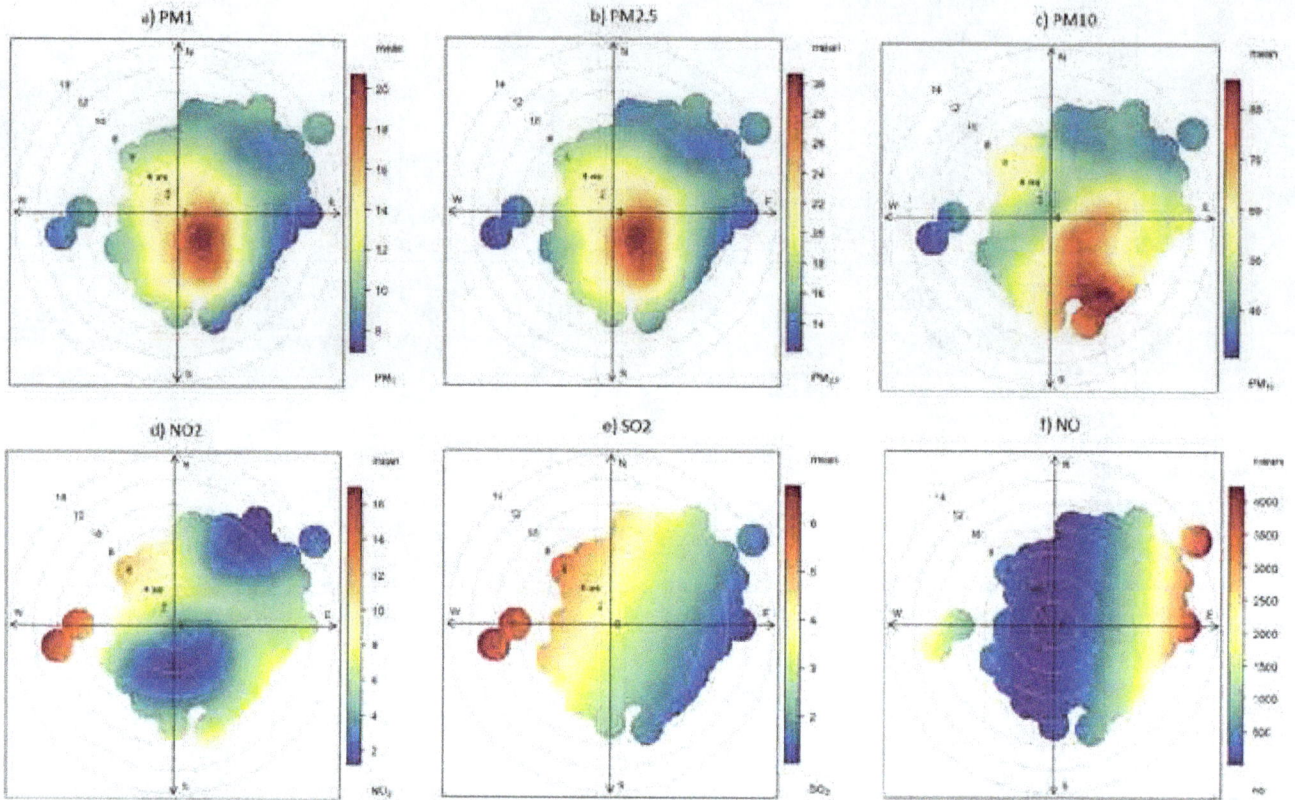

Figure 9: *Kibera Girls Soccer Academy. Bivariate plots of each pollutant (note even negative values of the gas pollutants were considered) with respect to wind speed and wind direction from May 5, 2016 to Jan11, 2017. The image at the bottom shows the site and the black arrow indicates the direction that the monitor.*

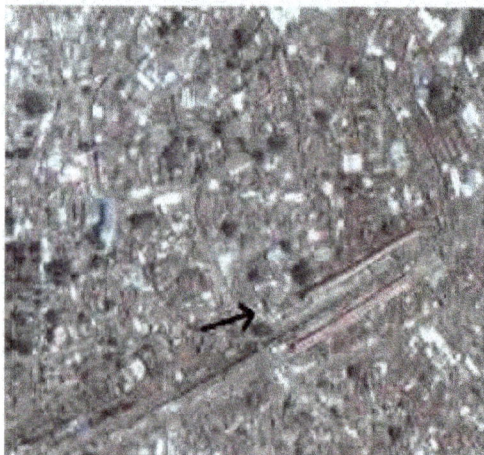

in the south east and north east for high wind speeds. There is a common source of SO_2 in the southeast. There are sources of NO in the north-north west and south-south-west of our site. There is a source of NO_2 and NO from the west. UN Avenue is located to the west of the monitor and is a potential source of vehicular pollution.

All Saint's Cathedral School

The particulate matter pollution at All Saint's Cathedral School is shown in Figure 16. Here as for the previous two sites, we see peaks of pollutants in the morning and in the evening corresponding to traffic patterns. Note that the values of PM registered at this site are in the same range as at UNEP. Here too, PM levels dip on Sunday but not on Saturday. This indicates that people come into the city on Saturdays but not Sunday .

From Figure 17, we see there is a common source of fine

particulates in the north west for low wind speeds. We see that there is also a source of SO_2 in the north west. There is a common source of SO_2 and coarse particulates in the east. Note there are several small industries and shops in this area which are potential sources of pollution. We did not have enough data to produce a similar plot of NO.

Alliance Girls School

We examined the data further to see why pollution was so high at Alliance Girls School. Figure 18 indicates that on some mornings, between midnight and 6 am, there is an immense spike in PM_{10} registered by the OPCs. When we examine the total PM time series plot in Figure 2, we see periodic spikes in PM_{10} as well. On speaking to the schoolchildren, we were told that boilers were lit using firewood at a site located very close to the deployed air quality monitor. Our monitors were thus able to highlight an important finding.

Figure 19 indicates NO$_2$ comes mainly from the south and west of the site at rather high wind-speeds. The Southern Bypass a major highway is in this direction, and it is possible that vehicular fumes from this road are a major source of NO$_2$. Figure 18 also indicates that there is a major common source of PM$_1$, PM$_{2.5}$ and SO$_2$ from the west as well which could also be due to traffic on the Southern Bypass. Trucks typically use the Southern Bypass. They burn diesel with high levels of sulphur, which could be the source of the SO$_2$ seen. However, there is a major source of PM$_{10}$ from the north-east as well. The burning of firewood takes place at the north of our site and thus it is highly likely that it is this that is the major source of the coarse particulates picked up by the monitor. Note we do not have enough data to plot NO for this site.

Discussion and Policy Implications

Even with technical limitations in both the study and the sensors, we were nevertheless able to glean a number of insights from the data. At a local level, the data we gathered led to new discussions about air pollution within the schools, which up to this point have not been sites for air quality measurements. The exception is the monitoring station at the University of Nairobi, which is primarily used for teaching and research on campus. This suggests that further experimenting with sensors through citizen science efforts can be a valuable way of spreading awareness and having public discussions, as long as the potential uncertainties in the data are also part of the conversation (Impressing on the communities the working of the optical particle counter that we used to measure particulate matter allowed them to understand the limitations of the instrument).

For example, identifying the peak in PM$_{10}$ at Alliance Girls School on Wednesday mornings was an important discovery- especially as the monitor was deployed on the wall of a dormitory. Conversations with the school led us to discover the burning of firewood to heat water as a source of this pollution. This allowed us to engage with the school and discuss with students and staff the hazards of air pollution, as well as ways to mitigate their particular source by using cleaner fuels or burning firewood in a different location far away from the students. Identifying that the school was in control of this burning allowed us to work with them to think through various possible pollution management plans. Continued monitoring will reveal if the measures the schools adopts are effective.

Conversations with students at the Kibera Girls Soccer Academy were more complex because the school is located in a large slum and faces a multi-faceted air pollution source problem. Thus, mitigation became part of the conversation – for example, whether planting trees might block the influx of particulates into the school premises from the south-east. This type of conversation around air pollution mitigation also came up in the conversations in Mathare slum (Ngo et al 2015b). More accurate measurements of local, canyon-influenced wind speed and wind direction over different seasons will be crucial to improving the efficacy of any interventions aimed at addressing sources. Given the poor services in these slum areas, waste burning is likely to be one source that needs addressing. However, without alternatives such as better solid waste disposal, mitigation techniques like tree planting or finding ways to avoiding the

worst sources where possible becomes important (Ngo et al. 2015b). Finally, our discussion with the community at the Viwandani Community Center led to the community leaders resolving to bring this issue up with the Nairobi City County, which is responsible for solid waste disposal, and also air quality along with the National Environmental Management Authority (NEMA).

It is important to note that the monitors were not stolen as many people had feared. Our discussions with the community led to them to appreciate the importance of our monitoring instruments. The Kibera Girls Soccer Academy even built a small gate to the alley on which our instrument was located, at their own cost to protect the instrument. However, the OPC at the Viwandani community center and at Alliance Girls School did lose power. Better understanding of the electric power situation and how it can be addressed at each location will be necessary for future deployments. This suggests overall, that more experiments with air quality sensors in collaboration with citizens are possible and provide a fruitful way to get some data and discussion on air quality in the absence of systematic air quality monitoring going on in the city. It is also a way to help citizens and entities like schools understand how they can play a role in improving air quality and ask more of their government.

Some broader conclusions can be drawn regarding air quality in the city of Nairobi. The pattern of peaks in data at most of our school sites indicates that vehicular emissions are a major source of pollution. Therefore, this implies that the city should prioritize a shift toward non-motorized transport, better fuel standards, and adopting cleaner vehicular technologies, as opposed to widening existing roads and building super highways. Another point of interest is that PM seems to peak in June over the roughly 8 months that the deployment took place. This needs to be examined in more detail. However, the policy implications could be that the Nairobi city council should focus especially on reducing vehicular traffic during this month. Another interesting observation is that the morning pollution peaks in the informal settlements occur earlier in the day than at the UNEP, St Scholastica and All Saints Cathedral sites. This is important to note, as it speaks to the way different groups of people use the city. Do people have to set off to work earlier in the informal settlements, as their workplaces are further, and transportation less convenient? This raises important questions around "spatial mismatch" in the city.

This study had many technical limitations. With sparse resources, we were not able to calibrate the gas sensors in the ambient conditions of Nairobi, which we know to be very important (Piedrahita et al. 2014). We therefore do not know how environmental factors and interference from other pollutants affected the gas sensors in the field. The interference from other pollutants could be large (Hasenfratz et al., 2012.; Popoola et al., 2012) . We also did not analyse the particle size distribution or the chemical composition of the particles sampled by the OPC, which could help determine the density of the particles sampled. In addition, the analysed data we obtained were noisy, and we were unable to determine which filter to apply to separate the signal from the noise without having access to any air quality measurements from a reference instrument.

We strongly recommend the calibration of low cost gas sensors

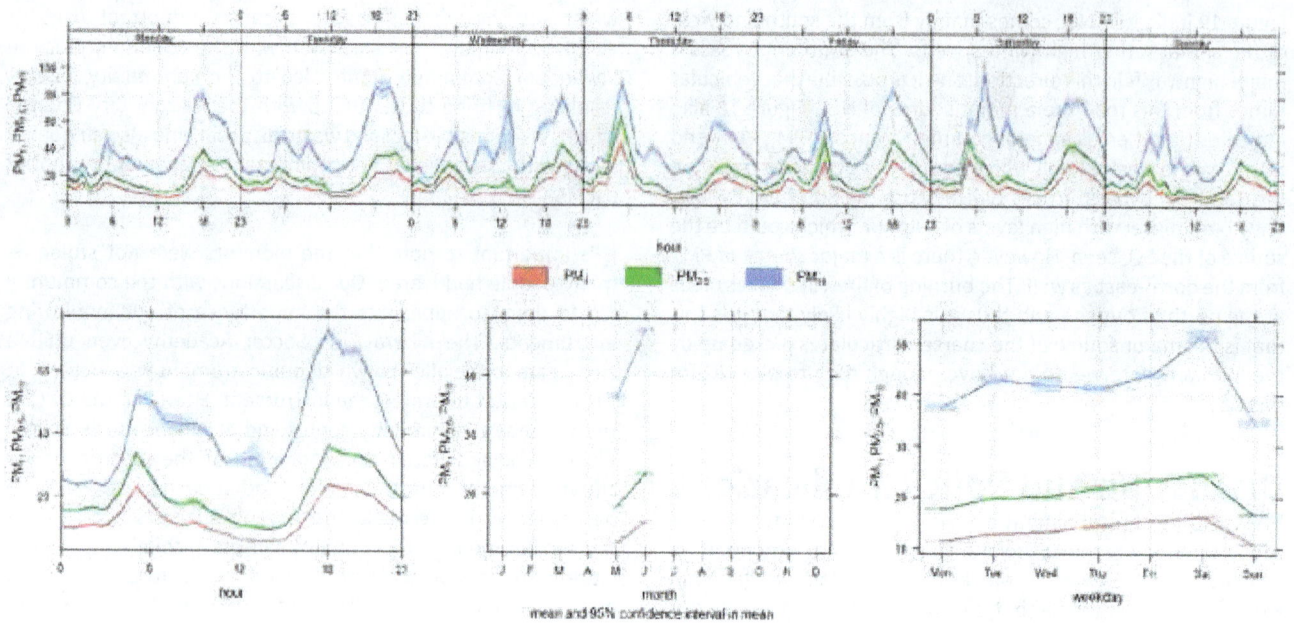

Figure 10: *Same as Figure 8 but for the Viwandani Community Center site for the period May 5, 2016-June 27, 2016.*

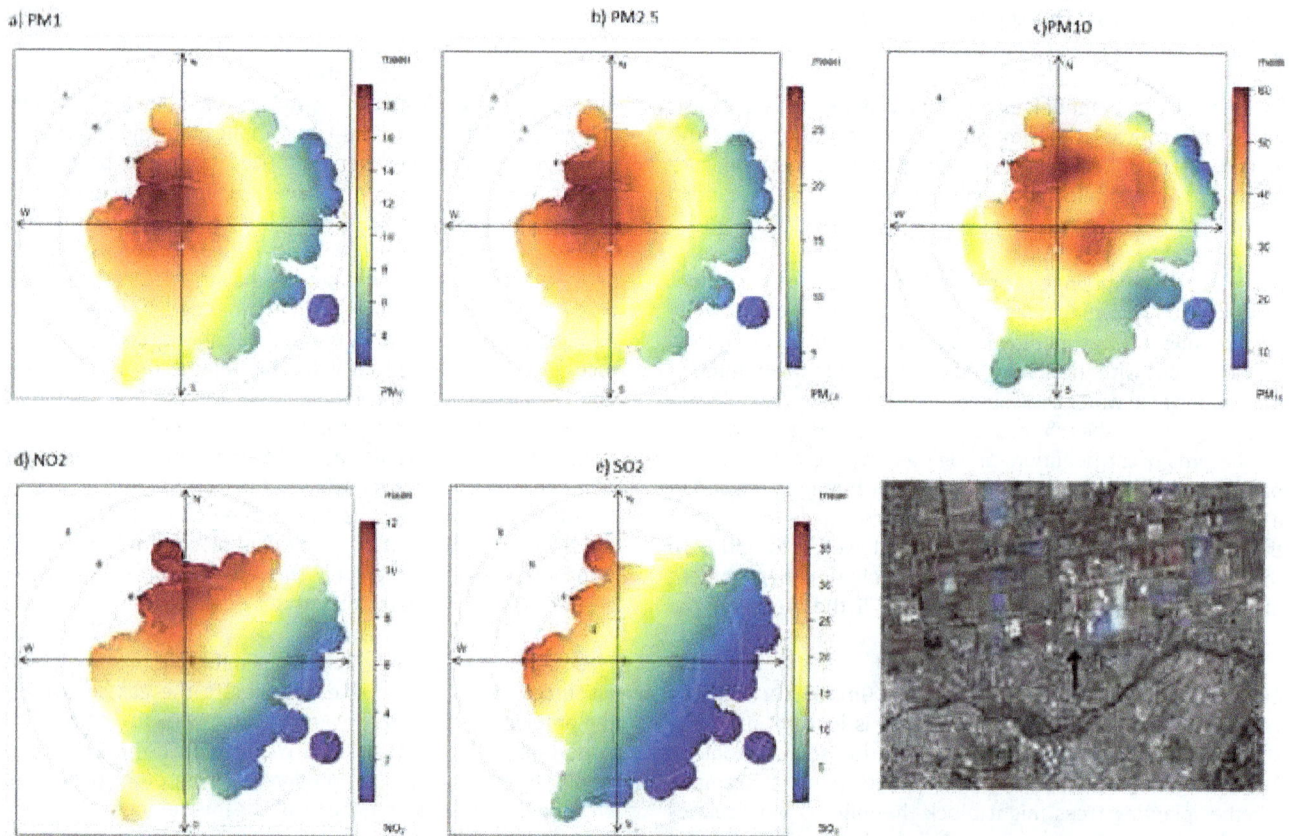

Figure 11: *Same as Figure 9 but for the Viwandani Community Center site for the period May 5, 2016-June 27, 2016. The image at the bottom shows the site and the black arrow indicates the direction that the monitor faces. The image has been taken such that the direction north in the image is towards the top of the page.*

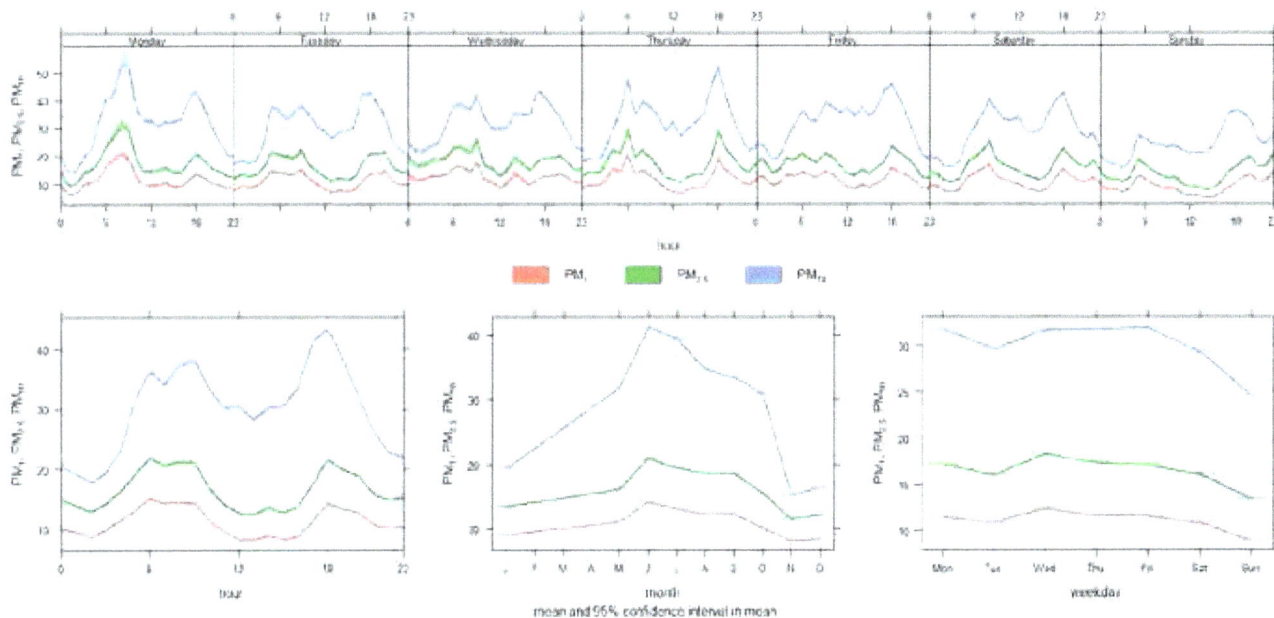

Figure 12: *Same as Figure 8, but for the St Scholastica site.*

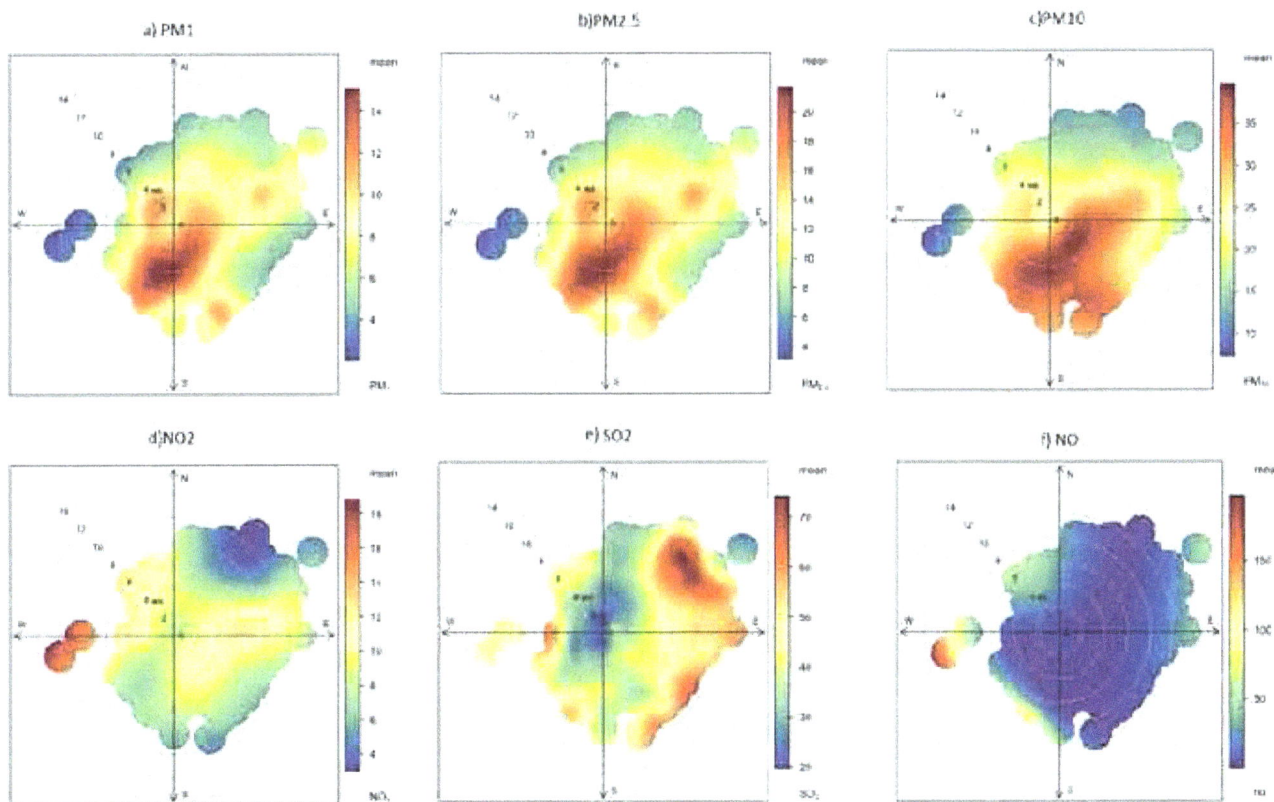

Figure 13: *Same as Figure 9 but for the St Scholastica site. The image at the bottom shows the site and the black arrow indicates the direction that the monitor faces. The image has been taken such that the direction north in the image is towards the top of the page.*

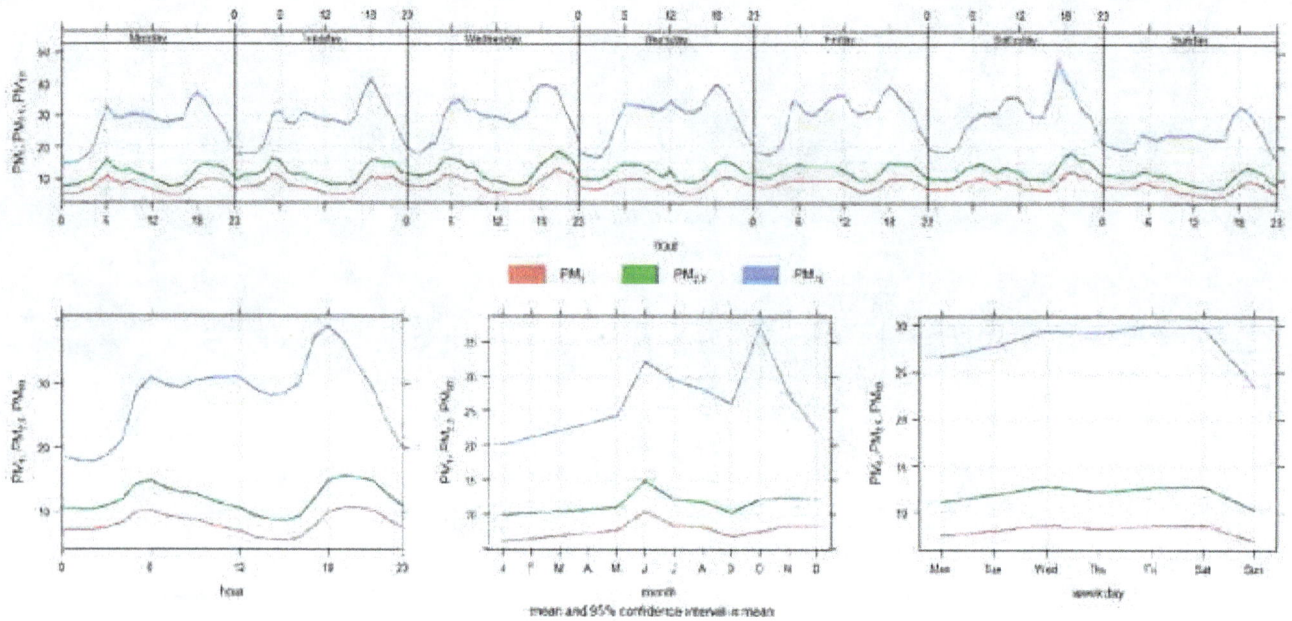

Figure 14: *Same as Figure 8, but for the UNEP site.*

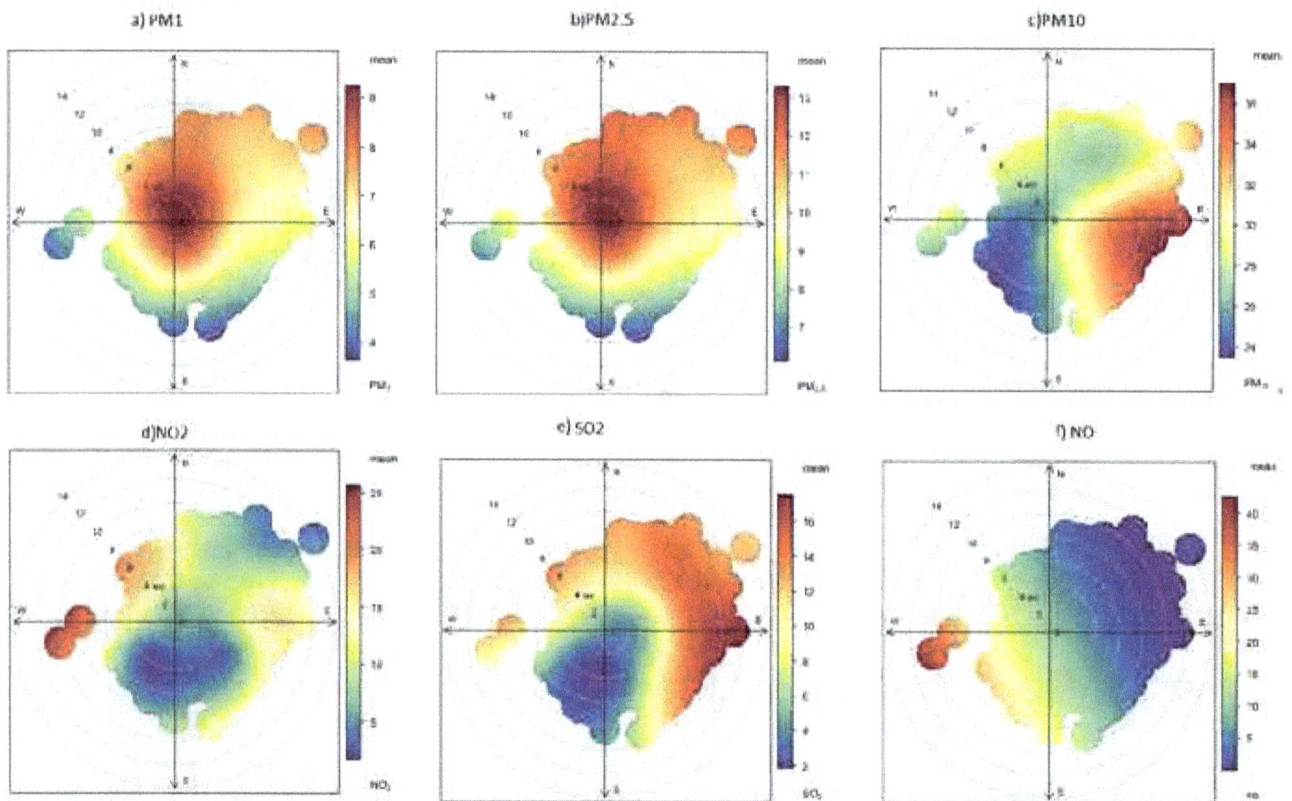

Figure 15: *Same as Figure 9 for the UNEP site. The image at the bottom shows the site and the black arrow indicates the direction that the monitor faces. The image has been taken such that the direction north in the image is towards the top of the page.*

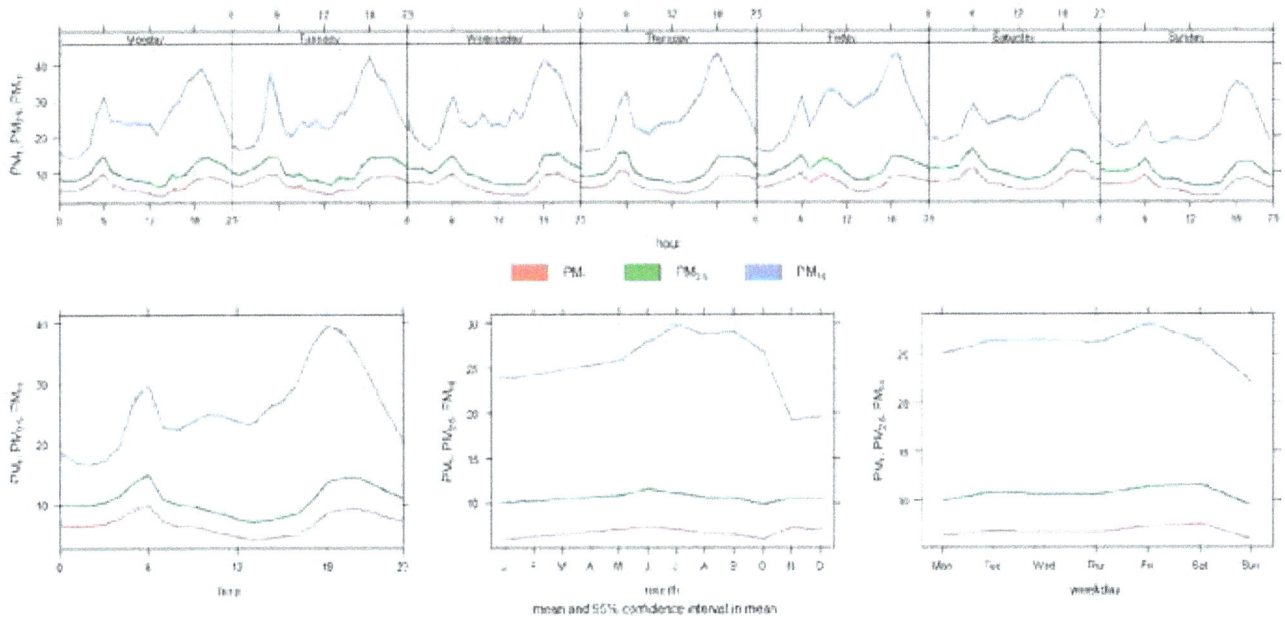

Figure 16: *Same as Figure 8 but for the All Saints.*

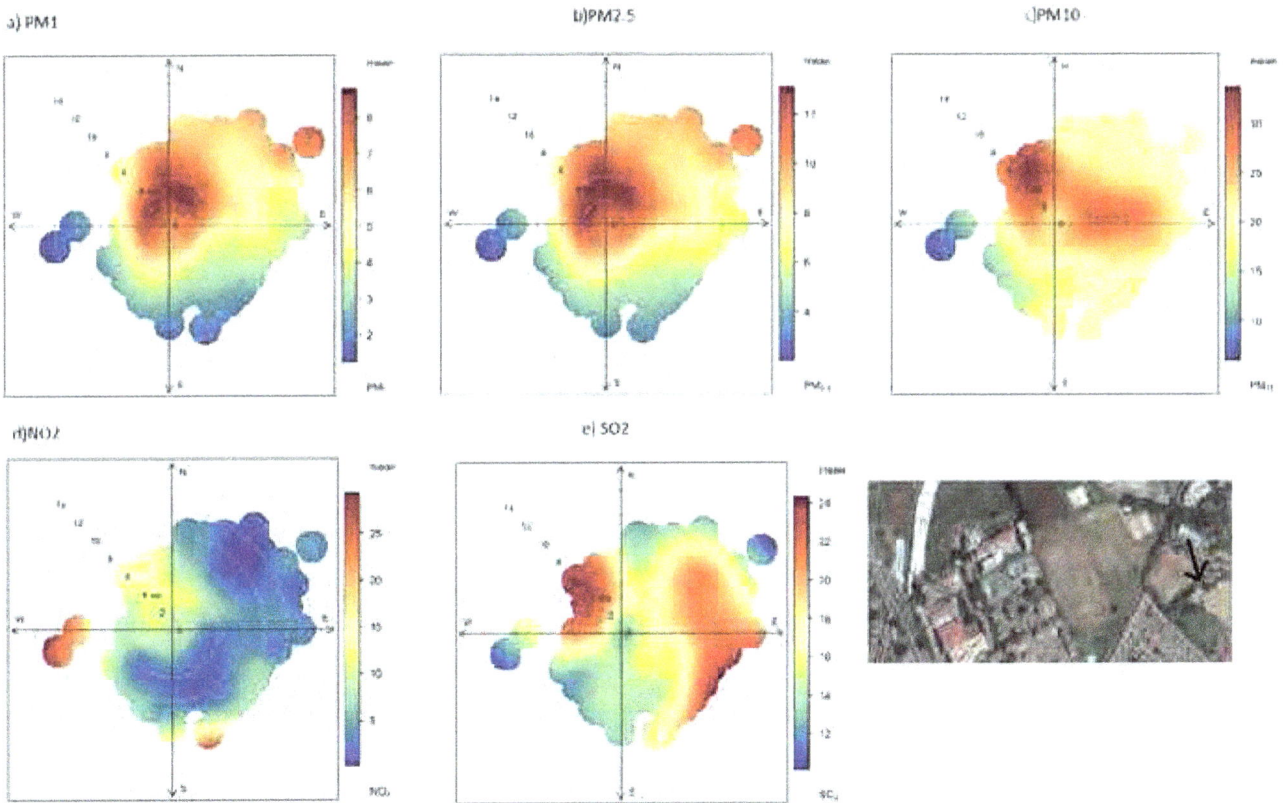

Figure 17: *Same as Figure 9 but for the All Saints Cathedral site. The image at the bottom shows the site and the black arrow indicates the direction that the monitor faces. The image has been taken such that the direction north in the image is towards the top of the page.*

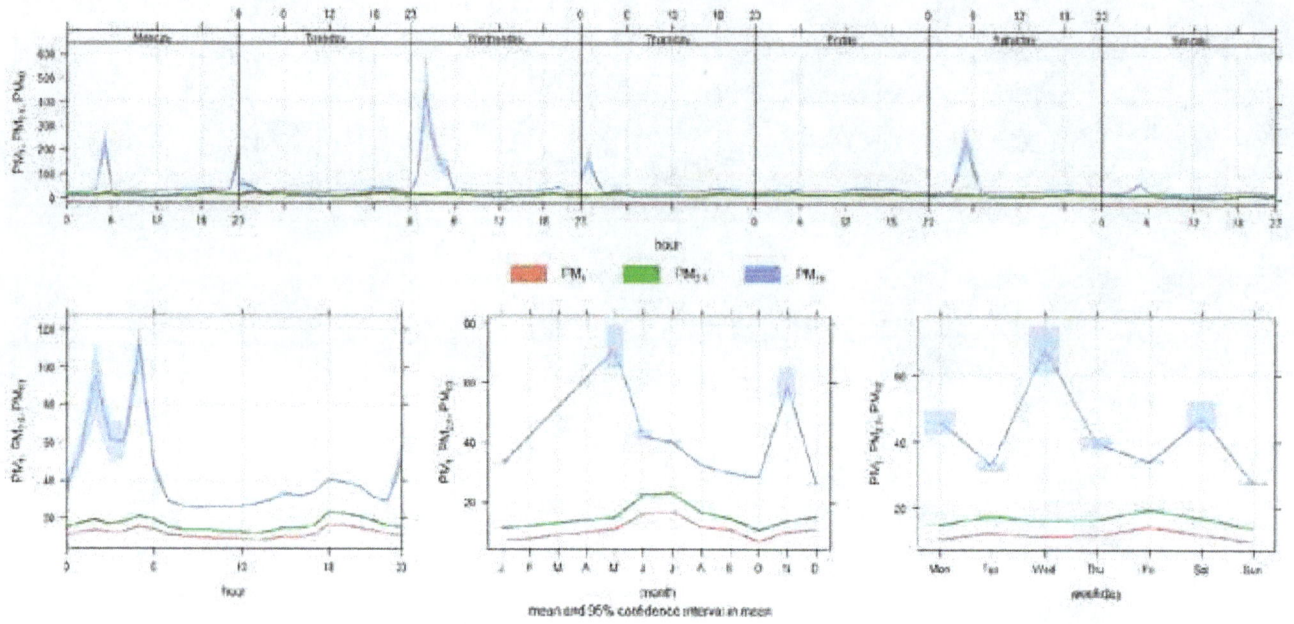

Figure 18: *Same as Figure 8 but for the Alliance Girls School site.*

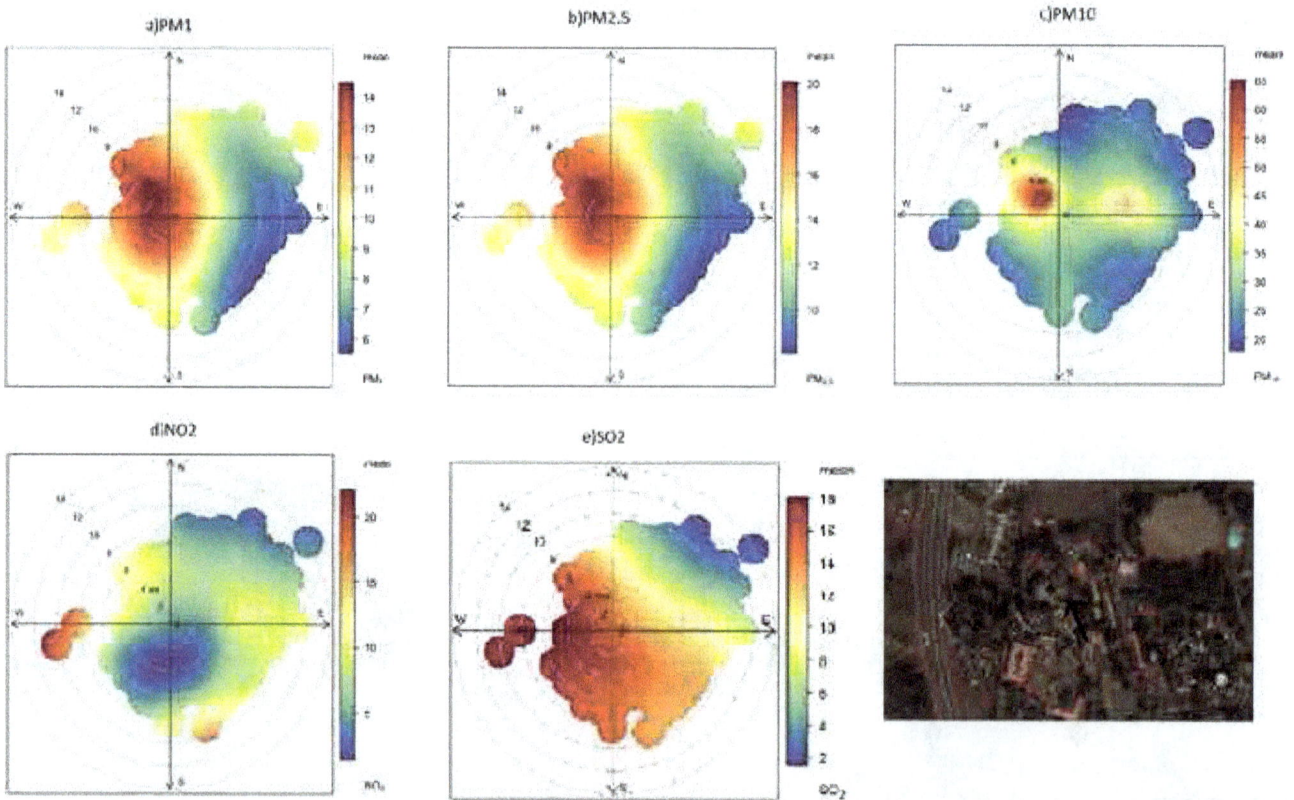

Figure 19: *Same as Figure 9 but for the Alliance Girls School site. The image at the bottom shows the site and the black arrow indicates the direction that the monitor faces. The image has been taken such that the direction north in the image is towards the top of the page.*

with reference air quality monitoring instruments in ambient conditions in order to determine the error in sensor readings due to interference effects of other pollutants and the effect of environmental conditions: specifically temperature and humidity. In addition, the rate of sensor drift depends on the season of the year so it is important to validate the network by regular calibrations of the sensors in each season. Work is underway on using machine learning algorithms to increase the accuracy of low cost sensors (Esposito et al., 2016), and we see this work as being very important for reducing calibration costs and improving data reliability.

We also strongly recommend obtaining a better understanding of the size distribution of particles collected by the OPC at each site over time. This is because the OPC does not function very well for counting particles of sizes < 380 nm. Thus, depending on the size distribution of particles (which varies over time), our measurements could have large errors, and understanding the extent of these errors is important for drawing inferences from the sources and type of particulate pollution. In addition, analysing the chemical composition of particulates at each site can help us develop a better understanding of sources of pollution, as well as help us in identifying correct value of density to use to convert the particle counts collected by the OPC to obtain particulate mass. This could also help us reduce the error in measurement.

Another limitation stemming from the proprietary nature of the technology is that we cannot report in detail on the performance of the electronics or the configuration of the device. These factors can affect results as well, and more research is required to identify standard configurations to facilitate comparisons of experiments. We believe that it is important to set a precedent for the reporting of the type of sensor used because, for example, it is unclear how the Alphasense A series gas sensors compare with the B series. No reports have been published examining this comparison. This makes comparing data from different low cost air quality sensor experiments difficult. In addition, components such as the analogue to digital converter (ADC) could add noise to the data, and thus reporting the kind of converter used could allow for a greater degree of comparison across networks. Over all, better reporting on the mechanics of low cost monitoring projects is needed moving forward.

Conclusions

Low cost sensors and apps that draw on their data to inform citizens about air pollution are becoming more and more prevalent. Given the magnitude of the data gaps in African cities, the growing availability of low cost sensors presents an important opportunity. This is especially the case as plans move forward to measure air pollutants for the Sustainable Development Goals and fight against climate change. However, much more research is needed on how well these new devices work under widely varied conditions, and whether the less accurate data these sensors generate is helpful or even harmful (Lewis and Edwards 2016, Kumar et al. 2015).

Our experiment using less expensive, lower-quality sensors in Nairobi schools contributes to this critical discussion. We did find significant technical limitations that need further work. However, we found that less accurate but carefully interpreted

data created by sensors within a citizen science initiative was clearly better than no data. Both the process of getting the data and the data itself, once carefully interpreted, helped to generate broader public understanding and interest in monitoring air quality and addressing likely sources of ambient outdoor air pollution. We also gained some idea of the air pollution problems affecting schoolchildren across class divides with more challenges clearly facing low income children in the slums.

The deployment and analysis of our network also showed that the cost of the sensors is only a small fraction of the total cost of network deployment. This is because maintenance of the network, calibration of the sensors and the analysis of the data is time consuming and therefore expensive. It is also abundantly clear that "low cost" sensors cannot obviate the need for stronger investment in high quality monitoring and related local scientific research around air quality in African cities. While low cost sensors can allow for more measurements and more civic engagement, this is ideally conducted in collaboration with local scientists who are well-equipped to ensure data are collected and interpreted accurately for the public. Lewis and Edwards (2016) suggest "well designed sensor experiments, that acknowledge the limitations of the technologies as well as the strengths, have the potential to simultaneously advance basic science, monitor air pollution — and bring the public along". We believe we have shown this to be the case for African cities like Nairobi that currently do not have an air quality monitoring system but do have a substantial air quality problem.

Acknowledgements

The authors would like to thank Dr. Jacqueline McGlade, United Nations Environment Program (UNEP), Sami Dimassi, United Nations Environment Program (UNEP), Valentin Foltescu, United Nations Environment Program (UNEP), the company Alphasense, the company Atmospheric Instruments, Charles Mwangi from the NASA GLOBE program, the Wajukuu Arts Collective and the Kibera Girls Soccer Academy, Alliance Girls School, St Scholastica, All Saint's Cathedral School and Viwandi Community Center. We also thank Dr Ralph Kahn for his valuable comments.

References

Amegah, A.K., Agyei-Mensah, S., n.d. Urban air pollution in Sub-Saharan Africa: Time for action. Environ. Pollut. doi:10.1016/j.envpol.2016.09.042.

AQ_GlobalReport_Summary.pdf, n.d.

Carslaw, D.C., Beevers, S.D., 2013. Characterising and understanding emission sources using bivariate polar plots and k-means clustering. Environ. Model. Softw. 40, 325–329. doi:10.1016/j.envsoft.2012.09.005.

Carslaw, D.C., Ropkins, K., 2012. openair — An R package for air quality data analysis. Environ. Model. Softw. 27–28, 52–61. doi:10.1016/j.envsoft.2011.09.008.

Development, O. of R.&, n.d. Sensor Evaluation Report [WWW Document]. URL https://cfpub.epa.gov/si/si_public_record_report.cfm?dirEntryId=277270 (accessed 10.17.16).

Dominici, F., Peng, R.D., Barr, C.D., Bell, M.L., 2010. Protecting Human Health from Air Pollution: Shifting from a Single-Pollutant to a Multi-pollutant Approach. Epidemiol. Camb. Mass 21, 187–194. doi:10.1097/EDE.0b013e3181cc86e8.

Egondi, T., Kyobutungi, C., Ng, N., Muindi, K., Oti, S., Vijver, S. van de, Ettarh, R., Rocklöv, J., 2013. Community Perceptions of Air Pollution and Related Health Risks in Nairobi Slums. Int. J. Environ. Res. Public. Health 10, 4851–4868. doi:10.3390/ijerph10104851.

Esposito, E., De Vito, S., Salvato, M., Bright, V., Jones, R.L., Popoola, O., 2016. Dynamic neural network architectures for on field stochastic calibration of indicative low cost air quality sensing systems. Sens. Actuators B Chem. 231, 701–713. doi:10.1016/j.snb.2016.03.038.

Gaita, S.M., Boman, J., Gatari, M.J., Pettersson, J.B.C., Janhäll, S., 2014. Source apportionment and seasonal variation of PM2.5 in a Sub-Saharan African city: Nairobi, Kenya. Atmos Chem Phys 14, 9977–9991. doi:10.5194/acp-14-9977-2014.

Gatari, M.J., Boman, J., 2003. Black carbon and total carbon measurements at urban and rural sites in Kenya, East Africa. Atmos. Environ. 37, 1149–1154. doi:10.1016/S1352-2310(02)01001-4.

Gatari, M.J., Boman, J., Wagner, A., 2009. Characterization of aerosol particles at an industrial background site in Nairobi, Kenya. X-Ray Spectrom. 38, 37–44. doi:10.1002/xrs.1097.

Hasenfratz, D., Saukh, O., Thiele, L., 2012. On-the-Fly Calibration of Low-Cost Gas Sensors, in: SpringerLink. Springer Berlin Heidelberg, pp. 228–244. doi:10.1007/978-3-642-28169-3_15.

Kinney, P.L., Gichuru, M.G., Volavka-Close, N., Ngo, N., Ndiba, P.K., Law, A., Gachanja, A., Gaita, S.M., Chillrud, S.N., Sclar, E., 2011. Traffic Impacts on PM2.5 Air Quality in Nairobi, Kenya. Environ. Sci. Policy 14, 369–378. doi:10.1016/j.envsci.2011.02.005.

Kumar, P., Morawska, L., Martani, C., Biskos, G., Neophytou, M., Di Sabatino, S., Bell, M., Norford, L., Britter, R., 2015. The rise of low-cost sensing for managing air pollution in cities. Environ. Int. 75, 199–205. doi:10.1016/j.envint.2014.11.019.

Lewis, A., Edwards, P., 2016. Validate personal air-pollution sensors. Nat. News 535, 29. doi:10.1038/535029a.

Lindén, J., Boman, J., Holmer, B., Thorsson, S., Eliasson, I., 2012. Intra-urban air pollution in a rapidly growing Sahelian city. Environ. Int. 40, 51–62. doi:10.1016/j.envint.2011.11.005.

Mead, M.I., Popoola, O.A.M., Stewart, G.B., Landshoff, P., Calleja, M., Hayes, M., Baldovi, J.J., McLeod, M.W., Hodgson, T.F., Dicks, J., Lewis, A., Cohen, J., Baron, R., Saffell, J.R., Jones, R.L., 2013. The use of electrochemical sensors for monitoring urban air quality in low-cost, high-density networks. Atmos. Environ. 70, 186–203. doi:10.1016/j.atmosenv.2012.11.060.

M. Fiore, A., Naik, V., V. Spracklen, D., Steiner, A., Unger, N., Prather, M., Bergmann, D., J. Cameron-Smith, P., Cionni, I., J. Collins, W., Dalsøren, S., Eyring, V., A. Folberth, G., Ginoux, P., W. Horowitz, L., Josse, B., Lamarque, J.-F., A. MacKenzie, I., Nagashima, T., M. O'Connor, F., Righi, M., T. Rumbold, S., T.

Shindell, D., B. Skeie, R., Sudo, K., Szopa, S., Takemura, T., Zeng, G., 2012. Global air quality and climate. Chem. Soc. Rev. 41, 6663–6683. doi:10.1039/C2CS35095E.

Muindi, K., Egondi, T., Kimani-Murage, E., Rocklov, J., Ng, N., 2014. "We are used to this": a qualitative assessment of the perceptions of and attitudes towards air pollution amongst slum residents in Nairobi. BMC Public Health 14, 226. doi:10.1186/1471-2458-14-226.

Ngo, N.S., Gatari, M., Yan, B., Chillrud, S.N., Bouhamam, K., Kinney, P.L., 2015. Occupational exposure to roadway emissions and inside informal settlements in sub-Saharan Africa: A pilot study in Nairobi, Kenya. Atmos. Environ. 111, 179–184. doi:10.1016/j.atmosenv.2015.04.008.

Ngo, N.S., Kokoyo, S., Klopp, J., 2017. Why participation matters for air quality studies: risk perceptions, understandings of air pollution and mobilization in a poor neighborhood in Nairobi, Kenya. Public Health 142, 177–185. doi:10.1016/j.puhe.2015.07.014.

Njee, R.M., Meliefste, K., Malebo, H.M., Hoek, G., 2016. Spatial Variability of Ambient Air Pollution Concentration in Dar es Salaam. J. Environ. Pollut. Hum. Health J. Environ. Pollut. Hum. Health 4, 83–90. doi:10.12691/jephh-4-4-2.

Petkova, E.P., Jack, D.W., Volavka-Close, N.H., Kinney, P.L., 2013. Particulate matter pollution in African cities. Air Qual. Atmosphere Health 6, 603–614. doi:10.1007/s11869-013-0199-6 Piedrahita, R., Xiang, Y., Masson, N., Ortega, J., Collier, A., Jiang, Y., Li, K., Dick, R.P., Lv, Q., Hannigan, M., Shang, L., 2014. The next generation of low-cost personal air quality sensors for quantitative exposure monitoring. Atmos Meas Tech 7, 3325–3336. doi:10.5194/amt-7-3325-2014.

Popoola, O., Stewart, G., Jones, R.L., Mead, I., Saffell, J., 2012. 7.4.4 Electrochemical Sensors for Environmental Monitoring in Cities. Proc. IMCS 2012 640–640. doi:http://dx.doi.org/10.5162/IMCS2012/7.4.4.

Schwela, D., 2012a. Review of urban air quality in Sub-Saharan Africa region - air quality profile of SSA countries (No. 67794). The World Bank.

Schwela, D., 2012b. Review of urban air quality in Sub-Saharan Africa region - air quality profile of SSA countries (No. 67794). The World Bank.

Sousan, S., Koehler, K., Hallett, L., Peters, T.M., 2016. Evaluation of the Alphasense optical particle counter (OPC-N2) and the Grimm portable aerosol spectrometer (PAS-1.108). Aerosol Sci. Technol. 50, 1352–1365. doi:10.1080/02786826.2016.1232859.

Vliet, E.D.S. van, Kinney, P.L., 2007. Impacts of roadway emissions on urban particulate matter concentrations in sub-Saharan Africa: new evidence from Nairobi, Kenya. Environ. Res. Lett. 2, 045028. doi:10.1088/1748-9326/2/4/045028.

World Health Organization (WHO) Burden of Disease from Ambient Air Pollution for 2012, WHO, Geneva (2014). Available: http://www.who.int/phe/health_topics/outdoorair/databases/AAP_BoD_results_March2014.pdf

Appendix

Figures 1A and 2A show the raw 1-minute data recorded of particulate pollutants and the gaseous pollutants, respectively.

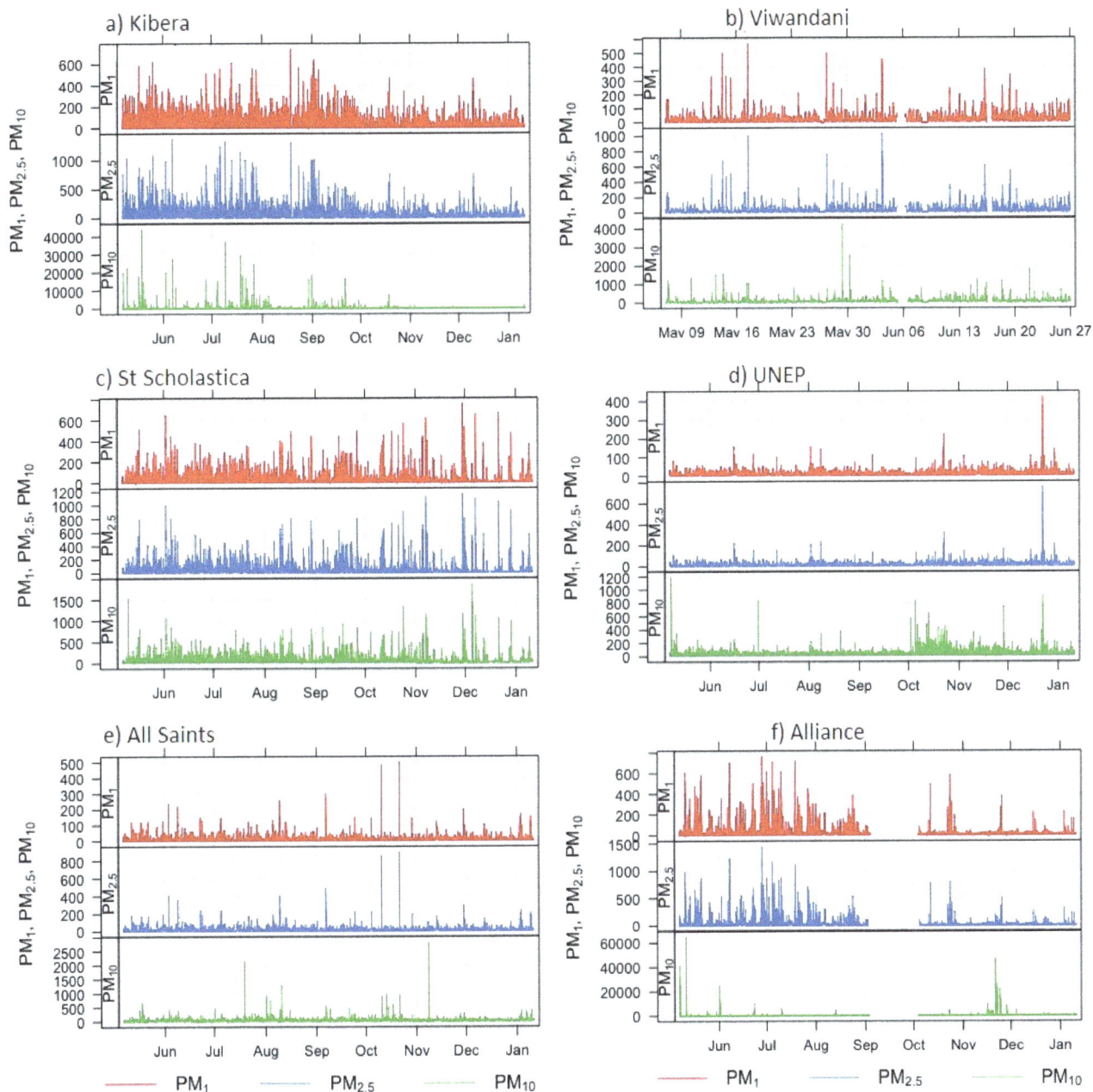

Figure 1A: *1-minute PM$_1$ (red), PM$_{2.5}$ (blue) and PM$_{10}$ (green) mass concentration (µg/m³) time series plots for each site; a) Kibera Girls Soccer Academy, b) Viwandani Community Center (note that due to an extended power outage this monitor stopped logging data after June 27, 2016), c) St Scholastics, d) UNEP, e) All Saints Cathedral School, f) Alliance Girls School from May 1, 2016 to January 11, 2017.*

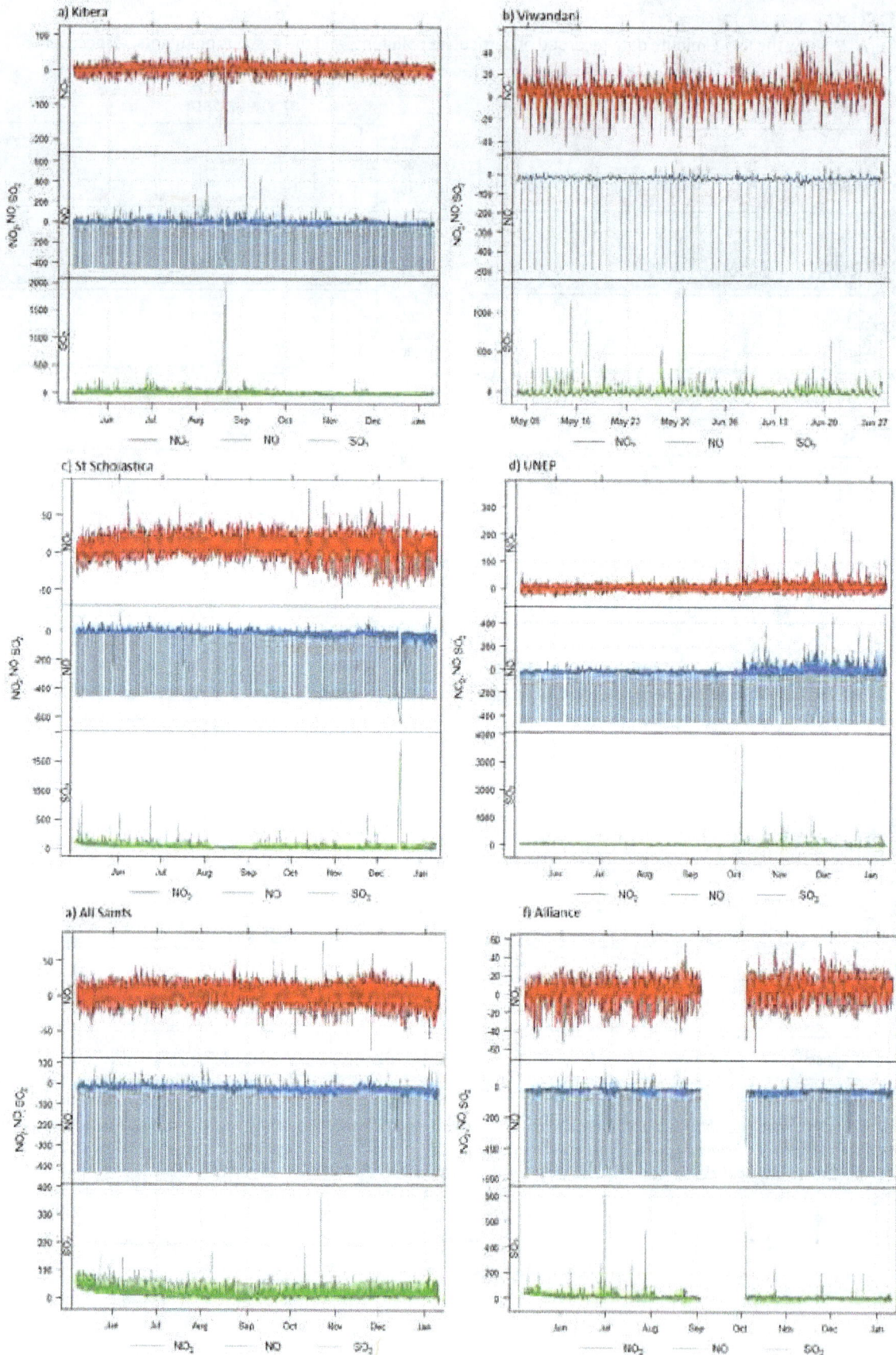

Figure 2A: : *1-minute NO$_2$ (red), NO (blue) and SO$_2$ (green) concentration (ppb) time series plots for each site a) Kibera Girls Soccer Academy, b) Viwandani Community Center (note that due to an extended power outage this monitor stopped logging data after June 27, 2016), c) St Scholastics, d) UNEP, e) All Saints Cathedral School, f) Alliance Girls School from May 1, 2016 to January 11, 2017.*

Table 1A shows the correlation between gaseous pollutant values > 0 and temperature/humidity and the other pollutants measured at each site. This table shows that for gaseous pollutants with values > 0, the correlation between temperature and humidity is low, and has the same sign across sites. This indicates that the signal registered is more likely to only be due to the pollutants and is not affected by environmental factors.

Table 1A: *Summary of the Pearson correlation coefficient (R) at each of the six sites for all gaseous pollutant observations greater than zero.*

	Kibera	Viwandani	St Scholastica	UNEP	All Saints	Alliance
Correlation of NO_2 with temperature	0.13	0.02	0.17	0.32	-0.045	0.38
Correlation of SO_2 with temperature	0.027	0.01	0.18	0.18	0.25	0.04
Correlation of NO with temperature	0.028	-0.12	-0.17	-0.04	-0.28	-0.12
Correlation of NO_2 with humidity	-0.13	-0.032	-0.27	-0.31	0.099	-0.39
Correlation of SO_2 with humidity	-0.056	-0.018	-0.16	-0.16	-0.11	0.013
Correlation of NO with humidity	-0.15	-0.09	0.26	0.06	0.058	-0.06
Correlation of NO with NO_2	0.32	0.11	0.098	0.26	-0.097	0.27
Correlation of NO with SO_2	0.55	0.24	0.31	0.33	0.47	0.36
Correlation of NO with PM_{10}	0.13	0.16	0.12	0.12	0.37	0.19
Correlation of NO with $PM_{2.5}$	0.16	0.13	0.09	0.06	0.27	0.16
Correlation of NO with PM_1	0.16	0.12	0.09	0.03	0.27	0.17
Correlation of NO_2 with SO_2	0.16	0.14	0.21	0.28	0.18	0.066
Correlation of NO_2 with $PM_{2.5}$	0.13	0.29	0.1	0.16	0.28	0.02
Correlation of NO_2 with PM_{10}	0.058	0.32	0.24	0.31	0.26	0
Correlation of NO_2 with PM_1	0.12	0.3	0.089	0.12	0.27	0.014
Correlation of SO_2 with $PM_{2.5}$	0.25	0.13	0.12	0.18	0.25	0.12
Correlation of SO_2 with PM_{10}	0.086	0.12	0.19	0.22	0.25	0.01
Correlation of SO_2 with PM_1	0.26	0.13	0.12	0.16	0.23	0.12

Figures 3A to 7A clearly show the variation of the gaseous pollutants with temperature. It is clear from these figures that for high temperatures (roughly > 200 C), negative values of pollutants ae registered. Co-location with a reference monitor is required in order to truly identify the ranges in which the values are correct. However, plotting these graphs is a rough way to identify temperature ranges in which the sensors clearly make incorrect measurements.

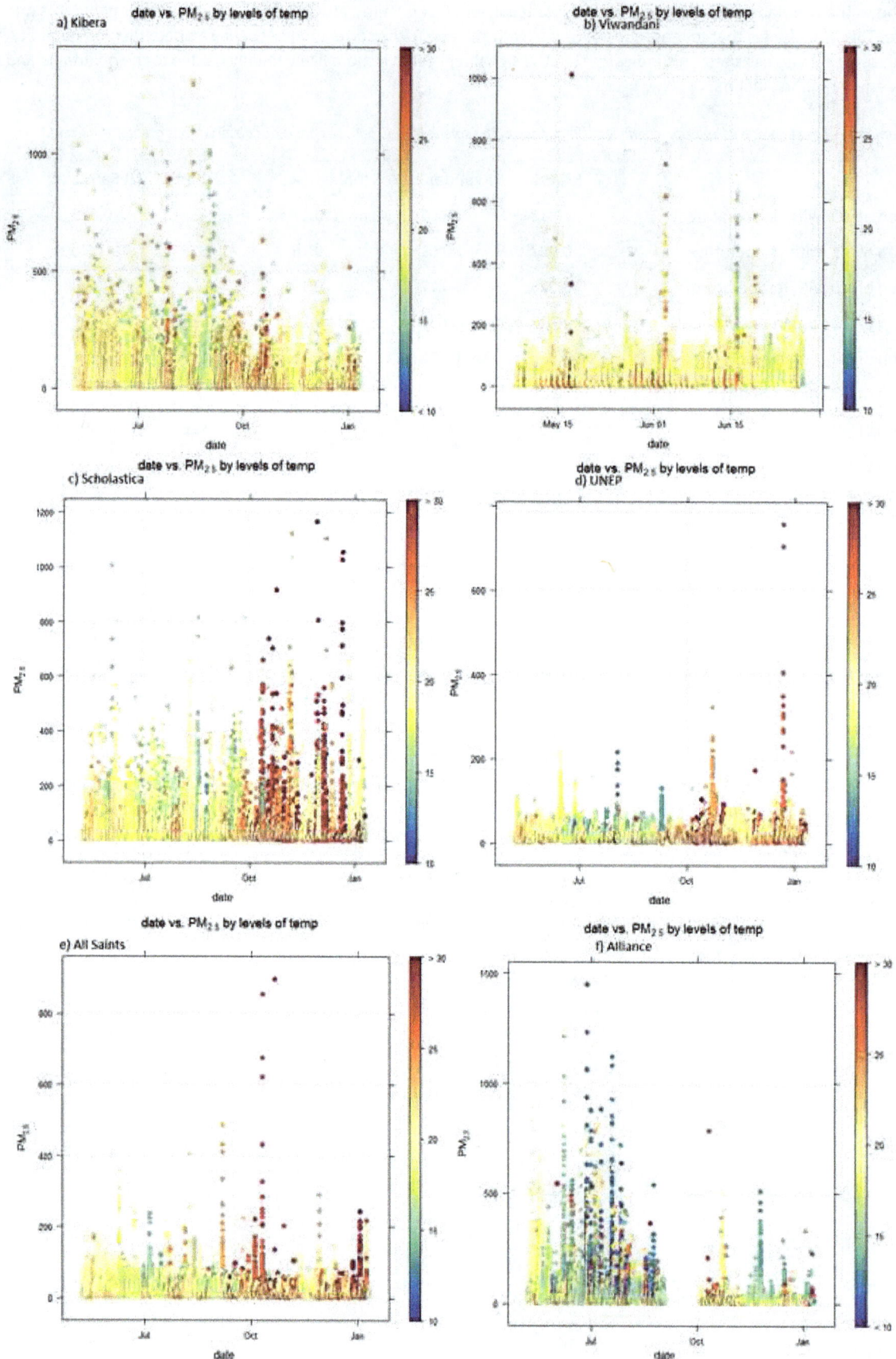

Figure 3A: *Time series of PM₂.₅ in units of μg/m³ with the color scale corresponding to temperature for the sites: a) Kibera Girls Soccer Academy, b) Viwandani Community Center (note that due to an extended power outage this monitor stopped logging data after June 27, 2016), c) St Scholastics, d) UNEP, e) All Saints Cathedral School, f) Alliance Girls School from May 5, 2016 to January 11, 2017.*

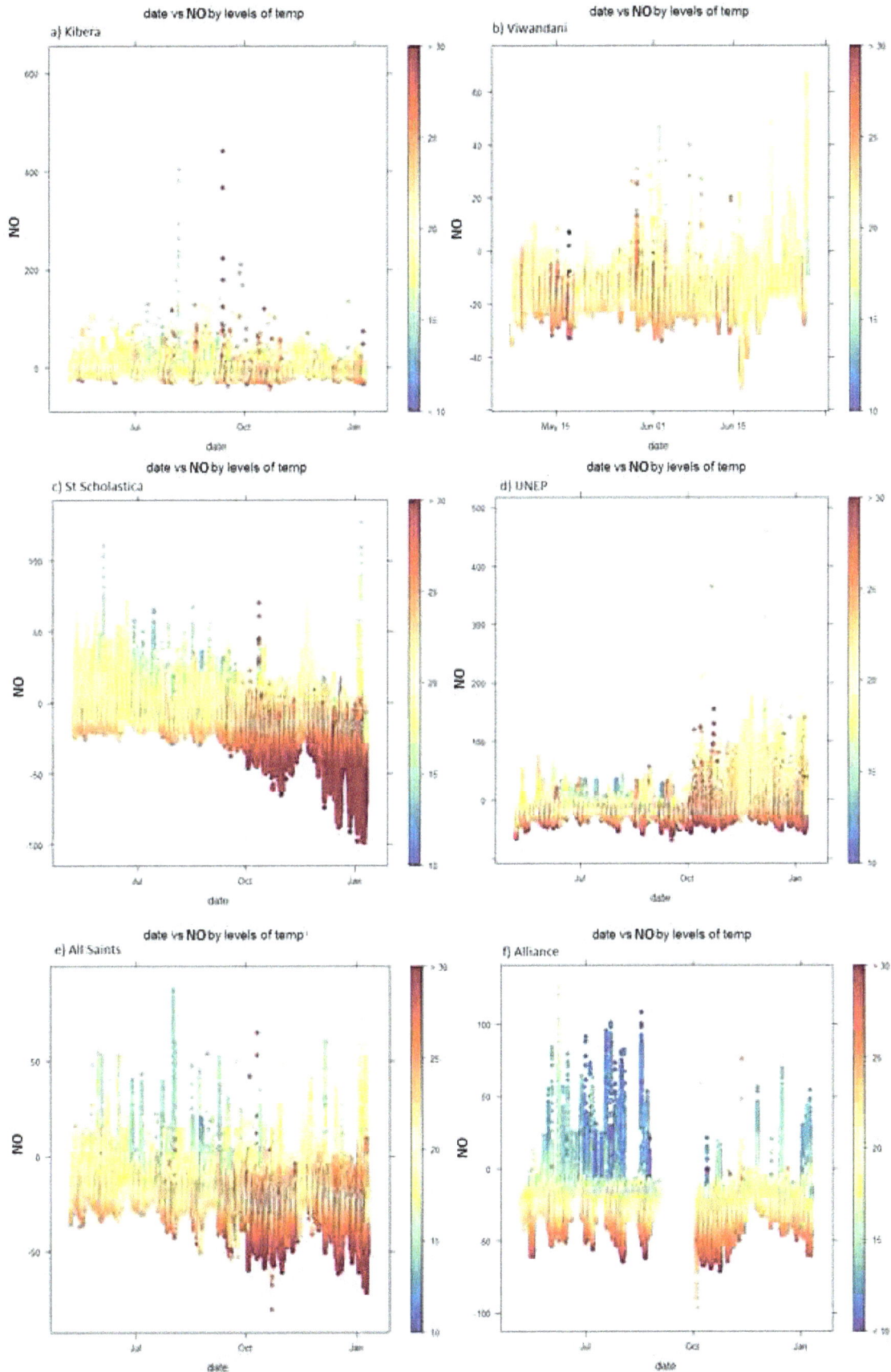

Figure 4A: *Time series of NO for recordings >-100ppb in units of ppb with the color scale corresponding to temperature for the sites: a) Kibera Girls Soccer Academy, b) Viwandani Community Center (note that due to an extended power outage this monitor stopped logging data after June 27, 2016), c) St Scholastics, d) UNEP, e) All Saints Cathedral School, f) Alliance Girls School from May 5, 2016 to January 11, 2017.*

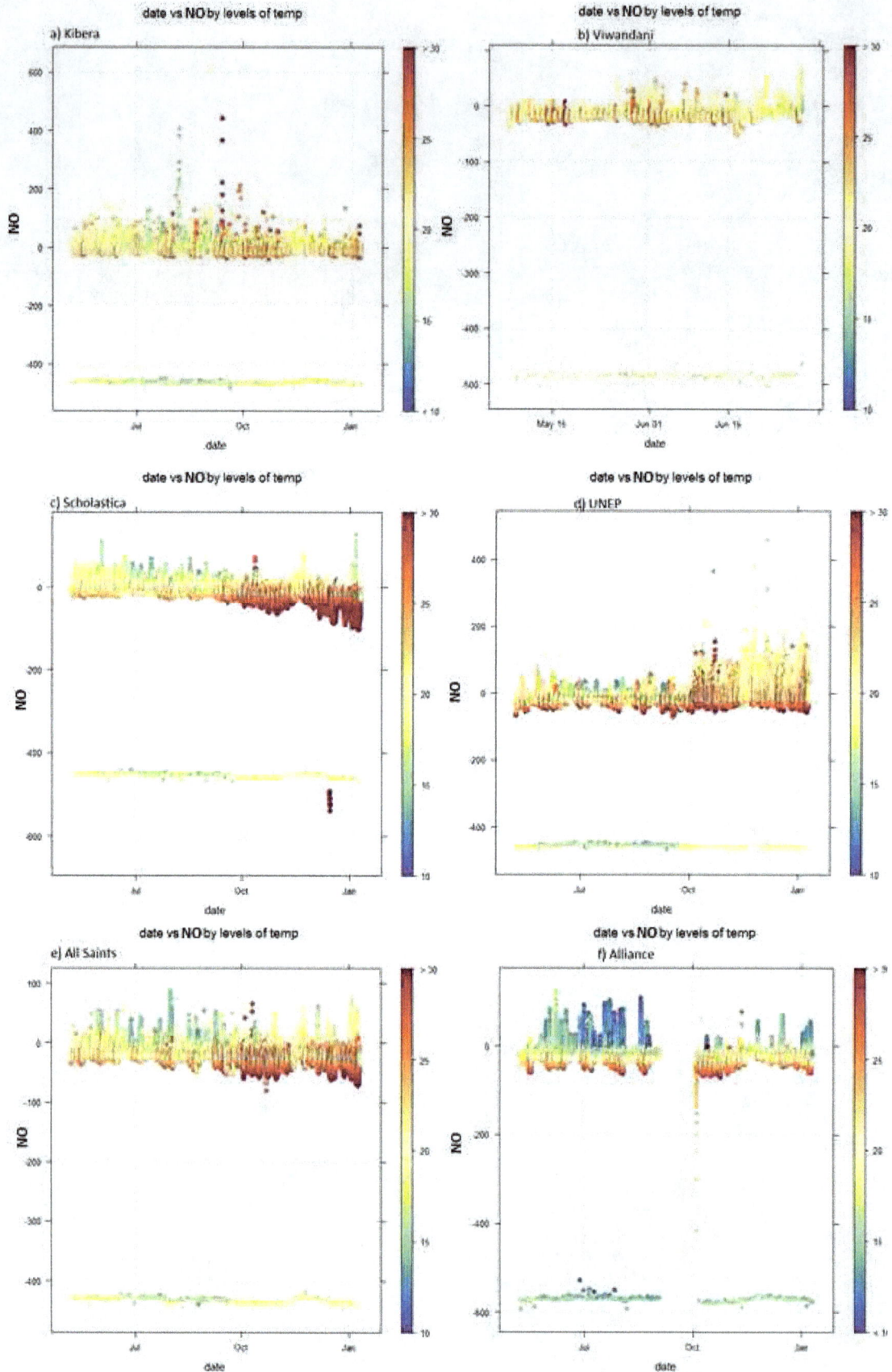

Figure 5A: *Time series of NO in units of ppb with the color scale corresponding to temperature for the sites. No filter was applied to the NO data: a) Kibera Girls Soccer Academy, b) Viwandani Community Center (note that due to an extended power outage this monitor stopped logging data after June 27, 2016), c) St Scholastics, d) UNEP, e) All Saints Cathedral School, f) Alliance Girls School from May 5, 2016 to January 11, 2017.*

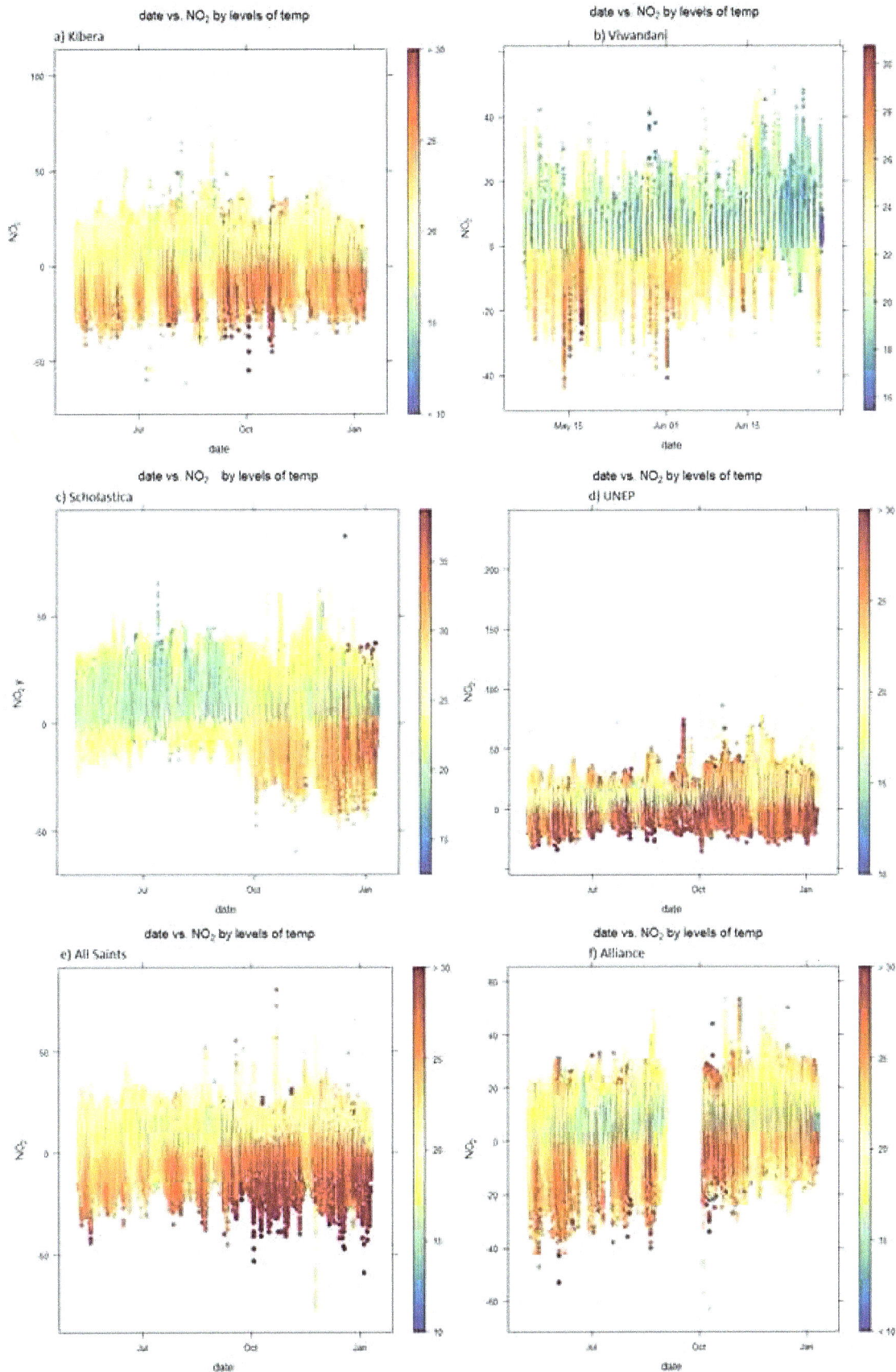

Figure 6A: *Time series of NO₂ in units of ppb with the color scale corresponding to temperature for the sites: a) Kibera Girls Soccer Academy, b) Viwandani Community Center (note that due to an extended power outage this monitor stopped logging data after June 27, 2016), c) St Scholastics, d) UNEP, e) All Saints Cathedral School, f) Alliance Girls School from May 5, 2016 to January 11, 2017.*

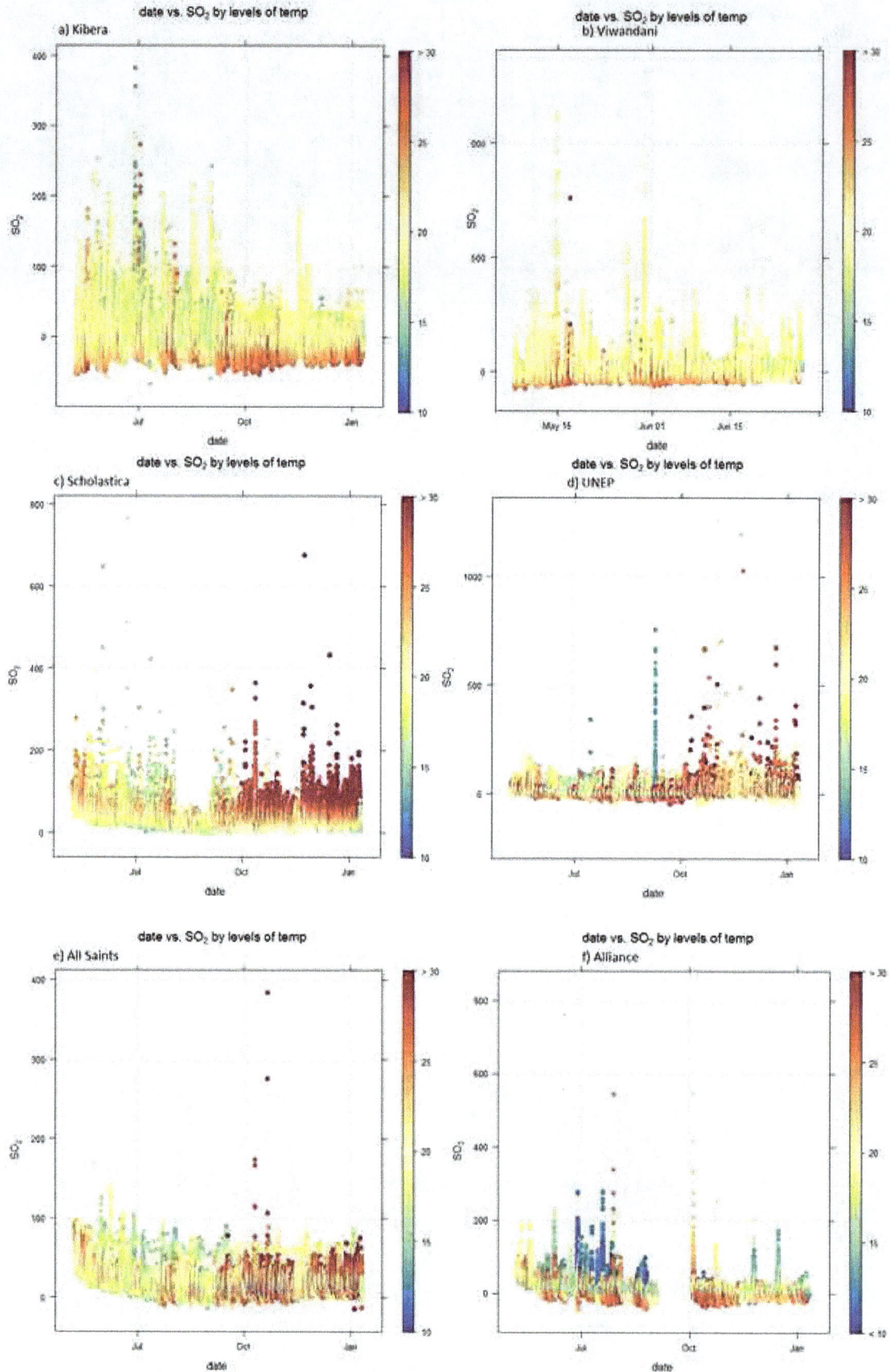

Figure 7A: *Time series of SO$_2$ in units of ppb with the color scale corresponding to temperature for the sites: a) Kibera Girls Soccer Academy, b) Viwandani Community Center (note that due to an extended power outage this monitor stopped logging data after June 27, 2016), c) St Scholastics, d) UNEP, e) All Saints Cathedral School, f) Alliance Girls School from May 5, 2016 to January 11, 2017.*

18

Monitoring the contribution of desert dust intrusion to PM$_{10}$ concentration in Northern Cyprus

Hassan Y. Sulaiman[*1] and Sedef Çakir[2]

[1]Yusuf Maitama Sule University Kano, Nigeria, hassanyousouph@gmail.com
[2]Cyprus International University, Mersin 10, Nicosia, Northern Cyprus, sedefcakir@ciu.edu.tr

Abstract

Air quality in the Mediterranean basin has been affected by PM$_{10}$ pollution induced by transported desert dust and local emission. The study used PM$_{10}$ data from Nicosia, Kyrenia, Guzelyurt and Famagusta urban representatives, Kalecik rural background and Alevkayasi regional background. HYSPLIT model and satellite data were used to identify dust days and dust input was quantified using the method suggested by the European Commission. Anthropogenic background contribution of each site was then estimated by subtracting the regional background concentrations. A total of 35 dust days occurred on Cyprus island within the 3-years period; mostly during winter and spring. Daily PM$_{10}$ concentration on dust days can reach up to 400 µg/m^3. After removing dust background, annual PM$_{10}$ concentrations were 48-58 µg/m^3 in Nicosia, 42-47 µg/m^3 in Famagusta, 40-50 µg/m^3 in Kyrenia, 33-41 µg/m^3 in Guzelyurt, 21-28 µg/m^3 in Alevkayasi, and 32-38 µg/m^3 in Kalecik. PM$_{10}$ concentrations were higher during winters in the urban sites. Despite the high frequency of dust events, only a fraction of exceedances of the standard limit in the urban sites were attributable to dust. Anthropogenic background sources contributions were 12.3 µg/m^3 in Guzelyurt, 18 µg/m^3 in Kyrenia, 18.4 µg/m^3 in Famagusta, 27.8 µg/m^3 in Nicosia and 9.7 µg/m^3 in Kalecik. Effects of other natural sources that the study did not assess, such as sea salt and local soil resuspension, could be the reason for exceedances.

Keywords

PM$_{10}$, Dust, HYSPLIT, Aerosol Optical Depth, Northern Cyprus

Introduction

Particulate Matter (PM), also known as aerosols, are complex mixture of tiny particles and liquid droplets that are composed of number of substances which include organic chemicals, acids, metals and dust and soil particles. The United States Environment Protection Agency (USEPA) (2015) listed particulate matter as one of the most common pollutants in the atmosphere. PM in the atmosphere may originate from various natural and anthropogenic sources. Natural sources of PM include crustal dust, sea salts, pollen and volcanic ashes. Human sources include burning fossil fuels in power plants, domestic heating, combustion engines of motor vehicles, re-suspension of dust by road traffic, quarrying and agricultural activities. Dust is transported into the receptive atmosphere by lifting and advection of non-vegetative soil in the desert. Wind can lift dust at an atmospheric altitude ranging from 1500 m to as high as 8000 m above ground level into the receptive site which may persist for days to few weeks especially on cloud free days (Vautard et al. 2005).

The island of Cyprus experiences high PM concentration from anthropogenic sources such as vehicular emissions, burning of solid fuel, road side dust resuspension, local soil resuspension and emission from industrialized European countries (Kubilay et al. 2000, Querol et. al. 2009, Achilleos et al. 2014). Also due to its close proximity to the Sahara Desert and the Arabian Peninsula, PM$_{10}$ concentration over the island has been heavily impacted by desert dust storm (Achilleos et al. 2014). Epidemiological investigations have linked increased cardiovascular, respiratory diseases and mortality to the exposure of people to high PM$_{10}$ concentration during dust episodes in the region (Middleton et al. 2008, Neophytou et al. 2013). This is evident in the Turkish Republic of Northern Cyprus (TRNC) as there is an increasing number of mortality resulting from PM$_{10}$ related diseases such as asthma and cardiovascular disease (State Planning Organization Statistics and Research Department (SPOSRD), 2015).

Yearly, Saharan dust contributes 5-10 µg/m^3 to the annual mean PM$_{10}$ concentration over the Mediterranean, which is higher than the 0-3 µg/m^3 it contributes to Northern Europe (Vautard et al. 2005). Desert dust intrusion may result in extreme PM$_{10}$ concentration in the island which may persist for few days (Middleton et al. 2008, Querol et al. 2009, Achilleos et al. 2014). Even among the Mediterranean countries, Cyprus is characterised as an epitome area where Saharan dust episodes cause high PM$_{10}$ concentration. Average PM$_{10}$ concentrations on dust days, as measured in Southern Cyprus regional background, could reach up to of 1000 µg/m^3 (European Environment Agency

(EEA), 2012). Previous studies investigating PM_{10} concentration over the island and the contribution of local and foreign sources found that concentration of PM_{10} in the regional, rural and urban background exceed that of most Mediterranean sites of equal background characterization (Querol et al. 2009, Achilleos et al. 2014).

Research on air quality across Europe often do not consider investigating remote locations such as Cyprus (Karagoulian et al, 2015, Priemus and Postma, 2009). The few previous investigations done in the region; were mainly carried out to cover the southern part of the island (Querol et al. 2009, Mazouridez et al. 2015, Achilleos et al. 2014, Neophytou et al. 2013, Middleton et al, 2008). Mouzourides et al. (2015) investigated the role of trans-boundary sources of PM_{10} in South Cyprus as a representative of South Eastern Mediterranean using Dust Regional Atmospheric Model from Barcelona Super Computing Centre (BSC/DREAM) and meteorological parameters. Their findings revealed that PM_{10} concentration, in 19% of the days that they examined, exceeded the critical value of 50 µg/m³ and were highly related to westerly dust from Sahara desert. The simulations they used identified sources, mode of dispersion and intrusion of the dust into the island. Despite that, the study made no attempt to quantify the amount of the dust or the contribution of dust to the concentration of particulates. Achilleos et al. (2014) used a combination of satellite imageries, Aerosol Optical Depth (AOD) and Hybrid Single Particle Integrated Trajectory (HYSPLIT) to identify dust days, and regression analysis to estimate the contribution of dust in an urban site and a regional background in South Cyprus. Their computations revealed that the overall concentrations and exceedances were higher than most European sites.

However, all these investigations were restricted to the southern part of the island. Conclusions derived from these studies could not necessarily reflect PM_{10} situation in the northern part of the region as meteorological conditions, building characteristic, land use pattern; economic activity and other anthropogenic factors vary between the north and south. These variations influence air flow and hence a spatial variation in PM_{10} concentrations (Pandis et al. 2005, Neophytou et al. 2013). Clear differences in climatic condition can be observed between cities in the region as distances and elevation from sea, which are significant factors that reflect differences in relative humidity and temperature, are not uniform over the island. Similar investigations are therefore required for the TRNC.

A preliminary assessment of the ambient air quality in the TRNC conducted in 2002-2003, under the Air Quality Framework Directive, revealed that the level of PM_{10} concentration exceeded the EU air quality objective and are affected significantly by urban sources, traffic and desert dust (Environmental Protection Department (EPD), 2015). Presently Air Quality in the TRNC is measured and maintained using strictly data from ground based monitoring stations, making it nearly impossible to estimate the contributions of the various sources.

Globally, measurements and models have been the main methods used (often in combination) for assessing air quality

and determining sources of pollution (Priemus and Postma, 2009). However, due to mechanical failures experienced in ground based monitoring stations, at EU level; air quality standards should be assessed more in combination with models and satellite observations (Priemus and Postma, 2009, European Commission (EC), 2011). Atmospheric model calculations are mathematical computer simulations of the sources and dispersion of substances in the atmosphere.

HYSPLIT (Draxler and Rolph, 2015) is considered to be one of the most reliable atmospheric models at this stage. The model is a complete system for simulating simple air parcel trajectories to complex dispersion and deposition. It computes the advection and pathways of pollutant particle hence its effectiveness in tracking dusty wind sources. Several investigators have used HYSPLIT to accurately identify desert dust intrusions and track their source area, for example Ashrafi et al. (2014) simulated dust event over Iran using HYSPLIT, Escudero et al. (2011) apportioned dust outbreak in the Mediterranean via HYSPLIT application and Escudero et al. (2006) determined the contribution of Saharan dust source to PM_{10} concentration in the central Iberian Peninsula using HYSPLIT.

To identify and estimate dust contribution, satellite measurements and observations such as AOD and Angstrom Exponent Value (α) from Moderate Resolution Imaging and Spectra Radiometer (MODIS) are used alongside HYSPLIT. AOD is a quantitative measurement of the extinction of solar beam by haze or dust between the observation point and the top of the atmosphere. It is a dimensionless number that defines the quantity of particulates in the vertical column of the atmosphere over a particular location during observation. AOD is the easiest, most precise and unique parameter used with ground based measurement to determine PM load (Holben et al. 2001). Angstrom Exponent Value on the other hand is a qualitative indicator of the sizes of aerosols. The value has been used to characterize aerosols from biomass burning in South America and Africa (Eck et al. 2001, Reid et al. 1999), urban emission (Eck et al. 2001, Kaskaoutis and Kambezidis, 2006), desert-dust aerosol in Africa and Asia (Masmoudi et al. 2003). Ground based measurement, AOD and α are used in combination to study desert dust contribution (Achilleos et al. 2014, Mazouridez et al. 2015, Barnaba and Gobbi, 2004).

To address the air quality monitoring problems related with PM_{10} concentration in TRNC, as mentioned earlier, this study used a combination of HYSPLIT model, satellite imagery, MODIS products (AOD and α), ground based measurements and an EC proposed method based on Escudero et al. (2007) to monitor the level of concentration of PM_{10} in the ambient air and measure the contribution of dust and background emission to the PM_{10} concentration in the country. Based on the available literature, nearly no attempt has been made to assess the contemporary PM_{10} situation and quantify the contribution of natural and anthropogenic sources in the country.

The study aims at ascertaining the level of PM_{10} pollution in the ambient air for the period of 2012-2014 and the contribution of desert dust to this pollution. The following objectives were set for the study:

- Identifying dust storm days and quantifying the amount of dust deposited in the ambient air over the study area.
- Evaluating the PM_{10} concentration attributable to land use characteristic.
- Assess the level of compliance to EU PM_{10} concentration limit and to determine whether cases of exceedances were caused by desert dust intrusion.

The significance of this study is mainly to provide complementary data to the existing ground based measurements for a comprehensive air quality management. The study will also provide a framework for checking whether the EU PM_{10} concentration limit in the ambient air has been exceeded. Exceedances of PM_{10} concentration limit can only be determined after the removal of the amount contributed by natural sources. The result of such estimation is important for formulating policies and designing strategies on the amount and place where emission needs to be cut off.

Quantifying PM_{10} contribution by natural sources provides information required in estimating population exposure to PM_{10} pollution necessary for health impact assessment. It worthy of notice that this study did not consider estimating the contribution of other natural sources of PM_{10} such as sea salt, pollen and volcanoes. Therefore, the effect of desert dust investigated here should not be generalised as the effect of all natural sources.

Methodology

Description of the Study Area

TRNC is the northern part of the island of Cyprus which is located in the Eastern Mediterranean. The island covers an area of 9,251 km^2 and is located off the south coast of Turkey, west coast of Syria, west of Lebanon, northwest of Israel, north of Egypt and east of Greece. Figure 1 is a satellite image showing the location of the island in the Mediterranean.

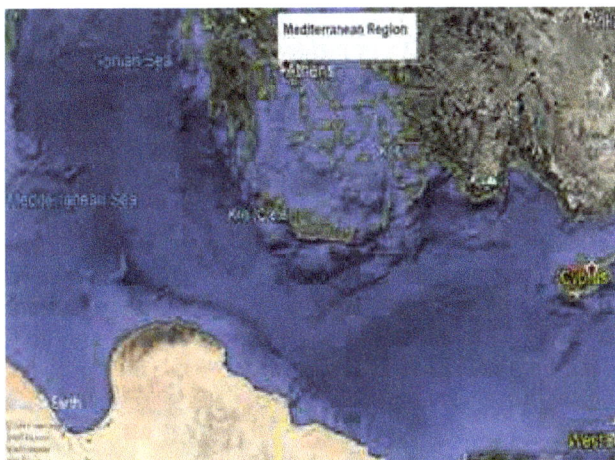

Figure 1: *Relative Location of Cyprus in the Mediterranean (Source: Google Earth 2017).*

Cyprus has a subtropical climate which is characterized as semi-arid with warm rainy winters around November to late March (average temperature of 17-18°C during the day and 8-10°C at night and average precipitation of 100 mm) and hot dry summers around June to late September (average temperature is around 33°C during the days and 23°C during the nights and average precipitation is around 4 mm). During winter, temperatures are higher over the inland than at coastal areas and vice versa in the summer. Spring season covers the period of April and May. Autumn is a short transition period to winter.

During winter westerly and south westerly surface winds prevail over the Eastern Mediterranean, while northerly and north-westerly winds prevail during summer periods. Variability in strength and direction in the wind blowing over Cyprus is influenced by the eastward moving cyclones crossing over the Mediterranean sea, sea and land breezes temperature differences, the continental anticyclone that stretch over Eurasia, the low pressure belt of North Africa, the monsoon low in summer, orographic factors and causes (Achilleos et al. 2014).

Description of the Monitoring Stations

This study used PM_{10} data obtained from the monitoring sites in TRNC which are monitored by the Air Quality Monitoring Network, Environmental Protection Department, Ministry of Environment and Culture, TRNC. A total of six monitoring stations which reflect some of the land uses in the region were selected. The positioning and characterization of these stations are within the EU framework legislation given in section C Annex III of CAFE-Directives 2008/50/EC on ambient air quality and cleaner air for Europe. Figure 2 shows the locations of the monitoring sites while Table 1 provides a summary of the characteristics of the monitoring sites.

Table 1: *Sampling sites and their characteristics*

Site	Type of site	Coordinates	Above Sea Level
Nicosia	Urban	35.20 N 33.35 E	108 m
Famagusta	Urban	35.13 N, 33.93 E	3 m
Kyrenia	Urban	35.33 N, 33.31 E	8 m
Guzelyurt	Urban	35.20 N, 33.00 E	51 m
Kalecik	Rural-Industrial	35.34 N, 34.00 E	10 m
Alevkayasi	Rural	35.30 N, 33.53 E	608 m

Method of Data Collection

24 hours PM_{10} concentrations were measured in the air quality monitoring stations and the average taken. These daily average PM_{10} data was collected from the EPD of TRNC. The data collected cover the three year study period (January 1st, 2012 through December 31st 2014) for all the monitoring sites. It was observed that within the period there were days with missing data coverage which were due to power failure or mechanical faults.

Figure 2: *Spatial arrangements of PM$_{10}$ monitoring sites in TRNC (Google Earth 2017).*

Desert Dust Storm Identification

Identification of dust events influencing particulate matter concentration is a multi-task which may demand the use of various tools such as ground PM concentration data, aerosol maps, receptor and dispersion modelling and back trajectory analysis. To identify dust event days in this study, the methods described by the 2011 Commission Staff Working Paper of the Council of EU with the reference "Establishing guideline for demonstration and subtraction of exceedances attributable to natural sources" under the Directive 2008/50/EC on ambient air quality and cleaner air for Europe was referred to. Combination of other methodologies used in previous researches was also employed. The procedures will be discussed in further detail in the following sections:

High PM$_{10}$ level identification

The occurrence of a dust storm was assumed to wholly affect the concentration of PM$_{10}$ in the region and cause high PM$_{10}$ level. High PM$_{10}$ level was defined as days with concentration above the 95th percentile value. Presence of possible dust episode was recognized by high PM$_{10}$ levels occurring in the same day at the sampled sites, especially a sudden high increase in the regional background station.

Identification using MODIS Products

Mean daily measurement MODIS AOD 550 nm and α from Giovanni Satellite Based Earth Science Online Data System were used to identify the particulate type. The data system was developed and managed by the National Aeronautic and Space Administration (NASA), Goddard Earth Sciences Data and Information Service Centre. The data are acquired by the MODIS sensor on both aqua and terra satellites on a spatial resolution of $1° \times 1°$.

AOD value of 0.01 corresponds to an extremely crystal clear atmosphere with little amount of aerosols, and a value of 0.4 and above means a very dusty aerosol dense atmosphere. While α is a qualitative measure of the sizes of aerosols. The value

is inversely related to the average size of the particles in the atmospheric aerosols; that is to say, the smaller the particle the higher α (NOOA, 2015).

Therefore, high AOD value reflects high total atmospheric concentration of particulates and a lower α indicates the particles are of coarser sizes. A combination of high AOD value of >0.3 and α <0.9 indicates the particulates are of desert dust origin. Barnaba and Gobbi (2004) recommended these indicator values for analysis of dust particles over the Mediterranean.

Identifying Dust using HYSPLIT Model

To verify a dust episode identified using the above method, a 5-day backward trajectories of air masses originating from the Sahara or Arabian Peninsula at three different altitudes of 750, 1500 and 2500 m above sea level were examined. HYSPLIT Model (Draxler and Rolph, 2015) which is provided by the National Oceanic and Atmospheric Administration (NOAA) Air Resource Laboratory (ARL), available at Real Time Environmental and Display System Website was used to compute the trajectories. Archive re-analysis meteorological data provided by the National Centre for Atmospheric Research (NCAR) was used as the composite data for this computation. HYSPLIT model is a computer based simulation model that is used to compute air parcel trajectories, dispersion and or deposition of atmospheric based pollutants (Draxler and Rolph, 2015). This model uses a combination of both Eulerian and Lagrangian approaches to track the source point of dust events using the u- and v-components of the wind, temperature, height and pressure at different level of the atmosphere and the resulting backward trajectory contains information about the origin and pathways of the dust transport (Banacos and Ekster, 2010). Air parcel trajectories that originate from the desert indicate possible presence of dust.

The identification of the dust event days largely depends on the availability of data for the days in focus. Dust events days include the days where the dusty wind from desert arrived the region and the subsequent days where significant amount of dust persists in the atmosphere. HYSPLIT model identified the arrival of the dust while the presence of dust in the atmosphere were basically identified using the combination of the PM$_{10}$ concentration data, AOD and α in accordance with the criteria mentioned in the methodology for determining possible dust in the atmosphere.

Lastly some days have missing AOD or α, or in some cases; both are missing. There are also days where ground based PM$_{10}$ concentration were not captured in some of the monitoring stations. Such missing data is expected to hinder the study from identifying some possible dust days and hence the calculated dust contribution may underestimate the actual contribution.

Identifying Dust Storm using Satellite Imageries

MODIS satellite images on air aqua and terra platform retrieved from NASA's website database were also used to support the identification.

Other methods that are recommended by the EU and applied by investigators to identify dust storm events include mineralogical analysis, and dispersion and receptor modelling. These methods were not applied in this study as the former requires PM samples and the later requires data on hourly concentrations which were not available for this investigation.

Quantifying Desert Dust

In order to measure the amount of PM_{10} concentration attributed to desert dust storm episodes, the Escudero et al. (2007) based conservative method suggested by EC, (2011) under the Directive 2008/50/EC was employed: A moving average of PM_{10} concentration of 15 days before and 15 days after the identified dust episode (excluding any dust day which may occur within the period) was calculated. The calculated value corresponds to a moving 50th percentile of 30 days. The calculated 30 days average is the supposed PM_{10} concentration assuming there was no dust intrusion. The net dust amount was calculated by subtracting the 30 days average value from the high PM_{10} concentration.

Escudero et al. (2007) based methods are scientifically validated (EC, 2011). The methods save time, cost and are simple to use. Unlike chemical analysis methods which may require long analysis time, expertise and laboratory costs. The methods were earlier applied by Querol et al. (2009) to the whole of the Mediterranean basin. The other Escudero et al. (2007) based method requires subtracting a monthly moving 40th percentile (excluding days with dust influence) from the bulk concentration of PM_{10} of the dust day in the regional background site, the returned value is the net dust PM_{10} concentration for the region; hence the net dust contribution for each site can be calculated by subtracting this net dust PM_{10} concentration. However, this method was not preferred here because the other sites (especially Kalecik) were found to reproduce a better PM_{10} concentration on some dust event days than the regional background site, thus making the method not applicable as the subtraction of net 40th percentile from these sites on such days would yield a negative PM_{10} concentration.

Estimating Background Sources Contribution

After removing the contribution of dust, the impact of local natural particulate sources such as soil and sea salt were assumed to be the same at all the sites. Then daily background concentration of each site was estimated as the difference between the daily concentration in the site and the daily concentration in the regional background. The new data base created became the "PM_{10} urban" or "industrial" contribution variables, depending on the dominant land-use in the site.

This method provides a good understanding of the level of emission from a background collectively where data for each emission sources is not available. Moreno et al. (2005) used this approach to assess the influence of urban sources in Spain. Achilleos et al. (2014) also used same method for similar investigation in South Cyprus.

Data Analysis

Microsoft Office Excel package was used to compute annual and monthly mean concentrations, percentages and frequency of exceedances of mean daily concentrations and dust contributions. Results were depicted using simple bar charts and tables.

To account for the loss of PM_{10} data capture during the estimation of exceedances of daily average concentration in calendar year, an adjustment was made to the available data. The adjustment assumes that the fraction of missing values that would have surpassed the EU limit is equivalent to the fraction of the available value that exceeded the limit. This approach was used on a quarterly basis, as EC (2013) suggested, and is computed as shown in equation 1:

$$Eq = \left(Vq \times \frac{Nq}{nq} \right)$$
Equation 1.

Eq in equation 1 refers to the estimated number of exceedances for the calendar quarter in question (qi). Vq refers to the observed number of exceedances for same calendar quarter. Nq is the number of days in the calendar quarter whereas nq is the number of days with valid daily values for the calendar quarter. q refers to the four calendar quarters in a year, that is to say; q= 1st, 2nd, 3rd, or 4th. The total number of exceedances (Ex) for the calendar year is then estimated as the summation of the estimated number of exceedances for all the calendar quarters of the years; as given in equation 2.

$$Ex = \sum_{q=1}^{4} eq$$
Equation 2.

Result and Discussion

The study identified dust storms, its frequency of occurrence and seasonal variability in the region. The study also estimated dust and local background contribution to PM_{10} concentration in the sampled sites.

Dust Storm Occurrences and Impact on Daily Concentrations

A total of 35 dust storm days were identified for the three years period in the region. The identified dust days and their daily PM_{10} average concentration are shown in Table 2 (see appendix), 4 dust days (less than 1% of the year) occurred in 2012, 8 dust days (2% days of the year) in 2012 and 24 dust days (7% days of the year) in 2013. These percentages correspond within the 1% (1993) - 9% (1998) dust days earlier estimated in the region (Achilleos et al. 2013). 75 percentile of the dust days in the urban areas fall within a mean daily concentration of ≥ 100 µg/m³, 75 percentile of the dust days in Guzelyurt have a mean daily concentration of ≥ 70 µg/m³, while in Alevkayasi and Kalecik; a 75 percentile of ≥ 50 µg/m³ was estimated.

On the other hand, dust storms were more frequent in 2013. Concentrations of PM_{10} in the sites range from 40 µg/m³ on minor dust days to 280 μg/m³ on intense dust days. The frequency of dust storm occurrence in 2013 was also high in the Mediterranean city of Athens, Greece, however maximum daily average PM$_{10}$ concentration on dust days were 125 μg/m³, less intense than those recorded in Cyprus (AIRUSE 2015). Frequent occurrence of dust in the Mediterranean is possibly as a result of drought and anthropogenic disturbances (such as clearing of vegetation cover) on the Saharan soil.

In 2014, daily average concentration of PM$_{10}$ range from 45 to 400 μg/m³. Estimated dust contribution range from 10 (06 May, 2014) to 360 μg/m³ (03 Mar 2014).

Dust storms in the three years were found to intrude the island in each of the four seasons in a year. However, only 3 dust days were experienced in the summer throughout the study period and 75% of the dust episodes occurred during the winter months especially March and November. A 14 day long dust episode was found to occur in November 2013. Similarly to this investigation, dust storms occurrences in neighbouring Israel for that year were also found to be in winters and spring, and summers were dust free (Krasnov et al. 2014). While contrary to this, AIRUSE (2015) found that dust storm that year in Athens were more frequent and pronounced during spring and summers.

Amount of dust deposited on dust days depends to an extent on the distance from the source region. Intrusion originating from closer proximity such as Egypt, Morocco and the Arabian Peninsula are found to be more pronounced than intrusions from far sources such as Mali, Niger Republic or Mauritania. Some amount of the dust may probably have been deposited elsewhere as they travel towards the Mediterranean, as dust are often deposited along the travel path to their destination (Moreno et al. 2005). Therefore, less deposition is expected when dust originates from far distance as the loss along travel path is likely to be more.

Analysis of the computed trajectories showed that about 65% of the dust intrusions into the island were from the Saharan desert, specifically the western part of the Sahara. The remaining dust storms originated from the Arabian Peninsula. Dust intrusions in 2014 reached the island in a lower vertical height than dust storms in the previous years. Dust arriving at lower vertical height implies more deposition at the ground especially on cloud free days.

Figure 3 shows an example of computed backward trajectories arriving Cyprus. The dusty wind arrived Cyprus on 02 March, 2014 from western part of Sahara, Egypt and the Arabia. Figure 4, which is a satellite view of the dust event, shows the dust originating from Sahara desert and covering the Mediterranean, which was cloudless at the time of deposition. In examining satellite views, cloud cover were seen to be associated with cyclonic conditions on some dust days, therefore obstructing clear view to the dust. However, the dust event on 02 March, 2017 was a good example of deposition on cloud free day. Concentration of PM$_{10}$ reached 280 μg/m³ in Kalecik and a shade above 200 μg/m³ in the other sites.

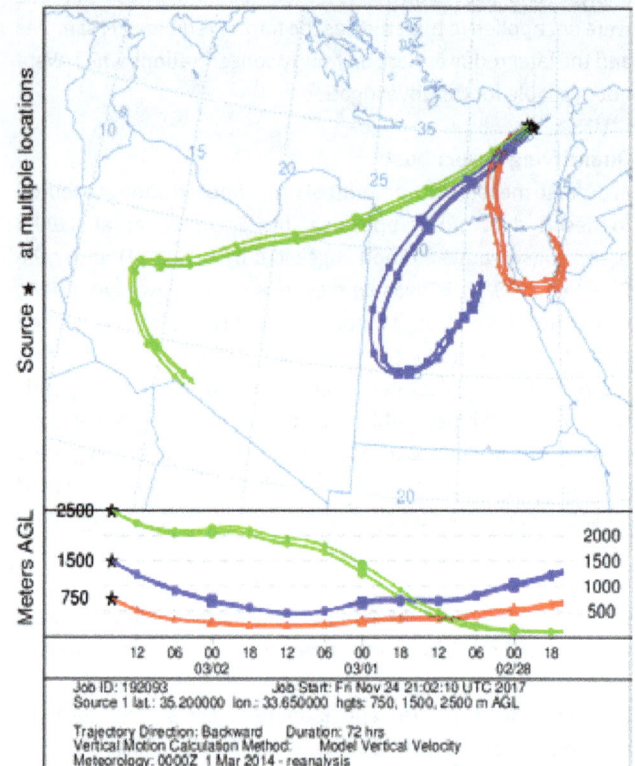

Figure 3: *Backward trajectories ending in Cyprus showing the arrival of dust on 02 March 2014.*

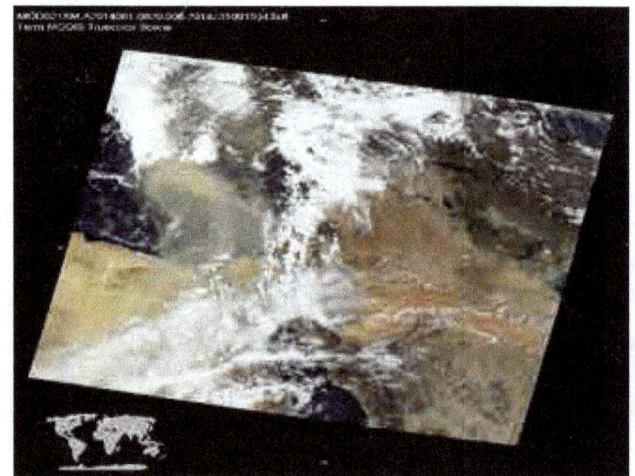

Figure 4: *MODIS satellite imagery over the Mediterranean on 02 Mar 2014.*

Contribution of Background Emission

Emission from urban background contribution was estimated for Nicosia, Kyrenia, Famagusta and Guzelyurt and industrial background contribution was estimated for Kalecik. As it can be seen in Figure 5, overall average urban background contribution for the three years period were 12.3 μg/m³ in Guzelyurt, 18 μg/m³ in Kyrenia, 18.4 μg/m³ in Famagusta and 27.8 μg/m³ in Nicosia. Industrial background contributed 9.7 μg/m³ to the overall average concentration of PM$_{10}$ in Kalecik. Kalecik is predominantly affected by emission from thermal station which after burning fossil fuel may give out high amount of coal fly-ash to the atmosphere.

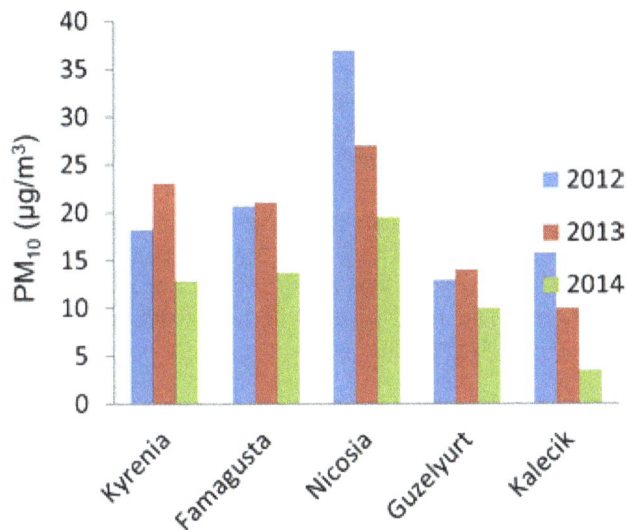

Figure 5: *Annual average contribution of background sources.*

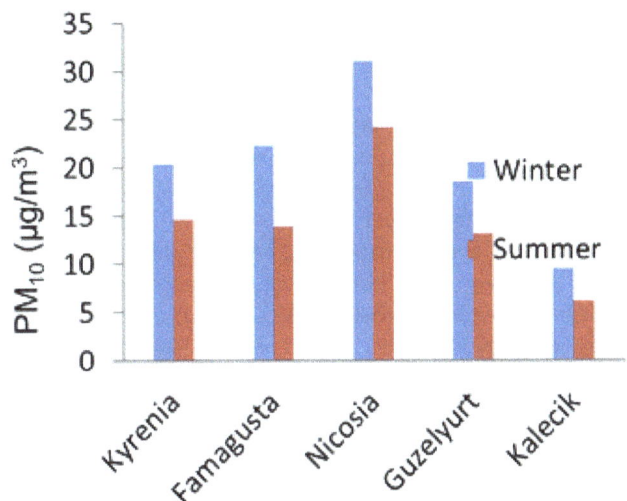

Figure 6: *Overall seasonal average contribution of local background emission.*

Effect of urban background to PM_{10} is spatially associated with traffic and population density. Nicosia among the urban sites, has the densest population and traffic hence the highest concentration while Guzelyurt has the lowest population and traffic density among the urban sites and hence the lowest concentration. Background sources of emission in the urban sites are majorly road dust resuspension and traffic emission, domestic heating and burning of solid fossil fuel, building construction and industries (especially in Nicosia). Shipping could contribute to the bulk of emission in Kyrenia and Famagusta. Current regulation regarding emission is not in compliance with the EU Directives. Field burning is still practiced in the country and these could be a major contributor of PM_{10} load in the affected sites (EPD 2015).

Figure 6 shows the average contribution of background emissions in relation to seasons. Overall background

contribution to PM_{10} concentration is more during the colder months than the warmer season, except in Teknecik where the background contribution is seen to be more in the summer. Usually urban sites experience winter sanding of roads.

Daily Average Concentration and Exceedance of the Limit Standard

The EU directive sets a daily average threshold value of 50 $\mu g/m^3$. This value is not permissible to be exceeded in 35 days (9.6% days) in a year. As summarized in Table 3, Nicosia, Kyrenia and Famagusta urban sites exceeded the daily average concentration of 50 $\mu g/m^3$ in more than 35 days in all the years (with and without dust effect). Exceedances of daily average limit solely as a result of dust events range from 4.4 to 10% in Kyrenia, 4.2 to 13.8% in Kyrenia and 0 to 6.3% in Nicosia.

Guzelyurt was within the 35 days limit in 2012. However, the limit was exceeded in the two subsequent years. The limit was exceeded in 2013 in Kalecik. Both exceedances of the 35 days limit in Guzelyurt and Kalecik were not as a result of dust effect. In the regional background, daily average concentrations were below the 35 Days limit.

Table 3: *Percentage days with daily mean concentration above the EU 50 $\mu g/m^3$ threshold value (A= with dust and B= without dust)*

Site	Year		
	2012	**2013**	**2014**
Kyrenia	A=18.6 B=17.7	A=37.5 B=34.2	A=16.4 B=14.7
Famagusta	A=20.4 B=18.6	A=32.2 B=26	A=18 B=15.3
Alevkayasi	A=1.6 B=0.5	A=6.8 B=2.2	A=7.3 B=5.7
Guzelyurt	A=22.7 B=7.4	A=16.9 B=13.6	A=11.7 B=11.2
Kalecik	A=51.1 B=22.2	A=16.7 B=10.1	A=8.5 B=1.9
Nicosia	A=51.1 B=51.1	A=45 B=43	A=35.1 B=43

Overall and Annual PM_{10} Concentrations

The overall average PM_{10} concentration was calculated as 54.7 $\mu g/m^3$ with a Standard Error ($\pm$) of 0.9 $\mu g/m^3$ for Nicosia and when the dust effect was removed the overall concentration decreased to 46.2 $\mu g/m^3$. The overall average concentration for Kyrenia and Famagusta was 43.2 $\mu g/m^3$ ±0.6 $\mu g/m^3$ and 43.6 $\mu g/m^3$ ±0.6 $\mu g/m^3$ respectively and when the dust effect was removed the concentration reduced to 41.4 $\mu g/m^3$ in Kyrenia and 41.8 $\mu g/m^3$ respectively. Guzelyurt has an overall average of 37.5 $\mu g/m^3$ ±0.58 $\mu g/m^3$ with dust effect and 35.6 $\mu g/m^3$ without the effect of dust. Kalecik has an overall concentration of 35.3 $\mu g/m^3$ ±0.6 and 32 $\mu g/m^3$ when the effect of dust intrusion

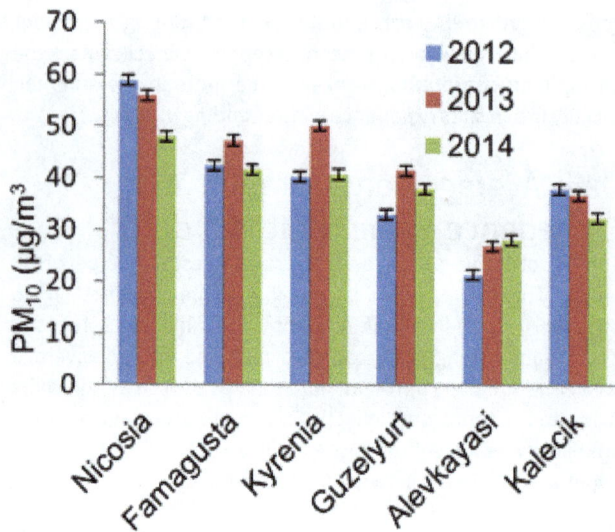

Figure 7: *Annual average concentration of PM$_{10}$ with dust effect inclusive.*

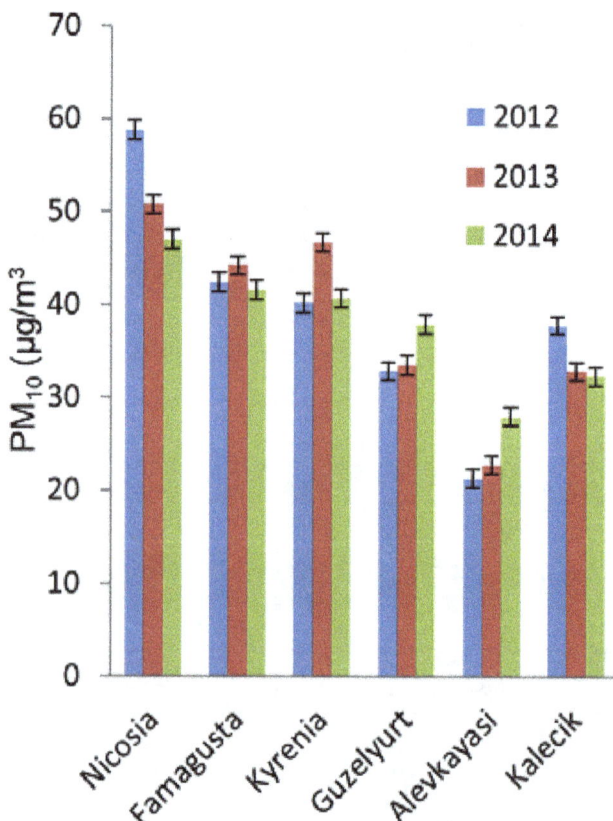

Figure 8: *Annual average concentration of PM$_{10}$ (without dust effect).*

was removed. Alevkayasi regional background has an overall average of 25.4 µg/m³ ±0.6 µg/m³ and an overall average of 23.1 µg/m³ when dust effect was removed.

The annual average PM$_{10}$ concentrations are shown in Figure 7 and 8. Annual average concentrations of PM$_{10}$ (with dust and without dust effect) in Nicosia were above the EU 2006 limit of 40 µg/m³ for all the years. Annual average concentration ranged from 48 µg/m³ (2014) to 59 µg/m³ (2012) and it was found to exhibit a reducing trend after every year. After removing the effect of dust; concentrations only reduced by 1-5%.

PM$_{10}$ Concentrations with dust effect were above the EU limit in all the years for Famagusta. Annual average concentrations were 42-47 µg/m³ with dust effect (Figure 7); however the limit was attained in 2014 after removing the effect of desert dust. Annual average concentration in Kyrenia were exceeded in 2013 and 2014, after removing dust effect the limit was achieved in 2014 (Figure 8). Annual average limit was exceeded in Guzelyurt in 2014 which after removing the effect of dust reduced to an acceptable value of 37 µg/m³.

Concentrations in the rural site and the rural industrial site were all within the EU annual mean limit. Concentrations were in the range 32-37.8 µg/m³ and 26-29 µg/m³ in Kalecik and Teknecik respectively. After removing dust effect, concentration reduced to 30-36 µg/m³ in Kalecik.

Annual average concentration in Alevkayasi ranges from 21-28 µg/m³ with dust effect. After removing effect of dust intrusion, it attained the EU 2010 target value of 20 µg/m³ in 2012. Annual concentrations exhibit an increasing trend Alevkayasi. Cristina et al. (2014) observed similar trend over the years in background sites among the EU sites.

Annual average PM$_{10}$ concentrations (without dust) in Guzelyurt can be compared with 31 µg/m³ in central Mediterranean city of Lampedusa, Italy (Marconi, et al. 2014), 24-30 µg/m³ in Athens suburban site, 26-27 µg/m³ in Milan and Porto urban sites (AIRUSE, 2015). Annual average concentrations in Famagusta and Kyrenia can be compared with 37-43 µg/m³ in Athens urban sites (AIRUSE, 2015), 29-42 µg/m³ in Berlin urban background (Langener, et al. 2011) and 37-43 µg/m³ in Madrid Spain (Salvador et. al, 2015). Annual Average concentration in Nicosia can be compared with those estimated in other Eastern Mediterranean urban sites such as 56 µg/m³ in Nicosia, South Cyprus, (Achilleos et al. 2014), 43-77 µg/m³ in Beer-Sheva, Israel, (Krasnov et al. 2014), 51 µg/m³ in Heraklion, Greece, 57 µg/m³ in Tel-Aviv, Israel, 47 µg/m³ in Istanbul, Turkey (Gerasoupoulos et al. 2006, Karaca et al. 2005). Annual average concentration in Alevkayasi corresponds with other rural sites in the EU such as Berlin rural background 19-25 µg/m³ (Langener et al. 2013), 20-25 µg/m³ in Campisabolos, Spain (Salvador et al. 2015).

Overall average in the regional background and Kalecik were lower than the estimated 32.1 µg/m³ in Agia Marina regional background, southern part of Cyprus (Achilleos et al. 2014), Annual dust contribution in TRNC can be compared with the 5 µg/m³ in Italy (Salvador et al. 2013). Vautard et al. (2005) also reported 5-10 µg/m³ as the annual average in the Mediterranean.

Conclusion and Recommendation

In this study, inter annual and annual PM$_{10}$ concentrations were analysed in Nicosia urban site, Famagusta urban site, Kyrenia urban site, Kalecik rural industrial background site and Alevkayasi rural background site. Dust episode and their contributions to daily, annual and seasonal PM$_{10}$ concentrations were also estimated as well as the contribution of collective anthropogenic background.

A total of 35 dust days occurred in the island within the 3 years period. Dust intrusion contributed more to PM_{10} concentration during winter and spring, daily concentration could reach as high as 400 µg/m³. Contribution of desert dust to PM_{10} concentration in the study area was averagely; 102 µg/m³ and can range from 22 to 183 µg/m³. Average contribution of dust to the annual average concentration were 8.5 µg/m³ in Nicosia, 2.2 µg/m³ in Kyrenia, 1.8 µg/m³ in Famagusta, 2 µg/m³ in Guzelyurt, 3.3 µg/m³ in Kalecik and 2.3 µg/m³ in Alevkayasi. Overall average urban contribution to PM_{10} concentration was 12.3 µg/m³ in Guzelyurt, 18 µg/m³ in Kyrenia, 18.4 µg/m³ in Famagusta and 27.8 µg/m³ in Nicosia. Average industrial background contribution was 9.7 µg/m³.

The study found that despite the high occurrence of dust events, desert dust was only responsible for exceedance of the 2006 EU mean annual PM_{10} concentration of 40 µg/m³ in Famagusta and Kyrenia in 2014, and Guzelyurt in 2013. However, no exceedance of the 35 days permissible daily average limit of 50 µg/m³ was attributed to dust storms in any of the site analysed/

It's worth reminding that the study only investigated one natural source of PM_{10} (dust storms). Impact of other natural sources such as pollen, sea salt and local soil resuspension were not assessed in this investigation. There is the likelihood that if impact of other natural sources were subtracted, exceedances may not be recorded in the sites. Therefore an investigation is required to ascertain the influences of other natural source of PM_{10}. A source apportionment that will include chemical or mineralogical analysis is also needed to evaluate the impact of each anthropogenic source (such as shipping, traffic, domestic burning of fuel, agriculture etc.) to the backgrounds.

Acknowledgement

We wish to thank the Environmental Department of Northern Cyprus for providing us with the ground monitoring station PM_{10} measurement data. We also thank the academics present at the Annual Nigerian National Association of Geographers Conference, 2017, for their viable contributions when the paper was presented.

References

Achilleos S., Evans J.S., Yiallorous P.K., Kleanthous S., Schwartz J. and Koutrakis P. 2014, 'PM_{10} concentration levels at an urban and background site in Cyprus: The impact of urban sources and dust storms, *Journal of the Air & Waste Management Association*, 64:1352-1360, doi: 1080/10962247.923061.

AIRUSE 2015, 'Contribution of natural sources to PM concentration levels', *AIRUSE LIFE11 ENV/ES/584*.

Banacos P.C. and Ekster M. 2010, 'The association of the elevated mixed layer with significant severe weather events in the Northeast United States, *WEA Forecasting*, 25:1082- 1102.

Barnaba F, and Gobbi G.P. 2004. Aerosol seasonal variability over the Mediterranean region and relative impact on maritime, continental and Saharan dust particle over the basin, *Atmos. Chem. Phys*. 4:2367-91. doi: 10.5194/acp-4-2367-2004.

Brook R.D., Rajagopalan S., Pope C.A., Brook J.R., Bhatnagar A., Diez-Roux A.V. and Peters A. 2010, 'Particulate matter air pollution and cardiovascular disease an update to the scientific statement from the American Heart Association', *Circulation*, 121(21), 2331-2378.

Cristina B.B.G., Foltesku V. and Leeuw F.D. 2014, 'Air quality status and trend in Europe', *Atmos.env*, 98:376, doi:10.1016.

Cristofanelli P., Marinoni A., Arduini J., Bonafe U., Calzolari F., Colombo T., Decesari S., Duchy R., Facchini M.C., Fierli F., Finesse E., Maione M., Chiari M., Calzolai G., Messina P., Orlandi E., Roccato F. and Bonasoni P. 2009, 'Significant variation of trace gas composition and aerosol properties at Mt. Cimone during air mass transport from North Africa-contribution from wildfire emissions and mineral dust'. *Atmos. Chem. Phys*. Discuss.9,7825-7872.

Dayan U., and Levy I. 2005, 'The influence of meteorological conditions and atmosphere circulation types on and visibility in Tel Aviv. J', *Appl. Meteor*. 44: 606-19. doi:10.1175/JAM2232.1.

De Gouw J.A., Brock C.A., Atlas E.L., Bates T.S., Fehsenfeld F.C., Goldan P.D. and Middlebrook A.M. 2008, 'Sources of particulate matter in the North-eastern United States in summer: Direct emissions and secondary formation of organic matter in urban plumes', *Journal of Geophysical Research: Atmospheres* 113(1984–2012).

Donaldson K., Stone V., Clouter A., Renwick L. and MacNee 2001, 'Ultra-fine particles', *Occup Environ Med*; 58:211-216 doi:10.1136/oem.58.3.211.

Donkelaar V., Martin A., Brauer V., Kahn M., Levy R. and Villeneuve C., 2010, 'Global estimates of ambient fine particulate matter concentrations from satellite-based aerosol optical depth: development and application'. *Environmental health perspectives*, 118(6), 847.

Draxler R.R., and Rolph G.D., 2015, 'HYSPLIT (Hybrid Single-Particle Langrangian Integrated Trajectory) Model, NOAA Air Resource Laboratory', College Park, MD: NOAA ARL READY website. N (Retrieved December 26, 2015).

EC (European Commission) 2011. 'Commission Staff Working Paper: establishing guideline for demonstration and subtraction of exceedances attributable to natural sources under the Directive 2008/50/EC on ambient air quality and cleaner air for Europe', *SEC(2011)208 final*.

EC (European Commission) 2013, 'Guidance on the commission implementing decision laying down rules for Directives 2004/107/EC and 2008/50/EC of the European parliament and of the council as regard the reciprocal exchange of information and reporting on ambient air' (*Decision 2011/850/EU*). *DG ENV*.

Eck T.F., Holben B.N., Reid J.S., Dubovik O., Smirnov A., O'Neill N.T., Kinne S. 1999, 'Wave length dependence of the optical depth of biomass burning, urban, and desert dust aerosols', *Journal of Geophysical Research: Atmospheres* (1984–2012), 104(D24), 31333-31349.

EEA (European Environment Agency) 2012, 'Technical Report: Particulate matter from natural sources and related reporting under the EU Air Quality Directive in 2008 and 2009', *Publication office of the European Union*, ISBN 978-9213-325-2. ISSN 1725-2237. doi:10.2800/55574.

EPD (Environmental Protection Department) 2015, 'Air Quality Monitoring Network', Ministry of Tourism Environment and Culture, Turkish Republic of Northern Cyprus, Website URL: http://www.turcyp.com/en/info/monitoring-information/index/html. (Retrieved 10 October 2015).

Escudero M., Querol X., Pey J., Alastuey A., Perez N., Ferreira F., Alonso S., Rodriguez S. and Cuevas E. 2007, 'A methodology for the quantification of African dust load in air quality monitoring network: Guidance to member states on PM_{10} monitoring and inter comparison with the reference method', *Atmospheric Environment* 41, 5516-5524.

Ganor E., Stupp A. and Alpert P. 2009, 'A method to determine the effect of mineral dust aerosols on air quality' *Atmospheric Environment*, 43(34), 5463-5468.

Gerasopoulos E., Kouvarakis G., Babasakalis P., Vrekoussis M., Putaud J.P. and Mihalopoulos N. 2006, 'Origin and variability of particulate matter (PM_{10}) mass concentration over the Eastern Mediterranean, Atmos. Environ. 40: 4679-90. doi:10.1016.

Giovanni 2015, 'The bridge between Data and Science. National Aeronautic and Space Administration, Goddard Earth Sciences Data and Information Service Centre', http://giovanni.gsfc.nasa.gov/giovanni. (Retrieved 26 December 2015).

Gobbi G.P., Barnaba F. and Ammannato L. 2007, 'Estimating the impact of Saharan dust on the year 2001 PM_{10} record of Rome, *Atmos. Environ*. 41,261e275.

Holben B.N., Smirnov A., Eck T.F., Slutsker I., Abuhassan N., Newcomb W.W. and Lavenu F. 2001, 'An emerging ground-based aerosol climatology- Aerosol optical depth from AERONET', *Journal of Geophysical Research*, 106(D11), 12067-12097.

Karaca F., Alagha O. and Erturk F. 2005, 'Statistical characterization of atmospheric PM_{10} and $PM_{2.5}$ concentrations at a non-impacted suburban site of Istanbul, Turkey', *Chemosphere* 59:1183-90. doi:10.1016/6.

Karagoulian F., Prus-Ustun A.M., Bounour S., Rohani H.A. and Aman M. 2015, 'Contributions to cities' ambient particulate matter: A systematic review of local source contributions at global level' *Atmos. Env*. 120, 475 483.

Kaskaoutis D.G., Kambezidis H.D., Adamopoulos A.D. and

Kassomenos P.A. 2006, ' The characterization of aerosols using the Angstrom exponent in the Athens area', *Journal of atmospheric and solar-terrestrial physics*, 68(18), 2147-2163.

Krasnov H., Katra I., Koutrakis P. and Friger M.D. 2014, 'Contribution of dust storms to PM_{10} level in an urban arid environment' *Journal of the air and waste management association*, 64:1. 89-94, doi: 10.1080/10962247.

Kubilay N., Nickovic S., Moulin C. and Dulac F. 2000, 'An illustration of the transport and deposition of mineral dust onto the Eastern Mediterranean' *Atmos.env* 34(8), 1293-1303.

Kumar A.V., Patil R.S. and Nambi K.S.V. 2001, 'Source apportionment of suspended particulate matter at two traffic junctions in Mumbai, India', *Atmospheric Environment*, 35(25), 4245-4251.

Laden F., Neas L.M., Dockery D.W. and Schwartz J. 2000, 'Association of fine particulate matter from different sources with daily mortality in six US cities', *Environmental health perspectives*, 108(10), 941.

Langner M., Draheim T. and Eindlicher W. 2011, 'Particulate matter in the urban atmosphere: concentration, distribution, reduction-result of studies in the Berlin Metropolitan Area', *Perspective in urban ecology*, 3-642-17731-6_2. Springer-Verlag Berlin Heidelberg.

Marconi M., Sferlazzo D.M., Becagli S., Bommarito C., Calzolai G., Chiari M. and Meloni D. 2014. 'Saharan dust aerosol over the central Mediterranean Sea: PM_{10} chemical composition and concentration versus optical columnar measurements', *Atmospheric Chemistry and Physics*, 14(4), 2039-2054.

Masmoudi M., Chaabane M., Tanré D., Gouloup P., Blarel L. and Elleuch F. 2003, 'Spatial and temporal variability of aerosol: size distribution and optical properties'. *Atmospheric Research* 66(1), 1-19.

Mazzei F., D'alessandro A., Lucarelli F., Nava S., Prati P., Valli G and Vecchi R. 2008, 'Characterization of particulate matter sources in an urban environment. Science of the Total Environment' 401(1), 81-89.

Middleton N., Yiallouros P., Kleanthous S., Kolokotroni O., Schwartz J., Dockery W.D., Demokritou P. and Koutrakisi P. 2008, 'A 10-year time series analysis of respiratory and cardiovascular morbidity in Nicosia, Cyprus: the effect of short-term changes in air pollution and dust storms', *Env. Health*, 7:39. doi: 10.1186/1476-069x-7-39.

Moreno T., Querol X., Alastuey A., Viana M. and Gibbon W., 2005, 'Exotic dust incursions into Central Spain, Implication for legislative controls on atmospheric particulate', *Atmos. Env*. 39:6109–20.

Mouzourides P., Kumar P., and Neophytou M.K. A. 2015, 'Assessment of long-term measurements of particulate matter and gaseous pollutants in South-East Mediterranean', *Atmos. Env*. 107, 148-165.

NASA (National Aeronautic and Space Administration) 2015, *Global browse image for MODIS Atmosphere*. NASA website. URL: http://modis atmos.gsfc.nasa.gov/IMAGES/index_L2Mosaics. html (Retrieved 30 December 2015).

Neophytou A., Yiallouros P., Coul B.A., Kleanthous S., Pavlou P., Pashiardis S., Dockery D.W., Koutrakis P. and Laden F. 2013, 'Particulate matter concentrations during desert dust outbreaks and daily mortality in Nicosia, Cyprus', *Journal of Exposure Science and Environmental Epidemiology* 23:275-280.

Nicolás J., Chiari M., Crespo J., Orellana I.G., Lucarelli F., Nava S. and Yubero E. 2008, 'Quantification of Saharan and local dust impact in an arid Mediterranean area by the Positive Matrix Factorization (PMF) technique', *Atmospheric Environment*, 42:8872-8882.

NOAA (National Oceanic and Atmospheric Administration) 2015, 'MODIS atmosphere', URL: http://modis-atmos.gsfc.nasa.gov/ pubs_main.html. (Retrieved 30 December 2015).

NOAA (National Oceanic and Atmospheric Administration) 2015, 'SURFRAD AEROSOL Optical Depth', *Earth System Research Laboratory, Global Monitoring Division*. NOAA Website URL: http://www.esrl.noaa.gov/gmd/grad/surfrad/aod.

Pandis S., Wexler A. and Seinfeld J. 1995, 'Dynamics of tropospheric aerosol', *J. Phys. Chem*. 99:9646-9659.

Priemus H., and Postma E.S. 2009, 'Notes on the particulate matter standards in the European Union and the Netherlands', *Int. J. Environ. Res. Public Health* 6(3): 1155–1173. doi:10.3390/ ijerph6031155 PMCID: PMC2672387.

Querol X., Pey J., Pandolfi M., Alastuey A., Cusack M., Perez N., Moreno T., Viana M., Mihalopoulos N., Kallos G. and Kleanthous S. 2009, 'African dust contribution to mean ambient PM$_{10}$ mass levels across the Mediterranean Basin', *J. Atmos. Environ*. 43:4266-77.

Salvador P., Artíñano B., Viana M., Alastuey A., and Querol X. 2015, 'Multi-criteria approach to interpret the variability of the levels of particulate matter and gaseous pollutants in the Madrid metropolitan area, during the 1999–2012 period', *Atmos. Env*. 109, 205-216.

Sharratt B.S. and Lauer D. 2006, 'Particulate matter concentration and air quality affected by windblown dust in the Columbia Plateau'. *Journal of environmental quality*, 35(6):2011-2016.

Slater J.F., Dibb J.E., Campbell J.W., Moore T.S. 2004, 'Physical and chemical properties of surface and column aerosols at a rural New England site during MODIS overpass', *Remote sensing of environment*, 92(2), 173-180.

SPOSRD (State Planning Organization Statistics and Research Department) 2015. *Statistical year book of 2012*, Turkish Republic of Northern Cyprus, Nicosia Mersin 10 Turkey.

Vautard R., Bessagnet B., Chin M. and Menut L. 2005, 'On the contribution of natural Aeolian sources to particulate matter concentrations in Europe: testing hypotheses with a modelling approach'. *Atmospheric Environment*, 39(18), 3291-3303.

Viana M., Kuhlbusch A.J., Querol X., Alastuey A., Harrison R.M., Hopke P.K. and Hueglin C. 2008, 'Source apportionment of particulate matter in Europe: a review of methods and results', *Journal of Aerosol Science*, 39(10), 827-849.

Wang T., Poon C.N., Kwok Y.H., Li Y.S. 2003. 'Characterizing the temporal variability and emission pattern of pollution plumes in the Pearl River Delta of China'. *Atmospheric Chemistry and Physics* 37, 3539-3550.

Wang Y.Q., Zhang X.Y. and Arimoto R. 2006, The contribution from distant dust sources to the atmospheric particulate matter loading at Xian, China during spring', *Science of the Total Environment*, 368, 875-883.

Wang Y.Q., Zhanq X.Y., Arimoto R., Cao J.J., and Shen Z.X. 2004, 'The transport pathways and sources of PM$_{10}$ pollution in Beijing during spring 2001, 2002 and 2003', *Geophys Res let*; 31:L14110.

WHO (World Health Organization) 2003, 'Wealth Aspect of Air Pollution with PM, Ozone and Nitrogen Dioxide', Report on *Working Group 2003*. Regional Office for Europe, Copenhagen.

WHO (World Health Organization) 2005, '*Air Quality Guidelines: Global Update (2005). Particulate matter, ozone, nitrogen dioxide and sulphur dioxide*', World Health Organization, 2006.

Appendix

		2012				
Date D/M	Nicosia (µg/m³)	Famagusta (µg/m)	Kyrenia (µg/m³)	Guzelyurt (µg/m³)	Alevkayasi (µg/m³)	Kalecik (µg/m³)
12/03	222.7	66.0	86.0	77.9	60.0	
13/03	167.1	81.5	107.6	103.2	62.4	
21/10	176.0	198.3	162.7			279.0
22/10	176.7	198.3	162.7			279.0

		2013				
Date D/M	Nicosia (µg/m³)	Famagusta (µg/m)	Kyrenia (µg/m³)	Guzelyurt (µg/m³)	Alevkayasi (µg/m³)	Kalecik (µg/m³)
18/01	222.3		87.6	44.1	126.2	154.7
19/01	80.0			126.2	62.3	75.0
23/02	132.1	95.2		106.8	89.8	116.5
24/02	77.7	78.8		107.2	44.4	89.5
11/03	138.4	90.1		138.4	101.1	119.9
12/03	108.8	60.8		130.2	236	76.4
13/03	276.8	216.6		286.4	118.6	229.5
1/04	275.8	126.7	267.2	153.8	272.6	149.4
2/02	112.2	88.7	125.9	239.6	94.6	87.6
9/04	228.8	129.5	133.8	119.5	110.0	101.5
31/05	141.3	88.7	132.8	110.9	77.0	99.3
31/10	128.3	64.4	78.8	72.6	53.2	58.5
01/11	136.2	89.9	75.8	86.9	53.7	63.0
02/11	114.3	69.0	74.5	75.1	57.0	74.1
03/11	70.1	83.1	61.0	64.8	56.5	79.9
04/11	104.3	73.9	76.3	54.6	51.0	60.5
05/11	102.7	84.9		57.6	38.4	52.9
06/11	82.7	77.2	109.3	62.3	52.1	71.2
07/11	77.3	102.3	91.1	71.1	48.0	84.1
08/11	91.7	78.8	105	62.2	60.6	84.8
09/11	99.9	79.3	61.9	80.2	30.6	51.2
10/11	57.6	84.2	47.4	53.3	32.9	51.7
05/11	102.7	84.9		57.6	38.4	52.9
06/11	82.7	77.2	109.3	62.3	52.1	71.2
07/11	77.3	102.3	91.1	71.1	48.0	84.1
08/11	91.7	78.8	105	62.2	60.6	84.8
09/11	99.9	79.3	61.9	80.2	30.6	51.2
10/11	57.6	84.2	47.4	53.3	32.9	51.7
11/11	102.3	62.5	84.1	55.9	47.5	56.7
12/11	116.7	58.3	87.6	77.4	44.1	63.6

		2014				
Date D/M	Nicosia (µg/m³)	Famagusta (µg/m)	Kyrenia (µg/m³)	Guzelyurt (µg/m³)	Alevkayasi (µg/m³)	Kalecik (µg/m³)
24/02	65.4	60.2	79.6	66.1	54.9	115.1
2/03	228.2	228.8	77.2	76.7	103.6	293.4
3/03	203	203.8	402.1	153.9	196.3	402.1
5/03		116.0	119.1	98.3	79.1	
6/06		97.3	108.5	95.0	89.3	
28/06	86.5	64.8	71.6	74.3	65.8	67.7

19

Understanding the atmospheric circulations that lead to high particulate matter concentrations on the west coast of Namibia

Hanlie Liebenberg-Enslin[1*], Hannes Rauntenbach[2,3], Reneé von Gruenewaldt[1], and Lucian Burger[1]

[1]Airshed Planning Professionals, Midrand, South Africa
[2]Climate Change and Variability, South African Weather Service, Pretoria, South Africa
[3]School of Health Systems and Public Health, University of Pretoria, South Africa

Abstract

Atmospheric circulations play a significant role in determining the extent and impact of local and regional air pollution. The Erongo Region, located in the western part of Namibia, falls within the west coast arid zone of southern Africa, and is characterised by low rainfall, extreme temperatures and unique climatic factors influencing the natural environment and biodiversity. Episodic dust storms, associated with easterly wind conditions, are common during austral autumn and winter months. During these events, dust is transported westwards over long distances across the Namibian continent towards the Atlantic Ocean. During 2017, such easterly wind conditions appeared to occur earlier and more frequently than in previous years. Of interest is that high PM_{10} concentrations (particulate matter with aerodynamic diameters of less than or equal to 10 micron) measured at the coastal towns of Swakopmund and Walvis Bay in the Erongo Region during 2017 were found to also coincide with south-westerly to north-westerly winds from the ocean during prevailing easterly wind events. In this study, the easterly wind events that occurred on 19 March 2017 and 6 July 2017 were assessed to investigate how local-scale coastal atmospheric circulation changes could have developed from the easterly wind conditions, and how such development could have contributed to wind direction deviations and the high PM_{10} concentrations measured at Swakopmund and Walvis Bay. It was found that in addition to the westward transport of PM_{10} from inland sources during easterly wind events, higher coastal concentrations of PM_{10} can also develop as a result of north-easterly / south-westerly wind conversion lines and the cyclonic circulation enhancement associated with easterly wind induced coastal troughs and coastal lows.

Keywords

Namibia, Erongo Region, air quality, particulate matter, wind patterns

Introduction

The Erongo Region is located in the western part of Namibia and is bounded by the Atlantic Ocean to the west and the continental escarpment to the east (approximately 180 km inland). From a hydrological perspective, the Erongo Region is drained in the central part by the deeply-incised Swakop and Khan Rivers, with the Kuiseb River separating the stony desert from the Namib sand dunes in the south (Tyson and Seely, 1980).

The Erongo Region falls within the west coast arid zone of southern Africa, and is characterised by low rainfall with extreme temperature ranges and unique climatic factors influencing the natural environment and biodiversity (Goudie, 2009). Episodic dust storms, associated with strong easterly wind conditions, are common during austral autumn and winter months. Associated dust is derived primarily from intermittent natural sources, giving rise to dust emissions only under conditions of high wind speeds. Windblown dust from natural sources is estimated to account between 75% (Ginoux et al., 2012) and 89% (Satheesh & Moorthy, 2005) of the global aerosol load, of which 25%

(Ginoux et al., 2012) to 50% (Tegen & Fung, 1995) is attributed to disturbed soil surfaces and the rest to natural soil surfaces. In Africa, approximately 54% of the dust is from desert and sparsely vegetated soils (Tegen & Fung, 1995). Anthropogenic sources account for 25% of global dust emissions (Ginoux et al., 2012). In the Erongo Region, anthropogenic sources of dust, such as unpaved roads, mining and exploration operations (primarily uranium prospecting), continuously contribute to atmospheric dust loads (Liebenberg-Enslin et al., 2010).

High concentrations of particulates in the air pose a risk to human health and welfare (Rashki et al., 2012; Rashki et al., 2013(a); Rashki et al., 2013(b)). Various studies have found a link between increased morbidity and mortality, especially amongst children and the elderly, and dust storm events (Ginoux et al., 2004; Karanasiou et al., 2012; De Longueville et al., 2013). In the Erongo region, radioactive dust associated with uranium mining and prospecting adds to the public concern (Liebenberg-Enslin et al., 2010).

Easterly wind events in the Erongo Region

During austral autumn and winter seasons, African continental anti-cyclonic circulation occasionally allows for easterly to north-easterly winds to descend along the downward slopes of the Namibian continent towards the Atlantic Ocean (Figure 1). This descend of air leads to a drop in air pressure as a result of vertical air column expansion, and the development of warm berg-wind conditions as a result of adiabatic heating. Although strong, hot and often uncomfortable for people, easterly wind conditions are usually relatively short lived[1].

During 2017, episodic dust storms associated with easterly wind events appeared to occur earlier and more frequently than in previous years. This heightened the public concern for high levels of radioactive dust from uranium prospecting and mining reaching the coastal towns of Swakopmund and Walvis Bay.

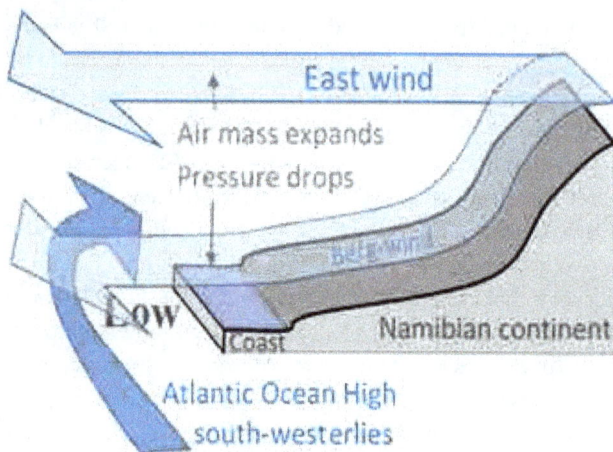

Figure 1: *Typical atmospheric flow characteristics during east wind conditions in Namibia. As berg-winds descends towards the coast, Mean Sea Level Pressures (MSLPs) drops to form either low pressure troughs or cut-off lows along the coastline. Because of the Coriolis force, such lows are associated with cyclonic (clockwise) circulation forcing.*

Although it is expected that fine particulates get lifted and carried across the interior of the Erongo Region towards the coast by easterly winds, easterly wind episodes that were monitored in 2017 indicated that high PM_{10} (Particulate Matter (PM) with an aerodynamic diameter of less or equal to 10 micron) concentrations recorded at Swakomund and Walvis Bay were also caused by PM_{10} that approached the towns from other directions than the prevailing easterly flow, which included onshore flow from south-westerly to north-westerly winds.

Of particular interest is that the high PM_{10} concentrations that were measured at the coastal towns of Swakopmund and Walvis Bay in the Erongo Region between 18 and 20 March 2017 were found to also coincided with south-westerly to north-westerly orientated winds during the prevailing easterly wind event, while wind direction deviations also occurred during the high PM_{10} concentration easterly wind event of 6 July 2017. While easterly wind conditions can explain the westward transport of PM_{10} from inland sources towards the Namibian coast, it is not clear why high PM_{10} concentrations were measured at the same

time under conditions of local-scale south-westerly to north-westerly wind directions, indicative of onshore flow from the Atlantic Ocean.

The purpose of this paper is to investigate how the local-scale atmospheric circulation deviations (south-westerly to north-westerly winds, in contrast to the large-scale prevailing easterly winds) observed on 19 March 2017, as well as 6 July 2017, could have developed from easterly wind conditions, and how these could have contributed to the high PM_{10} concentrations measured at Swakopmund and Walvis Bay.

Ambient air quality monitoring in the Erongo Region

Monitoring stations and parameters recorded

As part of the Strategic Environmental Management Plan (SEMP) initiated by the Namibian Ministry of Mines and Energy, an ambient monitoring network was established in the Erongo Region at the end of 2016 with the objective to measure PM_{10}, $PM_{2.5}$ and Radon concentrations. In addition, meteorological variables such as wind direction and speed, temperature, Relative Humidity (RH), solar radiation, barometric pressure and rainfall are also recorded at selected stations (Table 1). Monitoring locations were chosen based on the most populated areas in the region (i.e. towns, except for Jakalswater which serves as a background station). The towns identified were Swakopmund, Walvis Bay, Arandis and Henties Bay (Figure 2).

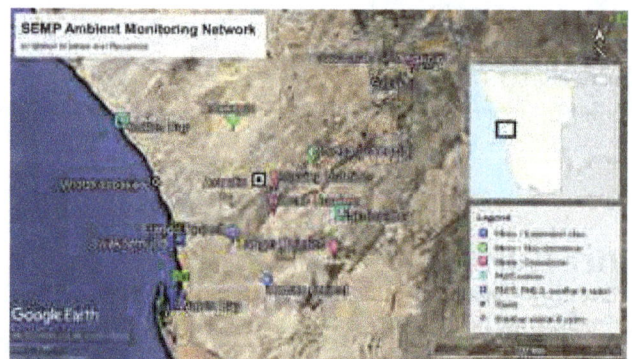

Figure 2: *Location of ambient air quality monitoring stations, meteorological stations and radon stations in relation to the mines and sensitive receptors.*

Ambient PM_{10} concentrations at Swakopmund and Walvis Bay are measured using Met One Instruments Model BAM 1020, designation for continuous PM monitoring. E-Samplers, a light-scatter Aerosol Monitor, are used to record PM_{10} concentrations at Henties Bay and Jakalswater and for $PM_{2.5}$ monitoring at Swakopmund and Walvis Bay. The E-Sampler at Jakalswater is fitted with a Met One Instrument measuring wind speed, wind direction and temperature. Met One weather stations at Swakopmund, Walvis Bay and Arandis are fitted with a wind speed sensor, wind direction vane, ambient air temperature

[1] http://www.raison.com.na

sensor, RH sensor, precipitation tipping bucket, as well as atmospheric pressure and solar radiation sensors. The Swakopmund and Walvis Bay stations are enclosed, while it should be noted that the Walvis Bay station is on top of the Walvis Bay Civic Centre (approximately 9 m above ground level), whereas the Swakopmund Station is on top of a 3m structure at the Swakopmund waste water works.

Table 1: Monitoring stations and parameters recorded

| Monitoring Location | Pollutant/ Parameter Measured | | | | | | | | | |
	PM_{10}	$PM_{2.5}$	Wind Speed	Wind Direction	Temperature	Relative Humidity	Solar Radiation	Barometric Pressure	Rainfall	Radon
Swakop-mund	X	X	X	X	X	X	X	X	X	X
Walvis Bay	X	X	X	X	X	X		X		X
Arandis			X	X	X	X	X	X	X	X
Henties Bay	X				X	X		X		
Jakalswa-ter	X		X	X	X	X		X		

Prevailing wind fields in the Erongo Region

The wind fields of the Erongo Region are influenced by a combination of synoptic and local scale circulations. Wind directions in the central-northern parts of the region are predominantly from the east, northeast and southwest, where the easterly and north-easterly winds (eastern wind conditions) are often being associated with high wind speeds. Along the coast, southerly to south-westerly wind directions are modulated by the Atlantic Ocean anti-cyclonic circulation as well as temperature and pressure gradients that are orientated parallel to the coastline between the upwelling Benguela ocean-current and the warm arid continent. During the period between November 2016 and April 2017, wind speeds in the region were found to be mostly between 0 m.s^{-1} to 10 m.s^{-1}. Inland, at Arandis, prevailing west-south-westerly winds dominated, while at Jakalswater, the prevailing winds were east-north-easterly (Figure 2). Easterly winds, associated with berg-wind conditions, were recorded for 22% of the period between November 2016 and April 2017 at Jakalswater, 16% at Arandis, 9% at Swakopmund and 10% at Walvis Bay. Easterly wind conditions were most prevalent during the month of March 2017, occurring for 32% of the time at Arandis and 41% at Jakalswater. At the coast, easterly flow was recorded at 9% and 10% of the time at Swakopmund and Walvis Bay, respectively.

During east wind conditions, high wind speeds of up to 22 m.s^{-1} were recorded at the Namibian coast. These strong winds are occasionally also being associated with south-westerly to north-westerly wind directions during east wind conditions at the towns of Swakopmund and Walvis Bay.

Case studies of easterly winds and high PM_{10} concentrations

Easterly wind event on 6 July 2017

Easterly berg-wind conditions prevailed across the Erongo Region on 6 July 2017. At Swakopmund wind directions were easterly from 00:00 to 09:00, from where it changed to north-easterly between 09:00 to 14:00 (Figure 4a). A significant change in wind direction appeared at 16:00 when a north-easterly wind developed. Shortly after this, the wind direction gradually returned to easterly winds in the later afternoon towards 24:00. On the same day, PM_{10} concentrations were mostly below 50 µg.m^{-3}. However, during the time interval when the winds turned to north-easterly (09:00-14:00), a significant increase in PM_{10} with a maximum of 312 µg.m^{-3} at 10:00 developed, most probably from sources to the north-east of Swakopmund. It is interesting to note that the PM_{10} concentration peak of 312 µg.m3 that occurred at Swakopmund is significantly higher than the six-month average (data for the period November 2016 to April 2017) which ranged between 30 µg.m^{-3} (Swakopmund) and 35 µg.m^{-3} (Walvis Bay).

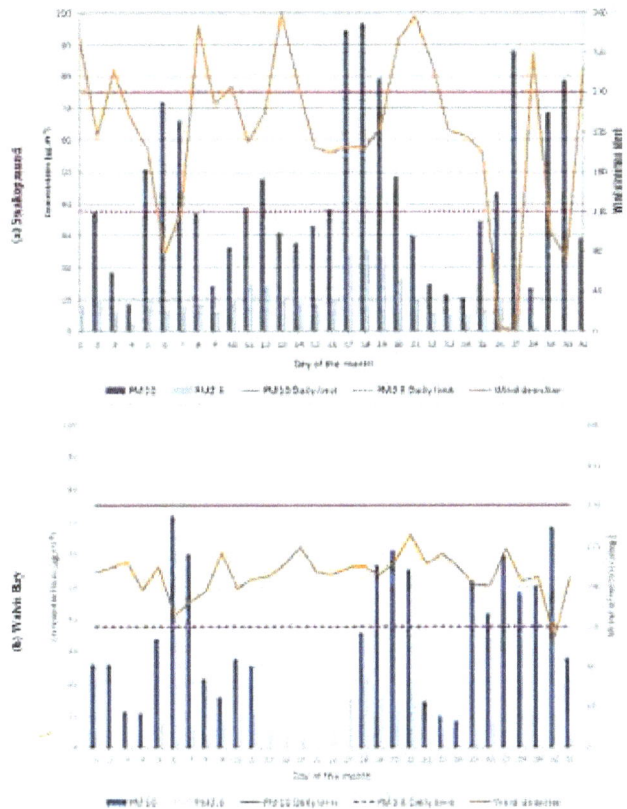

Figure 3: Daily PM_{10} and $PM_{2.5}$ concentrations at (a) Swakopmund and (b) Walvis Bay with average daily wind directions indicated for the month of July 2017.

At Walvis bay (Figure 4b), the wind direction pattern was very similar to what was recorded at Swakopmund, where the significant wind direction change at 16:00 was also captured, while PM_{10} concentrations were mostly around 10 $\mu g.m^{-3}$, but also slightly increased to 132 $\mu g.m^{-3}$ with the change in wind direction towards north-east when the concentration peak was recorded at Swakopmund.

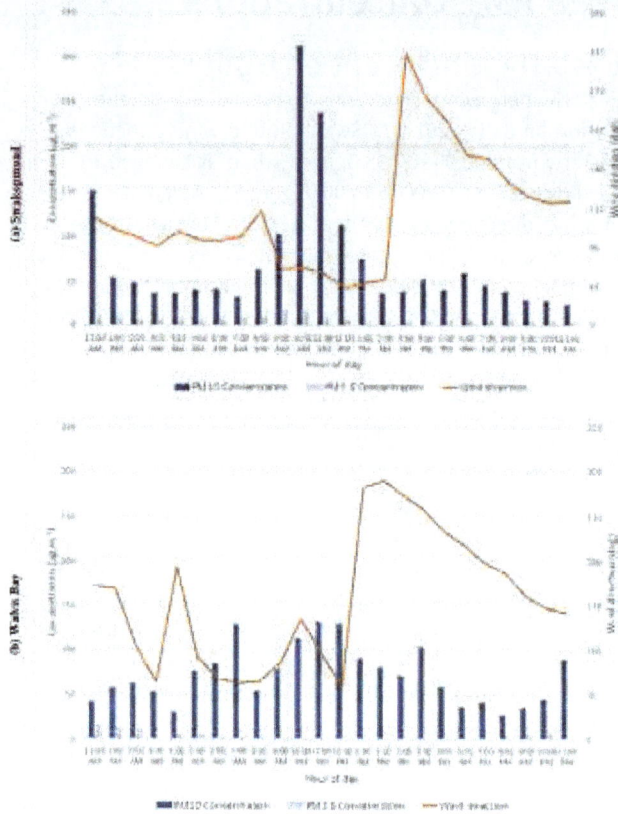

Figure 4: *Hourly PM_{10} and $PM_{2.5}$ concentrations at (a) Swakopmund and (b) Walvis Bay with average hourly wind directions indicated for 6 July 2017.*

Daily PM_{10} concentrations for the month July 2017 are presented as polar plots[2] in Figure 5. Polar plots represent the concentration in relation to the wind direction and wind speed from where it originated. At Swakopmund, high PM_{10} concentrations were recorded during higher wind speeds episodes (>8 $m.s^{-1}$) from the east-north-east, and during lower wind speeds (4-6 $m.s^{-1}$) from the east-south-east. Similarly, high PM_{10} concentrations at Walvis Bay occurred during strong east-north-easterly winds (>6 $m.s^{-1}$) with lower PM_{10} concentrations under higher wind speeds from the south and south-south-west. It therefore appears as if PM_{10} sources for Swakopmund are located to the north-west to south-south-west of the monitoring station, whereas for Walvis Bay sources might be located to the south-south-west to south-east of the station.

Polar plots for hourly PM_{10} concentrations on 6 July 2017 reflect the highest PM_{10} concentrations during strong east-north-easterly winds at both Swakopmund and Walvis Bay (Figure 6). Swakopmund recorded higher PM_{10} concentrations under

these higher wind speeds than at Walvis Bay (highest hourly PM_{10} concentration of 312 $\mu g.m^{-3}$ and maximum wind speed of 18.2 $m.s^{-1}$ at Swakopmund compared to the highest PM_{10} concentration of 132 $\mu g.m^{-3}$ and maximum wind speed of 7.6 $m.s^{-1}$ at Walvis Bay). Lower PM_{10} contributions were associated with south and south-westerly winds at the Swakopmund station, and south-west to north-westerly winds at the Walvis Bay station.

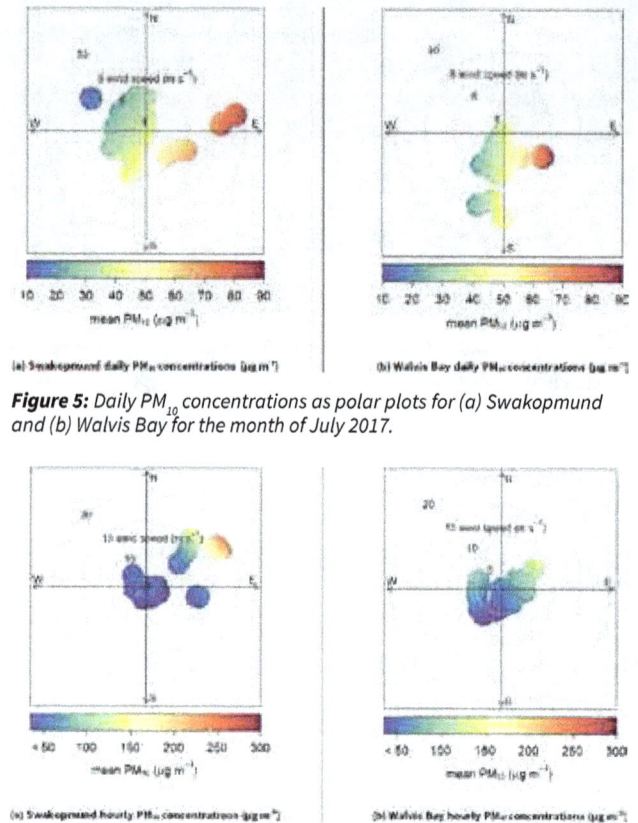

Figure 5: *Daily PM_{10} concentrations as polar plots for (a) Swakopmund and (b) Walvis Bay for the month of July 2017.*

Figure 6: *Hourly PM_{10} concentrations as polar plots for (a) Swakopmund and (b) Walvis Bay for 6 July 2017.*

European Reanalysis Interim (ERA-Interim) data (Dee at al., 2011) were downloaded to produce wind and sea-level pressure maps for 00:00, 06:00, 12:00 and 18:00 Greenwich Mean Time (GMT) on 6 July 2017 (Figures 7a, 7b, 7c and 7d). According to these maps, north-easterly winds were observed between the Namibian escarpment and coastline, with the Botswana anti-cyclone well-established to the east of the country.

As the east wind descended from the Namibian interior towards the coast, the vertical air mass expanded and air pressures decreased to form higher pressures along the escarpment to lower pressures along the coast (illustrated in Figure 1 and observed in Figures 7a, 7b, 7c and 7d). As a result, a trough associated with lower pressures developed along the Namibian coast line.

In this trough, Mean Sea Level Pressure (MSLP) values at Swakopmund and Walvis Bay were in the order of 1019 hPa at

---

[2] R package for air quality data analysis (Carslaw & Ropkins, 2012) and the Openair version 0.8 0 (Carslaw, 2013).

00:00 GMT. Pressures increased towards 06:00 to values of 1020 hPa. I noticeable drop in MSLP to 1017 hPa occurred between 06:00 GMT and 12:00 GMT, from where MSLPs stabilised towards 18:00 GMT at 1017 hPa.

While north-easterly winds prevailed at the two towns at 00:00 GMT, 06:00 GMT and 18:00 GMT, the deepening of the trough reflected for 12:00 GMT (Figure 7c) resulted in cyclonic (clockwise) circulation enhancement (north-eastern to the east of the trough and south-western to the west of the trough), which could explain the noticeable change in wind direction towards a north-westerly direction at 16:00 – as reflected in the weather station data (Figures 4a and 4b). The cyclonic circulation enhancement that appeared on 6 July 2017, however, did not result in exceptionally high PM_{10} concentrations. The high concentrations that were observed earlier in the day may rather be attributed to an upwind source towards the northeast.

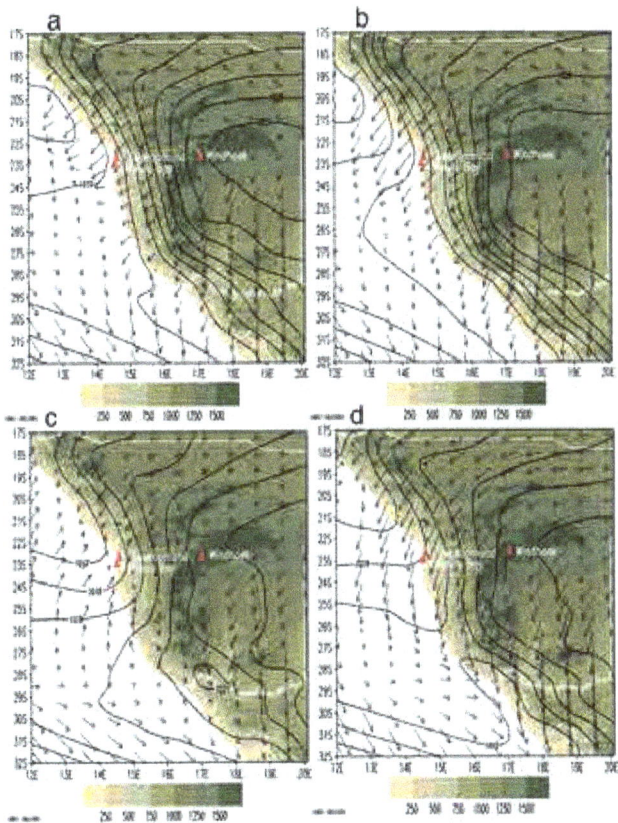

Figure 7: Map of Mean Sea Level Pressure (MSLP) (contours) and 10m wind vectors at 00:00, 06:00, 12:00 and 18:00 Greenwich Mean Time: GMT (a, b, c, d, respectively) on 6 July 2017. Topography (meters above mean sea level) is shaded, while the positions of Windhoek, Swakopmund and Walvis Bay are indicated by red triangles.

Easterly wind event on 19 March 2017

Easterly wind conditions were prevalent during March 2017, occurring for 32% of the time at inland stations of Arandis and 41% at Jakalswater. At the coast, easterly flow was recorded for 9% at Swakopmund and 10% at Walvis Bay. High PM_{10} concentrations were recorded on 18, 19, 20, 24, 25, 30 and 31 March 2017 at Swakopmund and Walvis Bay (Figure 8). On these days, easterly wind conditions prevailed inland at Jakalswater

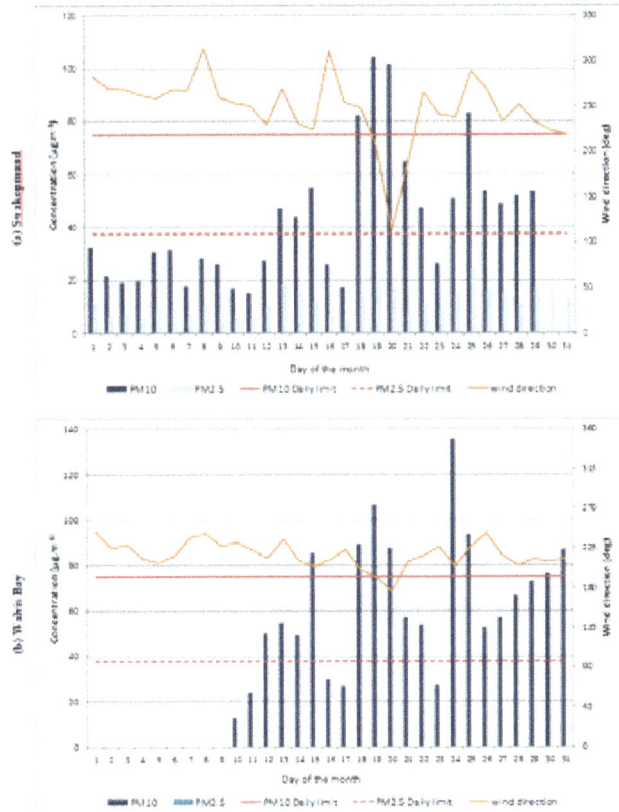

Figure 8: Daily PM_{10} and $PM_{2.5}$ concentrations for (a)Swakopmund and (b) Walvis Bay with average daily wind direction indicated for the month of March 2017.

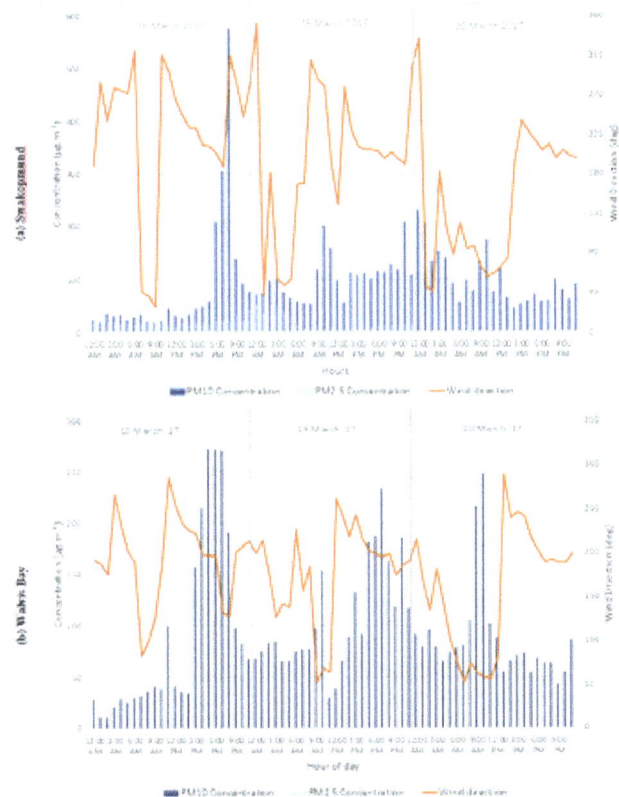

Figure 9: Hourly PM_{10} and $PM_{2.5}$ concentrations for (a) Swakopmund and (b) Walvis Bay with average hourly wind direction indicated for 19 and 20 March 2017.

and Arandis, but at Swakopmund and Walvis Bay high hourly PM$_{10}$ concentrations recorded between the late hours of 18 March and the early hours of 20 March were mainly associated with southerly to south-westerly winds. The highest and third highest hourly PM$_{10}$ concentrations recorded at Swakopmund of 575 µg.m^{-3} (20:00 on 18 March 2017) and 232 µg.m^{-3} (12:00 on 20 March 2017) were, however, recorded when the wind was from the north-west (Figure 9).

During the early hours of 19 March 2017 (the focus period for this study), wind directions at Swakopmund returned from north-west to easterly winds (Figure 9a). At 08:00 the wind direction turned to north-west in a very short time period, and maintained this direction until 13:00, from where south-westerly winds developed during most of the afternoon. Late at night, just before 12:00, the wind direction again changed to north-east over a relative short period. At Walvis Bay (Figure 9b), the changes in wind direction was not as extreme as observed at Swakopmund. Nevertheless, north-westerly winds with high PM$_{10}$ concentrations developed from just before 12:00. What is interesting is that the north-westerly winds were all associated with higher PM$_{10}$ concentrations at both Swakopmund and Walvis Bay.

Daily PM$_{10}$ concentrations, presented as polar plots for the month of March 2017, indicate windblown dust under high wind speeds of 8 m.s^{-1} (Swakopmund) and 8 to 10 m.s^{-1} (Walvis Bay). At Swakopmund, these concentrations were recorded during east-south-easterly and south to south-south-westerly winds, and south-south-westerly winds at Walvis Bay (Figure 10). The highest hourly concentrations at Walvis Bay were mainly associated with south-south-westerly winds (Figure 10). On average, higher hourly PM$_{10}$ concentrations and higher wind speeds were recorded at Swakopmund compared to Walvis Bay during 18 to 20 March 2017. From the highest PM$_{10}$ concentration measurements were made during onshore south-west to north-west winds (from the Atlantic Ocean) at both Swakopmund and Walvis Bay.

Highest hourly PM$_{10}$ concentrations for the period 18 to 20 March 2017, depicted as polar plots in Figure 11, show a combination of high PM$_{10}$ concentrations associated with east-north-easterly to south-south-westerly winds at both Swakopmund and Walvis Bay. Although the expected easterly wind signature is present, higher PM$_{10}$ concentrations resulted from winds from the Atlantic Ocean. The highest hourly PM$_{10}$ concentration (575 µg.m^{-3}) recorded at Swakopmund over these three days was under moderate winds (5 m.s^{-1}) from the north-west. The second highest PM$_{10}$ concentration (307 µg.m^{-3}) was during south-south-westerly winds at 6.4 m.s^{-1}. At Walvis Bay, the highest hourly PM$_{10}$ concentration recorded was 272 µg.m^{-3} over two (2) hours when the wind was from the south-south-west blowing at moderate velocities (9.4 and 7.6 m.s^{-1}).

European Reanalysis Interim (ERA-Interim) data (Dee at al., 2011) were downloaded to produce wind and sea-level pressure maps for 00:00, 06:00, 12:00 and 18:00 Greenwich Mean Time (GMT) on 19 March 2017 (Figures 12a, 12b, 12c and 12d). As

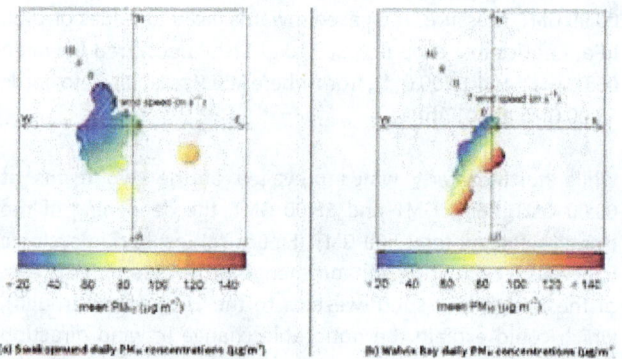

Figure 10: Daily PM$_{10}$ concentrations as polar plots for (a) Swakopmund and (b) Walvis Bay for the month of March 2017.

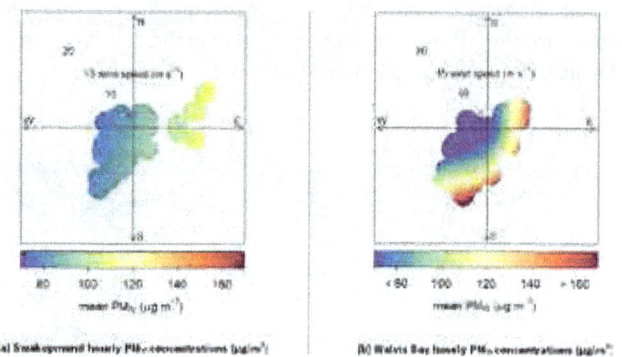

Figure 11: Hourly PM$_{10}$ concentrations as polar plots for (a) Swakopmund and (b) Walvis Bay for 18 to 20 March 2017.

the easterlies blows towards the Namibian coastline from the higher escarpment, air pressures dropped, as indicated by isobars that are almost parallel to the coastline in the area west of the high escarpment, which had led to the development of a coastline trough. In general, north-easterly conditions were observed in the Swakopmund / Walvis Bay area at 00:00 GMT. With these north-easterlies, strong Atlantic Ocean anti-cyclonic south-westerly winds appeared to allow for a north-easterly / south-westerly wind conversion line to develop along the coast of Namibia. A coastal-low feature which is an extension of the coastal through is also visible at 00:00 GMT (Figure 12a). Signs of cyclonic circulation is visible around the coastal-low, which developed even further towards 06:00 GMT as MSLPs dropped from 1013 hPa to 1011 hPa (Figure 12b). Wind directions also changed from northeast to southwest between 00:00 GMT and 06:00 GMT, which is also indicative of cyclonic circulation enhancement, which form part of the north-easterly / south-westerly wind conversion line. At 12:00 GMT, north-westerly winds develop with a significant increase in PM$_{10}$ concentrations, especially at Swakopmund (Figure 12c).

The change in wind direction from northeast to southwest to northwest can be interpreted as cyclonic circulation along the coast of Swakopmund and Walvis Bay, associated with the clockwise recirculation of PM$_{10}$. At the same time, the north-westerly / south-easterly wind conversion line would have prevented PM$_{10}$ to be dispersed westwards towards the deeper Atlantic Ocean, which could also have contributed to the accumulation of PM$_{10}$ along the coast that was recirculated

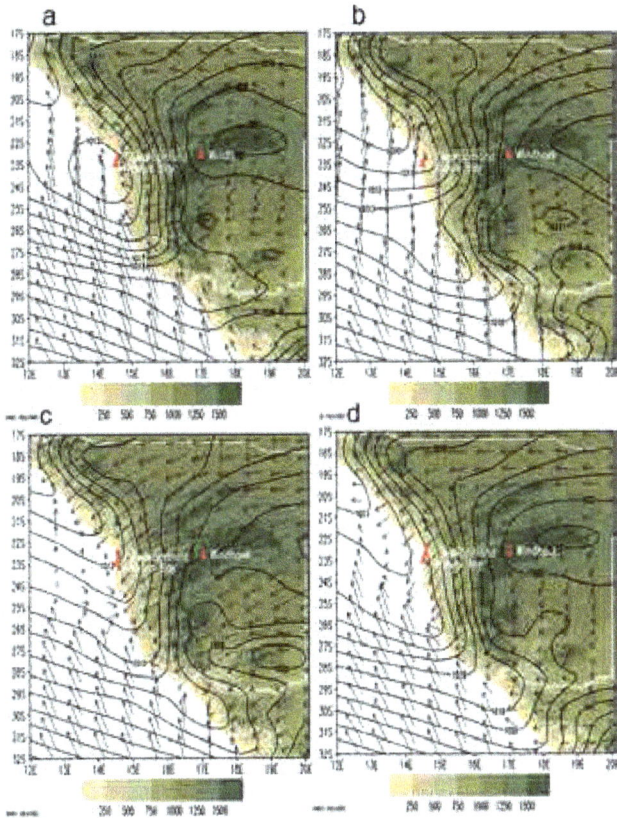

Figure 12: *Map of Mean Sea Level Pressure (MSLP) (contours) and 10m wind vectors at 00:00, 06:00, 12:00 and 18:00 Greenwich Mean Time: GMT (a, b, c, d, respectively) on 19 March 2017. Topography (meters above mean sea level) is shaded, while the positions of Windhoek, Swakopmund and Walvis Bay are indicated by red triangles.*

back towards the continent by north-westerly winds to generate conditions of exceptionally high PM_{10} concentrations, as indicated by the 12:00 GMT map (Figure 12c) as well as the high PM_{10} concentrations measured at Swakopmund (and Walvis Bay) (Figure 12a) between 08:00 and 13:00 on 19 March 2017.

Characterisation of likely dust sources

Filter tape from the PM_{10} BAM 1020 samplers at Swakopmund and Walvis Bay were analysed using Inductively Coupled Plasma - Mass Spectrometry (ICP-MS) for elemental content (43 elements) in an attempt to differentiate between the main sources contributing of high hourly PM_{10} concentrations at Swakopmund and Walvis Bay. Both towns border the Atlantic Ocean to the west and the desert gravel plains to the east, with the Kuiseb River mount and Namib Desert to the south of Walvis Bay. The main anthropogenic sources within the gravel plains of the Erongo Region are unpaved roads and mining and exploration activities (mainly uranium prospecting). Particulate matter from these areas reaching the coastal towns are wind dependent (windblown dust) whereas wind independent sources closer to the monitoring stations are likely to the associated with vehicle emissions, small boilers and hospital incinerators, and harbour activities at Walvis Bay.

ICP analysis was done for 43 elements on samples representing hours with high PM_{10} concentrations during the two case studies

– 6 July 2017 and 18 to 20 March 2017. Specific hours were selected reflecting the predominant wind direction during each case study as to eliminate dust from other directions during these periods.

For the 6th of July 2017 case study, 22 hours were selected from the Swakopmund station reflecting winds from the east-north-east to south-east (01:00 to 14:00 on 6 July and 04:00 to 11:00 on 7 July 2017). The main elements are presented in Table 2 and Figure 13. No filter tape was available from Walvis Bay for this period.

Table 2: *Elemental composition as percentages of the various samples*

Sample ID	Swakopmund		Swakopmund		Walvis Bay	
	06 July 2017		18-20 March 2018		18-20 March 2018	
Element	(mg/sample)	(%)	(mg/sample)	(%)	(mg/sample)	(%)
Arsenic, As	0.0007	1%	0.0010	1%	0.0000	0%
Iron, Fe	0.0122	26%	0.0122	16%	0.0177	38%
Magnesium, Mg	0.0012	3%	0.0074	10%	0.0083	18%
Manganese, Mn	0.0002	0%	0.0002	0%	0.0006	1%
Phosphorus, P	0.0010	2%	0.0019	3%	0.0000	0%
Selenium, Se	0.0001	0%	0.0004	1%	0.0002	0%
Sulphur, S	0.0165	35%	0.0524	69%	0.0195	41%
Titanium, Ti	0.0004	1%	0.0007	1%	0.0009	2%
Vanadium, V	0.0000	0%	0.0001	0%	0.0000	0%
	0.0322		0.0765		0.0471	

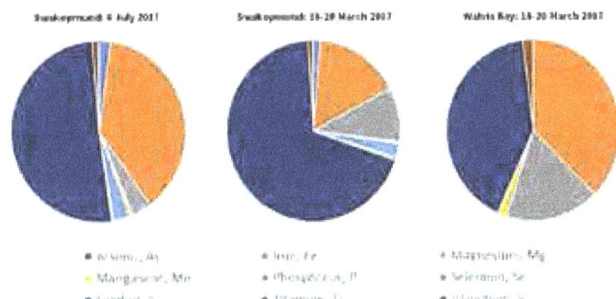

Figure 13: *Elemental composition of the samples from the case studies.*

Samples covering 25-hours from the Swakopmund station (06:00 to 23:00 on 18 March; 09:00 to 23:00 on 19 March and 00:00 to 04:00 on 20 March 2017) were analysed reflecting predominantly south-south-westerly to north-westerly winds during the period 18 to 20 March 2017. At Walvis Bay 15 hours (15:00 to 20:00 on 18 March and 15:00 to 23:00 on 19 March 2017) were selected reflecting similar wind directions. The main elements are presented in Table 2 and Figure 13.

The main elements from all three sample batches were arsenic (As), iron (Fe), magnesium (Mg), manganese (Mn), phosphorus (P), selenium (Se), sulphur (S), titanium (Ti) and vanadium (V). During the July east winds, sulphur (51%) and iron (38%) were more pronounced in the samples from Swakopmund, with lower fractions of 4% Mg and 3% P, respectively. For the two sites during the March case study, S and Fe remained the main elements (69% S from Swakopmund and 41% from Walvis Bay, and 16% Fe from Swakopmund and 38% from Walvis Bay). The Mg fraction was higher during these predominantly south-south-westerly to north-westerly winds (10% at Swakopmund and 18% at Walvis Bay). Elements associated with sea water are predominantly sodium (Na) and chloride (Cl), followed by Mg, V, S, calcium (Ca) and potassium (K) (https://en.wikipedia.org/wiki/Seawater). Elements associated with desert dust are mainly Ca, silica (Si), aluminium (Al), and some traces of Na and Mg (Rashki et al., 2012).

No clear distinction could be made in the chemical composition of the three sample batches from the two case studies where high PM_{10} concentrations were recorded. The expectation would have been that under predominant easterly winds (6th of July 2017 case study), the elements mostly associated with desert dust would dominate; whilst under prevailing south-westerly to north-westerly winds (between 18 and 20 March 2017), elements associated with the ocean would alternatively have dominated. Instead, the similar elemental composition of the PM_{10} samples indicates sources of possibly similar origin. All three samples show strong sea water signatures with some traces of desert dust. This is supported by the cyclonic circulation along the coast of Swakopmund and Walvis Bay, associated with the clockwise recirculation of PM_{10} back towards the continent by north-westerly winds.

Conclusions and recommendations

Two case studies, one between 18 and 20 March 2017 and the other on 6 July 2017, were considered to assess the atmospheric circulation patterns associated with the easterly wind conditions resulting in high PM_{10} concentrations at Swakopmund and Walvis Bay along west coast of Namibia. It was found that although easterly wind conditions prevailed over the Namibian continent, extensive temporal wind direction changes were observed at the Swakopmund and Walvis Bay coastline. Through further investigation it was concluded that in addition to the westward transport of PM_{10} from inland sources during easterly wind events, higher coastal concentrations of PM_{10} can also develop as a result north-easterly / south-westerly wind conversion lines as well as cyclonic circulation or cyclonic circulation enhancement, with the recirculation of PM_{10}, associated with coastal troughs and coastal lows. The study found a relationship between the (1) coastward decrease in continental air pressures due to easterly wind conditions, (2) the resultant development of coastal troughs (and sometimes coastal-lows), (3) the consequent development of conversion lines, (4) extensive changes in wind speed and direction, and (5) the high concentrations of PM_{10} concentrations (both as a result wind conversion line blocking and cyclonic recirculation) at Swakopmund and Walvis Bay. This is further confirmed by the similarity in elemental content of dust sources under various prevailing wind conditions.

However, these findings need further investigation, especially on the meso-scale, which will not only provide lead to an improved understanding of local wind behaviour, but also local PM_{10} variation characteristics along the coastline of Namibia and the main contributing sources.

Acknowledgements
The authors would like to acknowledge the Ministry of Mines and Energy, Geological Survey of Namibia and the Federal Institute for Geosciences and Natural Resources for making the data available for use in this study. The contribution of the South African Weather Service (SAWS) in providing advice on the ERA-Interim synoptic maps.

References
Carslaw, D. C. & Ropkins, K., 2012. Openair — an R package for air quality data analysis. Environmental Modelling & Software, 27 28, 52 61.

Carslaw, D. C., 2013. The openair manual — open source tools for analysing air pollution data. Manual for version 0.8 0, 14 February 2013, London: King's College CEPA/FPAC Working Group, 1998Colls, 2002.

De Longueville, F., Ozer, P., Doumbia, S. and Henry, S., 2013. Desert dust impacts on human health: an alarming worldwide reality and a need for studies in West Africa. International Journal of Biometeorology, 57, 1-19.

Dee, D. P., Uppala, S. M., Simmons, A. J., Berrisford, P., Poli, P., Kobayashi, S., Andrae, U., Balmaseda, M. A., Balsamo, G., Bauer, P., Bechtold, P., Beljaars, A. C. M., van der Berg, L., Bidlot, J., Bormann, N., Delsol, C., Dragani, R., Fuentes, M., Geer, A. J., Haimberger, L., Healy, S. B., Hersbach, H., Hólm, E. V., Isaksen, L., Kållberg, P., Köhler, M., Matricardi, M., McNally, A. P., Monge-Sanz, B. M., Morcrette, J.-J., Park, B.-K., Peubey, C., de Rosnay, P., Tavolato, C., Thépaut, J.-N. and Vitart, F. (2011), The ERA-Interim reanalysis: configuration and performance of the data assimilation system. Q.J.R. Meteorol. Soc., 137: 553–597. doi: 10.1002/qj.828

Goudie, A. S., 2009. Dust storms: Recent developments. Journal of Environmental Management, 90, 89–94.

Ginoux, P., Prospero, J. M., Torres, O. and Chin, M., 2004. Long term simulation of global dust distribution with the GOCART model: correlation with North Atlantic Oscillation. Environmental Modelling and Software, 19, 113 128.

Ginoux, P., Prospero, J. M., Gill, T. E., Hsu, N. C., and Zhao, M., 2012. Global-scale attribution of anthropogenic and natural dust sources and their emission rates based on MODIS Deep Blue aerosol products. Reviews of Geophysics, 50, RG3005, doi:10.1029/2012RG000388.

Karanasiou, A., Moreno, N., Moreno, T., Viana, M., de Leeuw, F., Querol, X., 2012. Health effects from Sahara dust episodes in Europe: literature review and research gaps. Environment International, 15, 107-114.

Liebenberg Enslin, H., Krause, N. & Breitenbach, N., 2010. Strategic Environmental Assessment for the Central Namib 'Uranium Rush Air Quality Specialist Report. Report APP/09/MME 02 Rev0 to the Ministry of Mines and Energy, Namibia.

 MME, 2017. Advanced Air Quality Management for the Strategic Environmental Management Plan for the Uranium and Other Industries in the Erongo Region: Ambient Air Quality Monitoring Report for the Period 1 November 2016 to 30 April 2017. Done on behalf of the Namibian Ministry of Mines and Energy. Report No. 16MME01-2.

Rashki, A., Eriksson, P.G., Rautenbach, C.J.deW., Kaskaoutis, D.G., Grote, W. and Dykstra, J., 2012. Assessment of chemical and mineralogical characteristics of airborne dust in the Sistan region, Iran. Chemosphere, 90, 227-236.

Rashki, A., Rautenbach, C.J.deW., Eriksson, P.G., Kaskaoutis, D.G. and Gupta, R., 2013(a). Temporal changes of particulate concentration in the ambient air over the city of Zahedan, Iran. Air Quality, Atmosphere & Health, 6, DOI 10.1007/s11869-011-0152-5, 123-135.

Rashki, A., Kaskaoutis, D.G., Eriksson, P.G., Rautenbach, C.J.deW, Flamant, C. and Abdi Vishkaee, F., 2013(b) Spatio-temporal variability of dust aerosols over the Sistan region in Iran based on satellite observations. Natural Hazards and Earth System Sciences, Volume 71, Issue 1, pp 563–585. DOI 10.1007/s11069-013-0927-0.

Satheesh, S. K. & Moorthy, K. K., 2005. Radiative effects of natural aerosols: a review. Atmospheric Environment, 39, 2089–110.

SAWS, 2017. Derived from Daily Synoptic Maps of the South African Weather Service (SAWS) http://www.weathersa.co.za/observations/synoptic-charts.

Tegen, I. & Fung, I., 1995. Contribution to the atmospheric mineral aerosol load from land surface modification. Journal of Geophysical Research, 100, 18707 18726.

Tyson, P.D., and Seely, M.K., 1980: Local winds over the Central Namib. The South African Geographical Journal, Vol. 62, No. 2, pp 135-150. September 1980.

20

Design considerations for a continuous emission measurement system for pressure type bag houses

Pierru Roberts[1], Luther Els[1] and Gerrit Kornelius[1,2]

[1] Resonant Environmental Technologies, Centurion, 0046, pierru@resonant.co.za, luther@resonant.co,za
[2] Department of Chemical Engineering, University of Pretoria, Pretoria, 0002, gerrit.kornelius@up.ac.za

Abstract

Measurement of the outlet particulate concentration on pressure-type bag filters, whether intermittent or continuous, has been avoided in South Africa as cumbersome and possibly inaccurate. A system based on the requirements of the US EPA's method 5D was however recently designed and installed on three pressure-type reverse air filters serving ferro-chrome electric arc furnaces in order to allow comparison with the legal emission concentration limits for this type of furnace. This paper reports on the design considerations and design process of this system.

Keywords

continuous sampling, pressure bag house, particulate matter

Introduction

Fabric filters are one of the most widely used devices for controlling emissions of particulate matter (PM). A fabric filter system typically consists of multiple filter elements, or bags, enclosed in a compartment, or housing. The process stream enters the housing and passes through the filter elements. Particulate matter (PM) accumulates as a dust cake on the surface of the bag. This dust layer effectively becomes the filtration medium. The filter elements are cleaned periodically to remove the collected dust (Cooper and Alley 2012).

Fabric filters generally are classified by cleaning method. The four types of cleaning method are reverse-air, shaker, pulse-jet, and sonic cleaning. Fabric filters can also be classified as either positive- or negative-pressure designs, depending upon the location of the fan(s) that provides the motive force for the exhaust stream through the unit. The fan is located upstream of the filter in a positive-pressure (forced-draft) unit, and downstream of the filter housing in a negative-pressure (induced-draft) unit.

Upstream location of the fan means that no duct work or exhaust stack is required on the outlet or downstream side. In fact, the gas after cleaning often exits the filter installation through openings similar to those used for roof ventilation in industrial buildings (called mono-vents). The customary measuring techniques for PM emissions cannot be used in this case, as there is no suitable measuring location that conforms to the flow conditions as required by the frequently used methods EPA 5A or ISO 9096, which are also the methods allowed by Schedule A of the s21 regulations under the South African Air Quality Act (Department of Environmental Affairs 2010, 2013).

Application description

Middelburg Ferrochrome (MFC), a subsidiary of Samancor Chrome, has three pressure-type reverse air bag filter installations serving their electric arc furnaces. MFC has taken the positive step of specifying a continuous emission sampling system. This enables them to not only report an annual emission result but also to continuously monitor their emissions in order to reduce emissions and the overall environmental impact.

Emission monitoring requirements

Fabric filters are capable of extremely high control efficiencies of both coarse and fine particles; outlet concentrations as low as 20-30 mg/Nm³ with conventional bags and emissions as low as 10 mg/Nm³ with membrane bags can be achieved. (Roberts et al 2013).

The South African emission limits as set out by the Department of Environmental Affairs (2010, 2013) in terms of regulations published under s21 of the National Environmental Management – Air Quality Act for an open furnace producing ferrochrome are given in Table 1.

Table 1: Emission limits - Ferro-alloy production

Common name	Plant status	mg/Nm³ *
Sulphur dioxide	New	500
	Existing	500
Oxides of Nitrogen	New	400
	Existing	750
Particular matter	New	30
	Existing	100

* Under normal conditions of 273 K and 101.3 kPa

Continuous emission monitoring (CEM) for particulate concentration is not currently a national requirement for ferro-alloy manufacturing under the above regulations. The local licensing authority has however specified continuous emission monitoring as a condition in the Atmospheric Emission License (AEL) issued to this specific installation.

CEM is gaining favour among major industry players and is increasingly added as a company policy decision to improve monitoring and reporting capability.

Methodology

The United States Environmental Protection Agency (EPA) method 5D (Federal Register 2000) specifies the identification of alternative locations and procedures for sampling the PM emissions from positive pressure fabric filters. The EPA methods as well as the international Standards Organisation (ISO) methods are both recognised according to the South African legislation for emission testing.

Sampling locations

The possible alternative measurement locations when an in-duct testing location is not available for peaked roof and ridge vent type fabric filters are:

- Sampling in multiple short stacks (if stacks are present), which do not meet the required location criteria for duct sampling.
- Sampling directly above the filter housing in the ventilator throat / roof mono-vent.
- Sampling direct from the compartment housing above the filter bags.

The second and third options are illustrated in Figure 1 below.

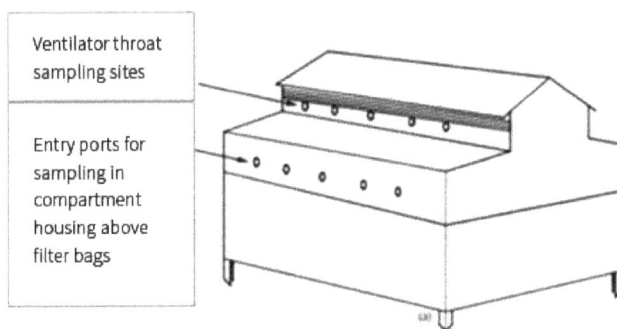

Figure 1: *Acceptable sampling locations for peaked roof fabric filters (Federal Register 2000)*

The alternative measurement locations need to be analysed according to structural suitability, measurement criteria and the choice between in-situ analysers versus an extraction system.

Multiple short stacks not meeting the location criteria for duct sampling

The filters at MFC have been designed with a mono-vent instead of multiple stacks to vent to atmosphere. It would however be

possible to construct multiple stacks over the mono-vent to conform to the EPA specifications.

A structural audit will then have to be performed on the filter supports to determine whether the structure can support the weight associated with multiple roof stacks. Maintenance access to the analysers would also have to be added to the structural load on the roof.

The audit, design and installation cost would be considerable and would probably not justify the cost for purpose, i.e. to determine continuous emission concentrations. For that reason this was not considered as a feasible option in this instance.

Roof mono-vent

The mono-vent located on top of the bag filter is open to atmosphere at the top, with a bottom section at a slightly lower level. The continuous sampling system in the roof or mono-vent will have to conform to the following EPA 5D requirements (Federal Register 2000):

- Measurement should take place at the bottom of the mono-vent.
- The measurement should take place above any upstream exhaust point.
- A minimum of 24 traverse points.

The minimum of 24 traverse points will have to be dealt with by either extraction at these points or by individual analysers for in situ measurement at each of the points. The reduced cross-sectional area of the vent compared to the compartment outlet will increase the velocities and will thereby increase the accuracy of the measurement.

In situ measurements will however require additional maintenance access at each of the analysers. The structure of the roof will have to be analysed but should be able to bear the weight of one person. Additional grating will have to be added on the roof and fitted with railings.

Compartment housing

Using the compartment housings will structurally be the optimum solution. Minimal structural changes will have to be made to allow for sampling.

A continuous sampling system in the compartment housing will have to conform to the following EPA 5D (Federal Register 2000) requirements:

- Testing should take place directly above the filter bags.
- A minimum of eight points or equivalent should be measured on each of the testing sites.
- A minimum of 50 % of the compartments needs to be sampled to obtain a representative sample.
- The sampling points need to be evenly distributed along the bag house.

In situ versus extraction system

Two options exist for measuring particulate matter concentration for the last two options described above. Measurements can either take place in situ meaning that analysers will be placed directly at each of the selected measurement location on the mono-vent, or by using an extractive system with a specified volume flow from each of the measurement points being sent to a separate combined measuring location.

In situ measurement

An in situ installation will require minimal structural changes to the bag filter housing. The large surface area of the bag filter outlet plenum and the limited range of the analysers will require a large number of analysers which will increase the installation as well as calibration and maintenance cost. In addition, the large number of analysers will each require an access point which will again increase the structural cost.

The expected velocity in the outlet of the bag compartments is in the order of 0.5 m/s to 2 m/s. Electrodynamic analysers are not designed for such low velocities. In situ measurement using an electrodynamic instrument will therefore not be possible.

Installing a dynamic opacity measuring system according to the EPA recommended standards will require a minimum of 8 transmitters and receivers per bag house.

The mono-vent of the filter housing is open to atmosphere. This would allow gas at temperatures above 100°C to be in contact with the analyser controller that is normally rated to approximately 50°C ambient operating temperatures. Additional controller cooling will then be required on each of the analysers adding to the already high analyser costs.

The only advantage of in-situ measurements would be the possibility of more accurately locating leaking bags.

Extraction system

An extraction system will transport the dust laden gas to a secondary location away from the harsh conditions at the top of the bag house. Only one analyser will be required per filter housing and no maintenance platforms will have to be constructed. This will off-set some of the construction cost associated with the installation of the extraction points.

Application

A continuous extraction system extracting from the mono-vent has therefore been selected. This method uses small extraction hoods to sample the gas from a number of locations and then transports the collective sample to a secondary measuring location. A structural layout indicating the hood locations as well as the duct routing is shown in Figure 2 below.

The rate of sampling cannot be determined by simply measuring the hood face velocity, as the velocities are low and measurement is therefore less accurate (Federal Register 2000). The EPA recognises the pitfalls involved in accurately measuring the velocities when dealing with low velocities varying across

the length of a mono-vent or filter housing and proposes an alternative calculation method.

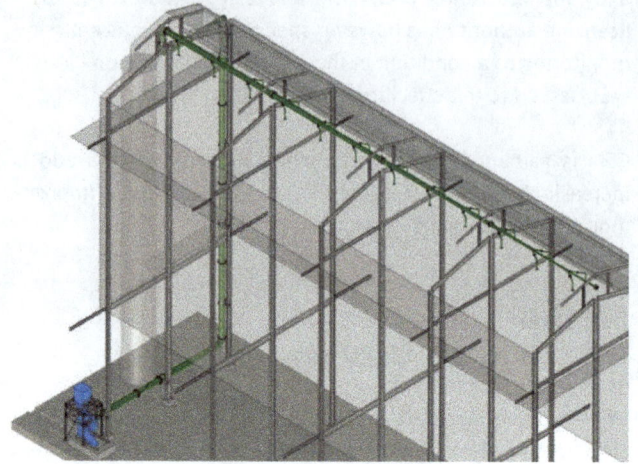

Figure 2: Structural layout
(Green: sampling ducting; Blue: Analyser location)

The prescribed calculation method is to measure the bag filter inlet volume (this would be measured for process reasons in any case) and calculate flow at the extraction points by using the temperature differences between the inlet and the sample extraction locations to adjust for dilution between the bags and the sampling location using a dilution factor. This is required because additional air is drawn in at the bottom of filter housing by the buoyancy of the hot outlet gas. The dilution factor is calculated according to Equation 1 (Federal Register 2000).

$$Q_D = \frac{Q_I(T_I - T_O)}{T_O - T_{AMB}} \quad (1)$$

Where
- Q_D is the dilution air volumetric flow rate (m³/s)
- Q_I is the filter inlet flow rate of the gas (m³/s)
- T_I is the filter inlet gas temperature (°C)
- T_O is the mono-vent outlet gas temperature (°C)
- T_{AMB} is the ambient temperature (°C)

The gas flow rate at the fabric filter inlet is measured for process purposes and therefore known. The required average velocity at the sample extraction location can therefore be calculated and approximate isokinetic sampling rates maintained at the measurement locations. An equal gas sampling rate at each of the sampling points is ensured by careful design of the diameter of the sampling header along its length, and could in principle be made adjustable by allowing for the installation of replicable orifices at each sampling point.

A process flow diagram illustrating the instrumentation required is shown in Figure 3 below.

A control system that includes volumetric measurements, thermocouples at three locations in the system as well as balancing orifice plates in each extraction line has been included into the design.

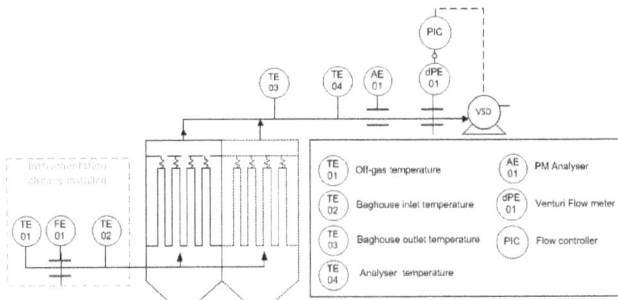

Figure 3: Process Flow Diagram

The sample flow will then be combined into a single duct and extracted with a fan. The electro-dynamic sampler was installed in the duct upstream of the extraction fan, together with suitable sampling ports for isokinetic testing to calibrate the instrument.

Conclusion

The following conclusions can be drawn from the continuous sampling design and installation project:

- Pressure-type reverse air bag houses do not conform to the laminar flow requirements normally required for accurate isokinetic sampling.
- The EPA method 5D prescribes alternative measurement locations and provides guidelines on the requirements of each of these locations.
- A solution for the specific application was designed.

References

Cooper, David C. & Alley F. C. 2012, *Air Pollution Control: A design Approach 4th ed.* Waveland Press, Long Grove Ill.

Department of Environmental Affairs 2010 'List of activities which result in atmospheric emissions which have or may have a significant detrimental effect on the environment, including health, social conditions, economic conditions, ecological conditions or cultural heritage' Government Notice 248, Government Gazette 33064, 31 Mar 2010. Pretoria

Department of Environmental Affairs 2013 'List of activities which result in atmospheric emissions which have or may have a significant detrimental effect on the environment, including health, social conditions, economic conditions, ecological conditions or cultural heritage' Government Notice 893, Government Gazette 37054, 22 Nov 2013. Pretoria

Federal Register 2000 Volume 65 Issue 201 'Amendments for Testing and Monitoring Provisions' Pages 61744 – 62273. Washington, DC

Roberts P., Els L., Noakes S. and de Montard B. 2013 'Ferroalloy off-gas systems- from design to implementation and design verification process' Proceedings, INFACON XIII, International Ferro-Alloy Conference Almaty, Kazakhstan.

21

The air quality perceptions of the residents of Bayview, Mossel Bay

Johann P. Schoeman[*1] and De Wet Schutte[2]

[1]28 Maroela Street, Mossel Bay, Western Cape, 6500, South Africa, jschoeman@edendm.co.za
[2]Cape Peninsula University of Technology, Post Office Box 652, Cape Town 8000, South Africa

Abstract

Background: In developing countries, it often occurs that little attention is given to air pollution emissions due to a lack of proper town planning, household combustion processes, energy production and the continuous growth in the transport sector (Norman *et al.*, 2007:783). There is an increase in urban air pollution in most of the major cities of developing countries which is amplified by population growth and industrialization (World Resource Institute, 1998, 1999:1). Air pollution studies are not complete, and may fail if the quality of life and the perceptions of the studied community are not taken into consideration. This paper investigates the air quality perceptions of a high income residency surrounded by industrial activities and Mossel Bay was rated as to have potentially poor air quality by the South African Department of Environmental Affairs and Tourism.

Methods: A cross-sectional survey was carried out in Bayview, Mossel Bay. The perceptions of the respondents were collected by a structured questionnaire. Components of perceptions that were tested included general opinion regarding air quality, visual perceptions of air quality, type of pollutants such as smoke and dust, perceptions regarding the source of air pollution, perceptions regarding the municipal health institution controlling air quality in Bayview, etc. These perceptions were investigated by age, gender, socio-economic status etc.

Conclusion: The findings of the study indicated that various factors, such as visual impacts, type of pollutants, role of the municipal health institution governing air quality, influence the air perceptions of the Bayview residents.

Keywords

air pollution, perceptions, air quality, monitoring, visual impacts, exposure, social status, local knowledge

Introduction

In developing countries it often occur that little attention is given to air pollution emissions due to a lack of proper town planning, household combustion processes, energy production and the continuous growth in the transport sector (Norman *et al.*, 2007). Most of the major cities of developing countries show an increase in urban air pollution which is amplified by population growth and industrialisation (World Resource Institute, 1998, 1999).

One of the major air quality problems in South Africa is poor regional and town planning, often driven by political ideologies where residential areas are developed adjacent or next to industrial areas (Matooane *et al.*, 2004). Mossel Bay was no exception to this rule and Bayview, one of the suburbs of this developing town is surrounded by industrial activities. The Mossel Bay region was also classified by the Department of Environmental Affairs as to have potentially poor air quality. One of the reasons for this classification was the petro-chemical industry situated 10km west of the Bayview residency. This industry triggered other industrial activities in the region. Considerable emphasis is placed on the monitoring of air pollutants that pose a risk and important aspects such as personal and community opinions on air quality and the effect on their health are often overlooked (Hunter *et al.*, 2003).

The development of perceptions about environmental pollution is a complex process due to the culmination of a wide spectrum of possible sensory inputs that could converge into perceptions Bickerstaff, 2003). In this respect it is clear that attitudes and concerns that an individual might have concerning the environment can play an important role in what is perceived as an environmental hazard (Stenlund *et al.*, 2009). Therefore, a general definition of risk perception is that it includes a person's attitude, behaviour, beliefs, opinions and concerns for the environmental hazard (Stenlund *et al.*, 2009). Bickerstaff (2003) takes this notion further as he argues that risk perceptions include the wider cultural and social disposition of people towards dangers and the disadvantages thereof. From this approach it is clear that perceptions regarding air pollution are predominantly based on the visual and chemo-sensory indicators, and showed as being useful indicators for possible air pollution when one has to mediate between environmental exposure and health (Stenlund *et al.*, 2009).

Hunter *et al.*, (2003) recorded that public stress related to the potential health effects of air pollution on individuals and their families increased over the last decades. Hunter *et al.*, (2003) furthermore mentioned various studies have been published on air pollution and the prevalence of respiratory diseases, the link between these two variables are still not clear, simply because of the lack of scientifically acceptable control over the variables in a real world setting. In this respect it was found that local knowledge could play an important role in shaping the people's perceptions of environmental risks (Scammel *et al.*, 2009). Research on perceptions of environmental risks showed that local context and experience plays an important role in the definition and perceptions of environmental risks (Bickerstaff, 2004). Added to this, it is also known that risk perceptions are also influenced by an individual's perception of the credibility of the environmental health institution that has to take care of his/her health (Scammel *et al.*, 2009).

Hyslop (2009) argued that the human physiology and psychology is particularly sensitive for visual inputs and that the observable, impacts to a large extent dictate how people perceive the world around them. Although man usually tends not to pay attention to his environment, it still consciously or subconsciously, influences how he sees life (Hyslop, 2009). However, since the 1960's there been a growing sensitivity for air pollution, mainly as a result of increased publicity, with the result of increased awareness regarding air quality (Hyslop, 2009).

This paper investigates the perceptions regarding the air quality of Bayview, middle to upper income residential suburb of Mossel Bay in South Africa. Being surrounded by industrial activities, Mossel Bay was rated by the South African Department of Environmental Affairs as having potentially poor air quality.

Methods

Bayview is middle to upper income residential suburb of Mossel Bay and is almost surrounded by industrial activities that include petrochemical storage tanks, a coal burning food industry, light industrial activities and also railway activities operating steam and diesel locomotives. A systematic sample of 114 plots (total 483) was selected from a cadastral map of Bayview. The sample size was calculated by means of Statcalc statistical software using a confidence interval of 5 points and a confidence level of 95% accuracy. A Politz frame was used to randomly select a respondent among the members of the selected household that qualifies for this study. Respondents that qualified for selection by means of the Politz frame had to be 18 years of age and older, usually sleep in the household and lived in the house since January 2009. The normal ethical conditions like anonymity, informed consent and voluntary participation applied during data collection. The research study and questionnaire was approved by the Cape Peninsula University of Technology Ethical Committee and Applied Sciences Higher degrees Committee.

The target population for this study consisted of all the residents that were 18 years and older and that have resided in Bayview for at least one year prior to the data collection for this research project. A questionnaire was designed to yield information regarding the perceptions of the residents regarding the air quality of the area. The questionnaire was based on literature research making use of the dendrogram-technique which is a conceptual framework highlighting the research questions on the lowest level of the dendrogram. The questionnaire

included sections on quality of life and perceptions. The section on perceptions included questions such as length of stay in Bayview, visibility of pollution and industrial activity, evaluating the general air quality of bayview, perception on industrial activity with biggest impact, perception on pollution control at industrial level and authority level, etc. Risk perceptions are also influenced by a person's perception of the worthiness of the health institution that has to protect that person's health (Scammel *et al.*, 2009). In order to test this theory the question of how the respondent perceived the health institution responsible for the health of the Bayview residents, was posed.

Independent variables in this evaluation were the respondent's age, gender, and period of residency, self reported health status, health consciousness, and exposure to polluted air in the workplace. Dependent variables in this study were their general perceptions of the air quality and visibility of air pollutants in Bayview. The results of the questionnaire were coded and analysed with the StatSoft Statistica version 9 software package. Chi square analysis was used to search for statistical significant relationships between independent and dependent variables.

Results and discussion

Table 1 shows the sample demographics for the 114 respondents.

Table 1: The age distribution of respondents

Age	n	%
18 - 23	5	4.4
24 - 30	3	2.6
31 - 45	20	17.5
46 - 60	41	36
61 +	45	39.5
Total	**114**	**100.0**

The age distribution indicated that the age group of 61+ was the highest age group in Bayview. Only 4.4% respondents were between 18 and 23 years old. The age profile thus indicated a high age profile.

Table 2: The comparison between male and female of the respondents

Sex	n	%
Male	52	45.6
Female	62	54.3
Total	**114**	**99.9**

Females were the predominant sex with a sex distribution of 54.3% female and 45.6% male respondents.

Table 3: The length of residency in Bayview

Length of residency in Bayview	n	%
0 - 5 years	32	28.1
6 - 10 years	19	16.7
11 - 20 years	44	38.6
21 years and longer	19	16.6
Total	**114**	**100.00**

Thirty-eight point six percent of the population resided in Bayview between 11-20 years and 16.6% for longer than 21 years. Based on this information it was derived that the population would have known the residency well and that their local knowledge would have been set.

Monitoring results

The study also took the air quality status of Bayview into account by studying the air quality data of the Department of Environmental Affairs and Development Planning ambient air quality station located in Bayview. The results indicated no exceedences of the South African National Ambient Air Quality Standards for the criteria pollutants. The annual average concentrations of the measured criteria pollutants can be observed in figure 1.

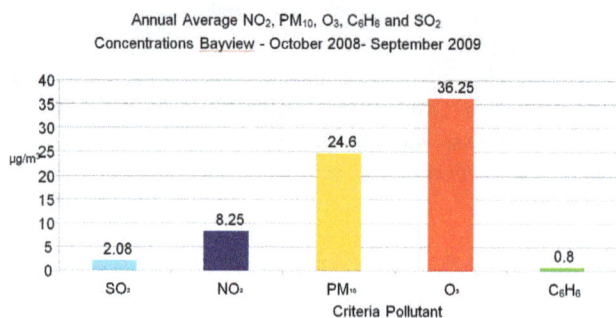

Figure 1: *The annual average concentrations of NO_2, PM_{10}, O_3, C_6H_6 and SO_2 for the Bayview residency from 1st October 2008 till 30 September 2009*

The data of the DEADP monitoring station were captured and compared to the 90% completeness against the USEPA and SANS1929 standards. See table 4.

Table 4: *The annual average of completeness of the captured data for the period 1 October 2009 to 30 September 2009 (South Africa. DEADP, 2009:10)*

Pollutant	SO_2	NO_2	O_3	PM_{10}
% collected	96.8%	98.9%	92.6%	97.6%

Perceptions

Almost two thirds 63.1% (n=72) of the respondents indicated that they perceived that the air quality was good, with a further 13.2% (n=15) who rated the air quality as excellent, whilst 19.3% (n=22) thought it was average (Table 5). The minority of respondents rated the air quality as poor to very poor (4.3%, n= 5).

Table 5: *The perceptions of the Bayview respondents on the status of air quality*

Opinion of the Bayview air quality	n	%
Excellent	15	13.2
Good	72	63.1
Average	22	19.3
Poor	4	3.5
Very poor	1	0.8
Total	**114**	**99.9**

Results of the logistic regression showed that the following factors may influence perceptions of air quality among the residents of Bayview, Mossel Bay. Although various influences

such as age, gender, length of stay, etc were tested, only the factors with statistical significance (P<0.05) are mentioned in this paper.

Visual impacts

Because visual impact plays an important role in the formation of perceptions regarding air quality, respondents were asked what they consider to be the main contributors to air pollution in Bayview. The results show that the perceived main contributor to pollution in the area is the Food factory (34.2%, n=39), followed by (19.3%, n=22) that perceive it is one of the local transport companies, followed by Petrochemical storage facility and the railway loco, both on (15.8%, n=18). (See Table 6)

Table 6: *Perceptions of Bayview respondents regarding the surrounding industries with the biggest air quality impact*

Industries with the major impact on Bayview air quality	n	%
Not answered	3	2.6
Petrochemical Storage facility	18	15.8
Local transport company	22	19.3
Railway loco	18	15.8
Tank farms	8	7.0
Oil tankers	4	3.5
Food Factory	39	34.2
Seal island	2	1.8
Total	**114**	**100**

Exposure to air pollution

The study found a statistically significant correlation (p=0.035) between the perception of the air quality of Bayview and the perception of the influence of the industries on the respondents' health (Table 7). It was found was that the respondents (80%, n=12) that perceived the air quality as excellent also indicated that the surrounding industries do not have a negative impact on their health. The same tendency was found with the respondents that perceived the air quality as good, when 61.1% (n=44) indicated that the surrounding industries do not have a negative influence on their health. This tendency was further confirmed when the majority (54.5%, n=12) among the 22 respondents (19.3%) that rated the air quality as average, indicated that they perceived that the industries do have a negative impact on their health.

Table 7: *The correlation between the perceptions of air quality and the perceptions of the impact of the industries on the health of the respondents*

Perceptions of air quality	Industries do influence my health		Industries does not influence my health		Unsure, if industries influence my health		Total	
	n	%	n	%	n	%	n	%
Excellent	1	6.7	12	80.0	2	13.3	15	100.0
Good	15	20.8	44	61.1	13	18.1	72	100.0
Average	12	54.5	6	27.3	4	18.2	22	100.0
Poor	3	75.0	0	0.0	1	25.0	4	100.0
Very poor	1	100.0	0	0.0	0	0.0	1	100.0
Total	**32**	**28.1**	**62**	**54.4**	**20**	**17.5**	**114**	**100.0**

Pearson Chi-square: 22.8375, df=8, p=.003582

Social status

Perceptions are also influenced by the environment, availability of information and socio-economic characteristics (MacKerron et al., 2009). Previous studies indicated that the more wealthy communities have the perception that air quality imposes a bigger threat to the less fortunate communities (Bickerstaff, 2004). The social status of the Bayview suburb was high. It was derived from their income level, work level and educational level. Thirty seven point eight percent, (n=43) respondents were professional, 39.4%, (n=45) had an annual income of R180 001 and higher and 43.8%, (n=50) had a degree or diploma. This research found a correlation between qualification and perceptions.

Table 8: The correlation between perceptions of air quality and the industrial emission control

Qualifica-tion	Yes, indus-tries emis-sion control		No, indus-tries emis-sion control		Unsure, industries emis-sion control		Total	
	n	%	n	%	n	%	n	%
< Grade 12	10	55.6	2	11.1	11.1	33.3	18	100.0
Grade 12	13	28.3	13	28.3	28.3	43.5	46	100.1
> Grade 12	8	16.0	17	34.0	34.0	50.0	50	100.0
Total	**31**	**27.2**	**32**	**28.0**	**51**	**44.8**	**114**	**100.0**

Pearson Chi-square: 10.9681, df=4, p=.026931

Eighteen of the respondents (15.7%) n=18 had qualifications lower than grade 12. Of those, 11.1% n=2 indicated that the industries don't do enough to control emissions, 33.3% n=6 was unsure and 55.6% n=10 indicated adequate emission control. Forty six (40.4%) respondents had grade 12. Of those, 28.3%, (n=13) indicated that industries don't do adequate emission control and almost half (43.5%, n=20) indicated that they don't know. Forty three point eight percent n=50 had a qualification higher than grade 12 (diploma or degree) of which 34%, (n=17) indicated that industries do not control emissions adequately and 50%, (n=25) indicated that they don't know. The perceptions of emission control fluctuated over educational qualification. The higher the qualification thus the higher the perception that industries do not implement adequate emission control. These groups were also unsure if the industries are doing emission control.

Risk perceptions and the perceptions of the air quality authority regulating air quality in Bayview

It was clear that the majority of the respondents that classified the air quality as excellent and good (n=15 and 72 respectively) did not know how to judge the health authority (46.7% and 36.1% respectively) (Table 9). Of these two groups, the second highest percentage (26.7%, n=4 and 34.7%, n=25) respectively also judged the health institution as good.

Table 9: The correlation between the perceptions of the Bayview respondents and the perception of the air quality authority

Perception air quality	Health institutions excellent		Health institutions good		Health institutions average		Health institutions poor		Health institutions don't know		Total	
	n	%	n	%	n	%	n	%	n	%	n	%
Excellent	3	20.0	4	26.7	1	6.7	0	0.0	7	46.7	15	100.1
Good	1	1.4	25	34.7	18	25.0	2	2.8	26	36.1	72	100.0
Average	0	0.0	5	22.7	6	27.3	4	18.2	7	31.8	22	100.0
Poor	0	0.0	1	25.0	0	0.0	2	50.0	1	25.0	4	100.0
Very poor	0	0.0	0	0.0	0	0.0	1	100.0	0	0.0	1	100.0
TOTAL	**4**	**3.5**	**35**	**30.7**	**25**	**22.0**	**9**	**7.9**	**41**	**36.0**	**114**	**100.1**

The influence of visibility on risk perceptions

On the question if any of the industries were visible from the respondents dwelling, 59.6%, (n=68) answered yes and 40.4%, (n=46) answered no. The majority respondents could see an industry from their home (Table 10).

Table 10: The visibility of the industries from the respondent's dwellings

Visibility of the industries	n	%
Yes	68	59.6
No	46	40.4
Total	**114**	**100.00**

There were 53.5%, (n=61) respondents that could see smoke from their dwellings (Table 11). More than two thirds 70.1%, n=80 of respondents experienced or saw dust from their dwellings.

Thirteen point two percent respondents (n=15) that perceived the air quality as excellent, the minority or 46.7%, (n=7) experienced or saw smoke from their houses. Almost two thirds or (63.2%, n=72) respondents that perceived the air quality to be good, the minority or 45.8%, n=33 could see smoke from their dwellings. Four respondents perceived the air quality as poor and all of them, n=4 could see smoke from their dwellings.

Table 11: The correlation between perceptions and the experiencing or visibility of smoke.

Perception of air quality	See smoke		Don't see smoke		Total	
	n	%	n	%	n	%
Excellent	7	46.7	8	53.3	15	100.0
Good	33	45.8	39	54.2	72	100.0
Average	17	77.3	5	22.7	22	100.0
Poor	4	100.0	0	0.0	4	100.0
Very poor	0	0.0	1	100.0	1	100.0
Total	**61**	**53.5**	**53**	**46.5**	**114**	**100.0**

Pearson Chi-square: 11.6079, df=4, p=.020523

Conclusion

This study found the higher the social status in terms of qualification, thus the lower the perception regarding the risk. Smoke posed a significant impact on the perceptions of the respondents. The importance of visibility was also emphasized as the industry that was most visible was also perceived as the industry with the biggest threat or impact. There is a correlation between general air quality perceptions and industrial air quality impact. There is a correlation between air quality perceptions and the perceptions of the air quality authority regulating air quality of that specific population.

References

Bickerstaff, K. 2004. *Risk Perception research: socio-cultural perspectives on the public experience of air pollution*. 2004. Environment International volume 30 (2004). Science direct.

www.elseriver.com/locate/envint   Pages 827-840. [11 August 2009].

Carter, S., Williams. M., Paterson, J., Iusitini, L. 2008. *Do perceptions of neighbourhood problems contribute to maternal health? Findings from the Pacific Islands Families study.* (PIF) Study, AUT University, Auckland, New Zealand. Health & Place. 2008. www.elserivier.com/locate/healthplace   Pages. 622-630. [15 May 2009].

Encyclopedia Britannica. 2009. Encyclopedia Britannica Online. 2009. http://www.search.eb.com/eb/article-9110465 Pages. 1-3. [22 May 2009].

Hunter, P., Davies, M., Hill, K., Whittaker, M., Sufi, F. 2003. The prevalence of self-reported symptoms of respiratory disease and community belief about the severity of pollution from various sources. September 2003. The International Journal of Environmental Health Research 13(3). Taylor and Francis Health Sciences. London UK. Pages 227-338.

Hyslop, N. 2009. *Impaired visibility: the air pollution people see.* Atmospheric Environment 43 (2009). Science Direct. www.elsvier.com/locate/atmosenv Pages 182-185. [22 May 2009].

MacKerron, G., Mourato, S. 2009. *Life satisfaction and air quality in London.* 2009. Ecological Economics Vol. 68 (2009) Science Direct. www.elserivier.com/locate/ecolecon Pages 1441-1450. [20 July 2009].

Matooane M., John, J., Oosthuizen, R en Binedell, M. 2004. Vulnerability of South African Communities to Air Pollution. Proceedings of the 2004 Conference of the South African Institute of Environmental Health, Durban, 22-27 February 2004.

Norman, R., Eugene Cairncross, Jongikhaya Witi, Debbie Bradshaw. 2007. *Estimating the burden of disease attributable to urban outdoor air pollution in South Africa in 2000.* South African Medical Journal, 97(7) August 2007. Pages 782-790.

Scammel, M., Senier, L., Darrah-Okike, J., Brown, P. 2009. Tangible evidence, trust and power: *Public perceptions of community environmental health studies.* Social Science & Medicine, 68(2009). Science Direct. www.elsevier.com/locate/socscimed Pages 143-146. [22 May 2009].

Stenlund, T., Lide, E., Andersson, K., Garvill, J., Nordin, S. 2009. *Annoyance and health symptoms and their influencing factors: A population-based air pollution intervention study.* Public Health. 2009. Science Direct. www.elsevierhealth.com/journals/pubh Pages 339-345. [15 May 2009].

World Resource Institute. Health and Environment,1998,1999. Health Effects of Air Pollution. http://www.wri.org/wr-98-99/airpoll.htm [11 May 2009]

22

Effectiveness of mediation in the resolution of environmental complaints against the activities of gold mining industries in the Witwatersrand region

Olusegun Oguntoke[*1,2] and Harold J. Annegarn[1]

[1]Department of Geography, Environmental Management and Energy studies, University of Johannesburg
PO Box 524 Auckland Park, 2006, Republic of South Africa, hannegarn@uj.ac.za
[2]Department of Environmental Management and Toxicology, University of Agriculture
PMB 2240 Abeokuta, Federal Republic of Nigeria, oluseguo@uj.ac.za; oguntokeo@funaab.edu.ng

Abstract

In the Witwatersrand gold mining area, there have been recurring public complaints about dust dispersed from gold tailings storage facilities (TSFs) that traverse the landscape. Although weather aggravates the frequency and intensity of dust emission from TSFs in the study area, the rapid conversion of buffer areas around the dumps to residential land–use is exposing more people to dust hazards. This study assessed the effectiveness of Crown Mines Dust Monitoring forum in Johannesburg as an alternative environmental dispute resolution mechanism. Records of complaints from 1995 to 2010 that were made available through the forum were collated and analysed with the aid of descriptive statistics. Within the study period, complaints about mine pollution were more frequent between August and October, i.e. the dry months. More than 70% of the complaints were made by companies whose properties, operations and employees were affected by dust emission from the TSFs. While 52% of the complainants reported pollution problems for the first time within the study period, other cases were follow-up to previous complaints. Mining companies responded to 31% of the public's grievances about dust pollution from their facilities within one week and another 12% in two weeks; response to the remaining complaints took much longer time. As part of mines' response to public complaints, site visits were organised to indicted facilities, and pollution control measures and mitigation plan adopted at sites were also explained. Moreover, additional control measures were installed in critical circumstances to ameliorate dust pollution. Only a few of the complaints reported to the forum escalated to litigation or issuance of penalty by government agency. Although, the forum provided an avenue for resolution of environmental conflicts in a pragmatic and mutually beneficial manner, the right of the public to a clean environment is still not being realised fully.

Keywords
environmental pollution, conflict resolution, public complaint, mine tailings, dust forum, environmental right sustainable mining

Introduction

Generally speaking, all informal channels of dispute resolution apart from full-scale court processes are regarded as alternative dispute resolution (ADR) mechanisms (Bingham 1986; the World Bank Group 2011). In ADR, disputants are generally encouraged to negotiate directly with each other through negotiation, conciliation, mediation or arbitration. The popular acceptability and use of ADR systems by the citizenry in many societies can be linked to their informal setting, flexibility, ease of access and, the acceptable and peaceful manner in which cases are handled. The World Bank Group (2011) noted that ADR increases access to justice for population groups by reducing cost and time, and also increases disputant satisfaction with outcomes. The mechanism is also capable of supporting disadvantaged people against the overriding influence of multinational companies.

Although the court of law is the last resort in the resolution of environmental conflicts in most civil societies, the use of alternative dispute resolution mechanisms is growing rapidly (Lee 2008). In China for instance, while the formal channels are dominant, Liang (2012) observed the active role of the informal channels in the resolution of environmental disputes. Similarly, public complaint is well established as a veritable means of seeking redress for perceived adverse impact of environmental pollution in Portugal and Brazil (Dong et al 2011; Carvalho and Fidélis 2011). This emerging significance of extra-litigation channels can be linked to a major setback suffered by public litigation against environmental pollution. In most cases when litigation is employed as a means of seeking redress for observed violation of the public's right to a harmless environment 'lack of sufficient evidence' thwarts such effort.

As a means of protecting the rights of the public, the South African constitution of 1996 (act 108) indicates that everyone has the right to an environment that is not harmful to their health and well-being; and to have the environment protected for the benefit of present and future generations, through reasonable legislative and other measures that prevent pollution and ecological degradation; promote conservation; and secure ecologically sustainable development and use of natural resources while promoting justifiable economic and social development.

In order to give effect to this constitutional provision, mandatory standards are set for selected air quality parameters that are considered critical to human health. These include; sulphur dioxide (48 ppb/24 hours), nitrogen dioxide (21 ppb/24 hours), carbon monoxide (26 ppm/1 hour), PM_{10} (120 µg/m³), ozone (61 ppb/8 hours), benzene (3.2 ppb/1 year), and lead (0.5 µg/m³/year) as published by South Africa Government Gazette (No. 32816, volume 534) of 2009. Moreover, environmental limits of dust-fall emanating from mining activities were set for residential (≤600 mg/m²/d) and non-residential areas (≤1200 mg/m²/d) (SANS 1929: 2005). In spite of these afore-mentioned constitutional and policy measures to safeguard the welfare of communities, complaints against dust dispersal into areas occupied by humans has been significant in the Witwatersrand areas.

Community complaints and protests in the Witwatersrand area stem from the irritation that is associated with human exposure to coarse dust, commonly referred to as nuisance dust. Additional worry is raised among exposed communities as mine dust is perceived to be hazardous (Wright et al 2014; Liu et al 2011), hence, capable of impairing their health and well-being. Among people that are generally aware of constitutional provision for human rights, environmental pollution by air-borne mine dust that affects residential areas is considered a denial of their right to a clean environment.

An analysis of frequency and intensity of dust episodes in the Witwatersrand environment showed a recurrent emission of dust well above set limits in some of the areas where tailings storage facilities (TSFs) are located (Oguntoke et al. 2013). This implies that despite constitutional and policy provisions for the rights of the public to an environment free of harm to their health, air pollution from mine facilities is frequent.

The Crown-Mines Dust Monitoring (CDM) forum that was inaugurated in the Witwatersrand region in 1992 is a proactive effort geared towards resolving environmental pollution complaints arising from mining activities. The aim was to provide information and resolve public complaints about the negative impact of mining operations in a practically feasible and peaceable manner.

CDM forum receives complaints against observed pollution from mines' operations, documents such and promptly addresses them. Once complaints are received, the owner of the indicted mine facility is identified and requested to respond to the allegation. For every case handled by the forum, the Secretary keeps a register of proceedings and circulates the minutes prior to the subsequent meeting. Follow-up actions to ensure a satisfactory mitigation of observed environmental pollution are monitored and reported to the forum.

Having operated for two decades, the current study aimed to assess the role of the CDM forum in resolving public complaints

against pollution arising from mining operations in the Witwatersrand region, South Africa. As a voluntary, non-formal and alternative environmental dispute resolution mechanism, the information about its operations, successes and defects are valuable for upgrading its operations so as to enhance the public's access to their right to a clean and healthy environment.

The study area

Mining activities and sites of dumps (TSFs) occupy approximately 12,200 km² of land area in the Witwatersrand basin of South Africa (Rossouw et al 2009). Coincidentally, this area serves as residence for a dense urban population in Gauteng province (Figure 1). In terms of the eco-climatic region, the Witwatersrand area is situated within the Highveld grassveld-savannah with summer rainfall ranging from 500–750 mm (O'Connor and Bredenkamp 1997). Daily temperature ranges between 2–12°C in the dry winter months, and from 14–30°C during the summer (Schulze 1997).

In Gauteng province, summer days are warm and winter days are crisp, clear and wind free. There are about six weeks of cold weather in mid-winter (from July to August) and summer offers warm sunshine followed by balmy nights (October to March). The rainy season is in summer rather than winter (June–August). Rainstorms are often harsh accompanied by much thunder and lightning and occasional hail, but they are brief and followed by warm sunshine. The moisture and wind characteristics of the seasons in the region have implications for dust storm frequency and severity during the year. Specifically, the absence of rainfall (dry conditions) coupled with high wind speed provide an enabling condition for dust dispersal from poorly vegetated TSFs.

From historical records, grasses, shrubs and trees were planted on some of the TSFs as early as 1969 and 1970 (GSG 1997). While additional re-grassing has been conducted, the general assessment of vegetation coverage of the TSFs showed varying degrees of degradation. These degraded sections of the TSFs have constituted a source of dust emission especially, during the dry and windy periods. The dust emission problem in the area is further worsened by the location of human settlements within the precinct of the TSFs. Some human residences are sited less than 100 m away from mine dumps in the area (Rossow et al 2009).

Figure 1: Location of the Witwatersrand Mega-dumps within residential land-use (Ojelede and Hannegarn, 2010)

Method

In this study, a retrospective study design was adopted for data collection (Mann 2003). Records of dust pollution complaints from 1995–2010 that were documented by the Crown-Gold Dust Monitoring forum (now, ERGO Gold Dust Monitoring forum) in Johannesburg were accessed. The forum has a chairperson, secretary, statutory member from relevant government departments, representatives of the mine companies, affected community members and interested individuals and organisations. Complaints from individuals, communities and companies that were communicated to the forum through email, telefax, letter, phone call and personal communication were printed on paper and compiled. From each of the cases, information such as date of complaint, nature of the complaint, the complainant, time between complaint and response, resolution of the complaint or otherwise, and whether the complaint escalated to legal action or issuance of penalty by a government agency or not, were extracted.

Registered cases that had insufficient information about the complainant, nature of complaint and relevant dates were excluded from the data analysis conducted in this paper. Similarly, six complaints about overgrown mine sites, noise emission from machinery and other complaints that are not directly related to dust pollution from mining activities were excluded. While it is practically impossible for all environmental complaints in the region to have been reported to the forum, those that were accessed provided valuable information about the forum's operations and mines' response to pollution complaints.

The attributes of each complaint registered with the forum were extracted and entered into an Excel™ spread-sheet. This data was exported into Statistical Package for the Social Sciences (IBM™ SPSS version 21) for statistical analysis. Descriptive tools such as frequency run, chi square and graphs were employed to analyse and summarise the data.

In order to complement the information retrieved from the CDM forum's register, participant observation was employed. The experience garnered from the scheduled meetings of the forum by the foundation chairperson and the participation of one of the authors between 2011 and 2013 were incorporated into this study. This additional information, unprinted though, formed a useful component of the discussion. It takes participation in the forum's proceedings for one to gain insight into the manner in which emotional out-bursts, intense arguments, and grievances of individuals are handled and amicably resolved in the most tactful and satisfactory manner.

Results

Within the study period, 75 complaints against dust pollution were documented by the CDM forum in Johannesburg. The annual trend of reported complaints showed the highest proportions in 1997 and 2006 (13.3% each). The second highest proportion (12.0%) of complaints was reported in 2009 (Figure 2). Although few new complaints about dust pollution were reported in 2007 and 2008, some of the complaints in the previous years were follow-up cases.

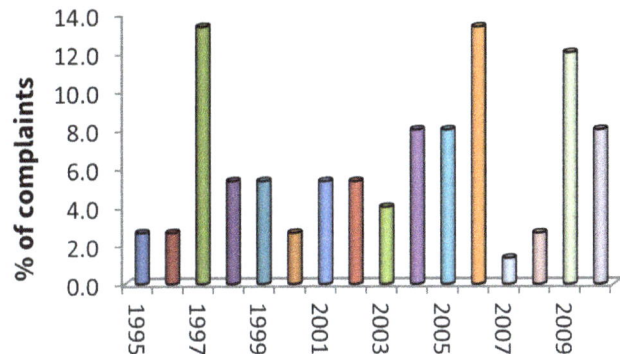

Figure 2: *Annual trend of reported dust pollution complaints*

Over the 16 year period considered in this study, reported complaints were more frequent from August to November (Figure 3). This period marks the transition from winter to spring, which is the dry and windy months in the study area, hence, the high frequency of complaints about mine dust pollution.

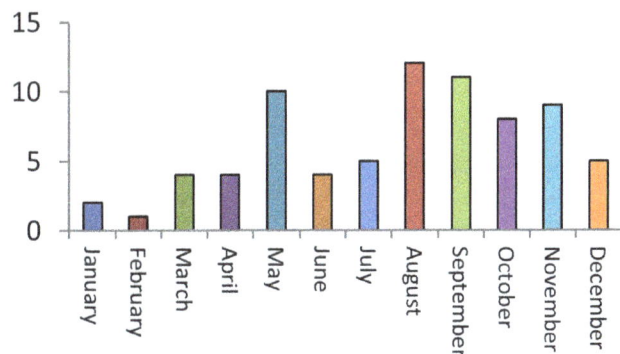

Figure 3: *Monthly frequency of dust pollution complaints (1995-2010)*

Seventy-two percent of the documented public complaints were reported by corporate organisations whose properties, operations and employees were affected by dust from mine facilities (Figure 4). Additionally, individuals, communities and government agencies brought 28% of the registered complaints to the CDM forum.

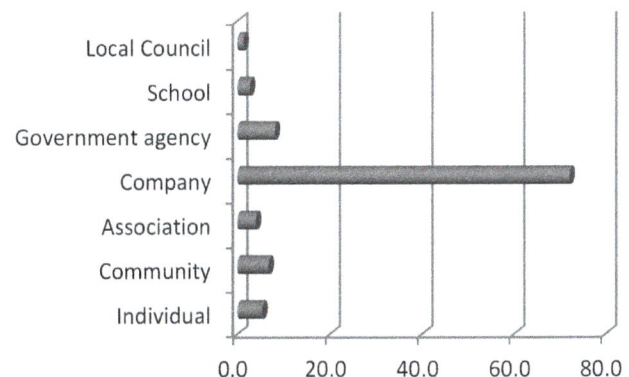

Figure 4: *Profile of registered complainants at the CDM forum*

While 52% of the complainants reported a pollution problem for the first time, other cases were follow-up to previous complaints already communicated to the concerned mines. Most complainants cited the nuisance caused by dust pollution (35%) and perceived health hazards (28%) as the motivation for registering their grievances against the mines (Table 1). Other reasons for filing cases against the mines included damage to factory equipment and property (17%), noise pollution (3%) and washing of mine materials into drains, roads and building premises (9%) among others. Whether real or perceived, mine dust pollution was indicated as the cause of irritation, discomfort, diseases and impaired human well-being.

Table 1: *Nature of complaints reported to the CDM forum*

Nature of the Complaint	Frequency	Percent
Dust deposition causing nuisance	26	34.7
Dust deposition and health concerns	21	28.0
Dust deposition, health hazard, property/equipment damaged	14	18.7
Dust deposition and Noise pollution	3	4.0
Mine materials washed on road, premises and drainage	7	9.3
Others	4	5.3
Total	**75**	**100**

The industrialists in areas affected by dust pollution attributed property damage and degradation, reduction in clients' patronage and the disruption of industrial processing to deposition of mine dust. They claimed that air-borne dust from mine tailings was deposited in their business premises, on vehicles parked within the premises and also, on offices' windowpanes. In the factories where manufacturing took place, damaged machines and contaminated production rooms were ascribed to mine dust pollution. Glass producing companies were at the fore-front of complaints against production room contamination. Dust pollution in the production environment is an inhibitory factor that could stall production. Hence, the limit of ultra-fine dust in the production room is set at 2930 microns (ISO Standard 14644).

Table 2: *Media used for registering air pollution complaints with the CDM forum*

Medium	First complaint	Follow-up	Issuance of ultimatum	Others	Total
Letter	16	6	1	2	25
Fax	11	11	0	1	23
Email	4	10	0	0	14
Telephone	6	4	0	0	10
Telephone / Email	0	1	0	0	1
Not mentioned	2	0	0	0	2
Total	39	32	1	3	75
Percent	52	43	1	4	100

About 65% of the registered complaints were communicated to the forum through physical letters and fax messages (Table 2). Other methods employed by the complainants to communicate their grievances against the infringement of their right to clean air were electronic mails (19%) and telephone calls (13%). These messages are normally printed and archived in the forum's register for discussion. After thorough consideration of each complaint, the final decision and follow-up activities are inscribed on hard-copies of the complaints.

In 32% of the reported cases, there were no clear indications of response time to complaints; the mines responded to 34% of the registered cases within a week (Figure 5). Twelve percent of the complaints received responses by indicted companies within two weeks and, another 12% between two to four weeks. In about 11% of the cases, concrete response to complaints took between one and two months. Considering the time lag between complaints and response, many complaints got delayed responses apart from 11 cases that received prompt attention from the mines.

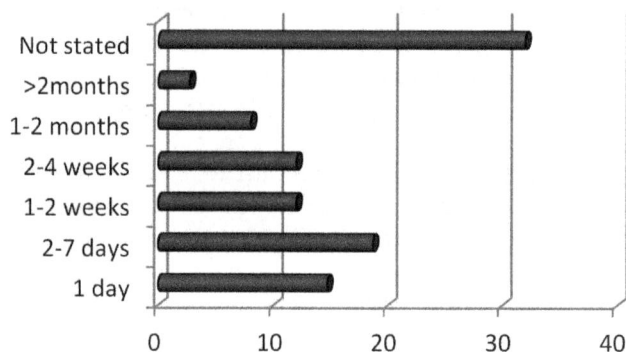

Figure 5: *Response time of mines to pollution complaints*

In addition to the time lag between complaint and response, the nature of the responses to pollution allegation is quite germane. For instance, it is pertinent to consider whether or not the responses were adequate to abate dust episodes and were satisfactory to the complainants. In 25% of the cases, the indicted mines explained to complainants the mitigation measures adopted for pollution control in their facilities (Table 3). Similarly, meetings were organised with affected and concerned communities in order to address their complaints and or resolved issues raised against the operations of the mining companies. Immediate intervention which entailed fixing malfunctioning dust suppressants, and installation of additional mitigation facilities were done only in a quarter of the registered cases. There were no specific actions taken with respect to 32% of the registered complaints, due to the unwillingness of mines to commit funds to solve the alleged pollution problem. In all cases where dust pollution allegations were traced by the mines to off-site activities such as open ground, road works and other non-mining sources, no specific actions were taken to control the pollution.

Only 3 of the registered cases escalated to litigation, threat of litigation and liability claims. However, it is possible that other cases might have gone to court for legal processes without the knowledge of the CDM forum. Since the forum is a voluntary conflict resolution medium and hence, has no legal power to enforce actions on mines, complainants are at liberty to litigate against mines if they are not satisfied by the response received. The exercise of constitutional right of individuals and communities was not precluded by registering their grievances with the CDM forum. The fact that cases are considered objectively, realistic mitigation plans and timelines indicated and sometimes, immediate action were engaged at no cost to the complainant, positioned the forum as a viable alternative for conflict resolution in the study area.

Table 3: Responses of the mining companies to public complaints

Response of mines to complaint	Frequency	Percent
Explanation of adopted mitigation plan	5	6.7
Arrangement of site visit with concerned individuals	14	18.7
Installation of additional mitigation measures	11	14.6
Meeting held with concerned individuals for resolution	12	16.0
Immediate response for intervention	5	6.7
No specific action or response	24	32.0
Others	4	5.3
Total	**75**	**100**

Considering the proportion of cases (62.7%) that were resolved by the forum over the years, some level of success was achieved. The *modus operand is* of the forum, transparency and openness of the arbitration process, dissemination of seemingly technical information in comprehensible manner, and the friendly atmosphere in which proceedings are conducted are commendable and perhaps, enhanced the level of success achieved on the cases handled. Equally commendable is the effort of the forum in arranging the meeting of complainants and mining companies for mutual discussion so as to arrive at a feasible action plan, which is monitored by the forum.

Discussion

The variation in the number of new cases reported to the forum annually can be associated with the prevalent weather conditions and possibly, public awareness of the forum's activities. According to the analysis of rainfall pattern over the study area between 1978 and 2009, 1997 and 2006 were identified as dry years with rainfall amount (526 mm and 364 mm) below the overall average in summer (Dyson 2009). On the other hand, 2000 and 2007 with few complaints had higher rainfall (793 mm and 708 mm) during summer.

The fact that August and September mark the transition from winter to spring, which is the dry months in the study area, explains the prevalence of complaints about mine dust pollution within the period. While it is true that operating mines employ mitigation measures to curtail dust emission from the tailing facilities, the dry conditions of spring make dust readily available for air-lifting that results in dust episodes (Annegarn and Sithole 2002, Preston-Whyte and Tyson 1988). Closely related to the paucity of rainfall between August and September is the increasing average monthly wind speed in the study area. From the weather data of Johannesburg (South Africa Weather Service, 2014) from 2001–2010, the average wind speed increased from 4.0 to 4.6 m/s within the dry months. As predicted by Annegarn *et al* (1991) and Oguntoke *et al* (2013), a wind speed ≥4 m/s was critical for dust episode incidence.

Moreover, mines re-processing old tailings for further gold extraction produce finer dust particles that are capable of being transported over longer distances (Ozkan and Ipekoglu 2002). Generally, residents reacted negatively to dust pollution once dust episodes formed plumes that caused irritation and concerns of possible negative health consequences. The mines will require greater effort in the dry and windy months to suppress dust emission from tailings by engaging integrated tailing management strategies.

High intensity of dust-fall in non-residential areas and the fact that companies, being corporate entities, are better informed and more capable to pursue legal issues may explain their higher visits to the CDM forum. The voice of a community or interest group is considered stronger that the single voice of an individual when seeking redress against the infringement of their rights by multi-national companies. Hence, individuals seek the support of their community, associations, and government agencies to register complaints of right abuses on their behalf. For example, Kerona, a Community-based Association and Redirile Development Project lodged a complaint against dust pollution from a mine dump in Diepkloof (May 2004), after harvesting signatures of more than 300 people in the community that supported the complaint. This observation agrees with the submission of Liu *et al* (2011) that a collective complaint is preferred by the citizens against environmental pollution.

The specific ailments attributed by the complainants to mine dust exposure included asthma, breathing difficulty, sore throat, cough, irritation, bronchitis, sneezing, tuberculosis, lung cancer, and eye, skin and hearing problems. Other issues raised as a basis of lodging complaints were vehicular accidents due to poor visibility caused by dust, fear of cyanide residues in the environment and general hazards. The fear of the complainants, whether real or perceived that mine dust is hazardous agrees with the submissions of Naicker *et al.* (2003), Bright (2007) and ATS (1997) that exposure to respirable mine dust that contains silica poses health hazard to humans living within the vicinity of TSFs.

A critical evaluation of the response time of the mines to pollution allegations showed unacceptable delay. A situation where more than half of the complaints did not receive immediate response is tantamount to serious violation of the right of affected individuals.

Conclusion

Over the years, the public living around mine dumps in the Witwatersrand area have been exposed to dust pollution from TSFs. In an effort to resolve the ensuing grievances and create easier access to their environmental rights, the CDM forum was inaugurated. Complaints against mine dust pollution were received by the forum, registered, processed and followed-up until some respite was received by the complainants. This enhanced public confidence in the forum particularly as litigations to redress environmental pollution are tricky and often fail due to lack of evidence.

Although, the forum provided an avenue for resolution of environmental conflicts in a pragmatic and mutually beneficial manner, the right of the public to a clean and healthy environment is not being realised fully. The gold mining companies operating in the area need to demonstrate greater commitment to pollution control so as to protect the public's environmental rights.

While the CDM forum is not an official or legal body, it should be accorded the support of the government in view of the valuable services it provides in respect of environmental conflict resolution. For instance, the agencies of government that are directly connected with mine operations and environmental monitoring should take special interest to follow-up cases that are reported to the forum. The forum is capable of providing first-hand information on areas that require immediate and urgent intervention for community protection and environmental preservation. Furthermore, the achievement of sustainable development in the mining industry, which centres on maximising net environmental and human welfare, is enhanced.

Acknowledgement

We sincerely appreciate the information made available to us by the Crown-Mines Dust Monitoring forum (now, ERGO Gold Dust Monitoring forum). Special thanks to the Secretary of the forum for her painstaking record keeping. The opinions expressed by various attendees over the study period have been of tremendous benefit to us.

References

American Thoracic Society-ATS 1997, Adverse effects of crystalline silica exposure: American Thoracic Society Committee of the Scientific Assembly on Environmental Occupational Health. *Am J Respir Crit Care Med*. 155: 761–765.

Annegarn H.J. and Sithole S.J. 2002, Dust Monitoring and Mitigation on Gold tailings reclamation. Paper presented at Mine Ventilation Society Symposium, 1-13.

Annegarn HJ, Surridge AD, Hlapolosa HS, Swanepoel DJ, Horne AR. A review of 10 years of environmental dust monitoring at Crown Mines. Journal of the Mine Ventilation Society of South Africa 1991; 43: 46–60.

Bingham G. 1986, Resolving Environmental Disputes: A decade of experience. The Conservation Foundation, Washington D.C. 1-12.

Carvalho D.S. and Fidélis T. 2011, 'Citizen complaints as a new source of information for local environmental governance', *Management of Environmental Quality: An International Journal*, 22(3): 386 – 400.

Dong Y., Ishikawa M., Hamori S. & Liu X. 2011, The determinants of citizen complaints on environmental pollution: An empirical study from China. Journal of Cleaner Production, 19: 1306-1314.

Dyson L.L. 2009, Heavy daily-rainfall characteristics over the Gauteng Province. Water SA (Online) 35(5), Available from http://www.scielo.org.za/scielo.php?pid=S1816-79502009 000500011&script=sciarttext#nt (accessed June 12 2012)

ISO Standard 14644-1, Clean room class limits for particulates. Available online at: http://www.engineeringtoolbox.com/clean-rooms-iso-d_933.html (accessed July 18 2012).

Lee J.C. 2008, 'Pollute first, control later' No more: Combating Environmental degradation in China through an approach based in Public interest litigation and public participation. *Pacific Rim Law & Policy Journal Association*, 7(3): 795–823.

Liang M. 2012, Pollution Severity, Alternative Judicial Channels, and Citizens' Environmental Complaints: Evidence from Chinese Provinces. Paper prepared for the 3rd Global Forum of Chinese Scholars in Public Administration, Shandong University, Jinan, June 1-3, 2012.

Liu X., Dong Y., Wang C., & Shishime T. 2011, Citizen Complaints about Environmental Pollution: A Survey Study in Suzhou, China. *Journal of Current Chinese Affairs*, 3: 193-219.

Mann C. J. 2003, Observational research methods. Research design II: cohort, cross sectional and case-control studies. *Emergency Medicine Journal* 2003; 20: 54-60

Naicker K., Cukrowska E. & McCarthy T. 2003, Acid mine drainage arising from gold mine activity in Johannesburg, South Africa and environ. *Environmental Pollution*, 122: 29–40.

O'Connor T.G. and Bredenkamp G.J. 1997, Grassland. In Vegetation of Southern Africa. Cowling, R.M., Richardson, D.M. and Pierce, S.M. (eds). Cambridge University Press, Cambridge. 215-257.

Oguntoke O., Ojelede E.M and Annegarn, J.H. 2013, Frequency of mine dust episodes and the influence of meteorological parameters in the Witwatersrand area, South Africa. International Journal of Atmospheric Sciences, 1: 1–19.

Ojelede M.E. and Annegarn H.J. 2010, The implications of aeolian emissions from gold mine tailings on ambient air quality: the Witwatersrand scenario. A paper presented at the Waste Revolution Seminar Series: Mining, Witbank, 4 November 2010. Available at http://alive2green.com/conference-presentations /MiningWaste2010/MatthewOjelede_the%20 implications%20 of%20aeolian%20emissions%20from%20gold%20mine%20 tailings%20on%20ambient%20air%20quality%20the%20 Witwatersrand%20scenerio.pdf

Ozkan S. and Ipekoglu B. 2002, Investigation of Environmental impacts of tailings dam. *Environmental Management and Health*, 13(3): 242–248.

Preston-Whyte R.A. and Tyson P.D. 1988: The Atmosphere and Weather of Southern Africa. Oxford: Cape Town.

Rossouw A.S., Furniss D.G., Annegarn H.J., Weiersby I.M., Ndolo U. & Cooper M. 2009. Evaluation of a 20-40 year old mine tailings rehabilitation project on the Witwatersrand, South Africa. Mine Closure (Fourie AB, Tibett M.: Eds.) Australian Centre for Geo-mechanics, Perth; 123-136.

SANS-1929. 2005, Ambient Air Quality – Limits for Common Pollutants, SANS 1929: 1.1, 13-14, Pretoria, South Africa.

Schulze, R.E. 1997, *South African Atlas of Agrohydrology and -climatology*. Water Research Commission, Pretoria, South Africa. Report TT82/96.

South Africa Government Gazette (No. 32816, volume 534) of 2009. As available form http://www.google.com/l?sa=t&rct=j&q =South+Africa+Government+Gazette+(No.+32816,+volume+534)+of+2009 (accessed May 27, 2013)

South African Weather Service 2011, Hourly Rainfall, Temperature, Wind Speed, Wind Direction, Humidity and Pressure Data (1985–2010). Weatherline: 082 162

The Constitution of the Republic of South Africa, Act 108 of 1996. As seen in Environmental Right available from http:// www.sahrc.org.za/home/21/files/Reports/4th_esr_chap_8.pdf (accessed May 7 2013)

The World Bank Group (WBG), 2011. Alternative Dispute Resolution Guidelines. Investment Climate Advisory Services of the World Bank Group, 1818 H Street, N.W., Washington D.C. 20433.

Wright C.Y., Matooane M., Oosthuizen M.A. & Phala N. 2014, Risk perceptions of dust and its impacts among communities living in a mining area of the Witwatersrand, South Africa. *Clean Air Journal*, 24(1): 22-27.

23

Aerosol particle morphology of residential coal combustion smoke

Tafadzwa Makonese[*1], Patricia Forbes[2], Lorraine Mudau[2] and Harold J. Annegarn[1, 3]

[1] University of Johannesburg, Dept. of Geography, Environmental Management & Energy Studies, PO Box 524, Auckland Park 2006, Johannesburg, South Africa. taffywandi@gmail.com
[2] Laboratory for Separation Science, Department of Chemistry, University of Pretoria, Pretoria, 0002.
[3] Energy Institute, Cape Peninsula University of Technology, Cape Town

Abstract

A study carried out at the University of Pretoria characterised aerosol particle morphology of residential coal combustion smoke. The general approach in this study was on individual particle conglomerations because the radiative, environmental, and health effects of particles may depend on specific properties of individual particles rather than on the averaged bulk composition properties. A novel, miniature denuder system, developed and tested at the University of Pretoria, was used to capture particle emissions from the coal fires. The denuder consists of two silicone rubber traps (for gas phase semi-volatile organic compound monitoring) in series separated by a quartz fibre filter (for particle collection). The denuders were positioned 1 m away from the fire and were connected to pumps that sampled ~5 litres of air over a 10 min sampling interval. A JSM 5800LV Scanning Electron Microscope with a Thermo Scientific EDS was used to analyse the structure and morphology of different aerosol samples from the quartz fibre filters. Eight samples from the different fire lighting methods were selected for SEM analysis. The punched samples were sputter coated with gold for ~15 minutes using a K550 Emitech Sputter Coater. Results show that apart from the fine and ultra-fine particles, coal smoke from domestic burning also contains aerosols greater than 5 µm in diameter. Consequently, we describe the potential for generation of 'giant' carbonaceous soot conglomerates with outer diameters of 5 to 100 µm. However, the exact mechanism for formation of such large soot conglomerates remains to be determined. We also describe the presence of spherules and solid 'melted toffee' irregular surfaces. Circumstantial evidence is used to postulate and discuss the possible modes of formation in terms of condensation, and partial melting. This work provides a description of the modes of formation and transformation of conglomerates originating from low temperature (<800ºC) coal combustion.

Keywords

Particle morphology, coal combustion, conglomerates, particulate matter, soot, smoke

Introduction

In many developing countries, such as South Africa, China and India, coal is used as a primary source of energy for generation of electricity, industrialisation and enhancement of standard of living for the increasing populations [Makonese et al., 2014; Finkelman et al., 2007]. In South Africa, although the majority of coal is used for electricity generation, high levels of coal fuel is burned in self-fabricated and inefficient cooking devices resulting in elevated emissions of products of incomplete combustion, which are released into the immediate environment. Particulate matter emissions from coal combustion are receiving significant attention from regulatory authorities and environmental scientists because of their effects on health.

Scanning electron microscopy with energy-dispersive X-ray analysis (SEM/EDS) has proved to be a valuable tool for analysing single particles from combustion processes [Li et al., 2010]. The tool provides useful information on the morphology, elemental composition and particle density of aerosols and also gives us a better insight about the origin of particles that whether emitted from anthropogenic or natural processes [Pachauri et al., 2013:523].

In this paper we describe the potential for generation of 'giant' carbonaceous soot conglomerates with outer diameters of 5 to 100 µm. However, the exact mechanism for formation of such large soot conglomerates remains to be determined. We also describe the presence of spherules and solid 'melted toffee' irregular surfaces. Circumstantial evidence is used to postulate and discuss the possible modes of formation in terms of condensation, and partial melting. This work is vital to understanding the modes of formation and transformation of conglomerates originating from low temperature (<800ºC) coal combustion.

Materials and methodology
Combustion devices

Fires were made in representative artisan–manufactured braziers, commonly known as *imbawulas*, purchased from users in residential areas of Johannesburg (Figure 1).

Figure 1: *Illustration of* imbawulas *used in the experiment: (a) high ventilation case; (b) medium ventilation case; and (c) low ventilation case.*

Fuel analysis

Experimental fuels were tested and analysed using appropriate standard methods at an independent laboratory. The fuel samples were analysed on an air dried basis. Results for the proximate and ultimate analysis for the coal grade used in this study are Table 1.

Table 1: *Proximate and Ultimate analysis results of the D-grade type coal used in the study.*

Parameter (Air Dried Basis)	Standard Method	Slater Coal D-Grade
Moisture content (%)	ISO 5925	3.5
Volatiles (%)	ISO 562	20.3
Ash (%)	ISO 1171	24.2
Fixed carbon (%)	By difference	52.0
Calorific value (MJ/kg)	ISO 1928	23.39
Calorific value (Kcal/kg)	ISO 1928	5588
Total sulphur (%)	ASTM D4239	0.63
Carbon (%)	ASTM D5373	62.56
Hydrogen (%)	ASTM D5373	2.72
Nitrogen (%)	ASTM D5373	1.43
Oxygen (%)	By difference	4.96
Total Silica as S_iO_2 (%)	ASTM D4326	58.6
Aluminium as Al_2O_3 (%)	ASTM D4326	27.6
Total Iron as Fe_2O_3 (%)	ASTM D4326	6.63
Titanium as TiO_2 (%)	ASTM D4326	0.82
Phosphorous as P_2O_5 (%)	ASTM D4326	0.55
Calcium as CaO (%)	ASTM D4326	2.30
Magnesium as MgO (%)	ASTM D4326	0.83
Sodium as Na_2O (%)	ASTM D4326	0.42
Potassium as K_2O (%)	ASTM D4326	0.79
Sulphur as SO_3 (%)	ASTM D4326	1.10
Manganese as MnO_2 (%)	ASTM D4326	0.12

Fire ignition method

Two methods of lighting a coal fire in an *imbawula* stove namely the conventional/traditional method, and the *Basa njengo Magogo (BnM)* method, were compared in this study. The traditional way of lighting a coal fire entails laying the fire in the following order: paper, wood, ignition, after which coal is added at an appropriate time after the wood fire is established. In our experiments, ~1 000 g of coal were placed onto a grate at the bottom of the brazier followed by 36 g of rolled paper and 360 g of pine wood chips. After ignition, about 2 000 g of coal was added on top of the already burning kindling.

In the BnM, the order of laying the fire is reversed – first coal, paper, and then wood, with a few lumps of coal added at an appropriate time after the fire has been lit. As such, 2 000 g of coal was added to the bottom of the brazier onto a fuel grate followed by 36 g of paper and 360 g of kindling. After lighting the kindling, about 1 000 g of coal was added to the brazier above the kindling.

Filter preparation

6 mm quartz fibre filter punches were prepared and then placed in a vial to which methanol was added and swirled for one minute before being decanted. The same procedure was repeated with dichloromethane. The cleaned filter punches were then placed in an oven to dry for 30 min at 100°C and were then stored in a desiccator [Makonese *et al.*, 2014]. The quartz filters are used to trap the particle phase emissions in the denuder configuration as shown in Figure 2.

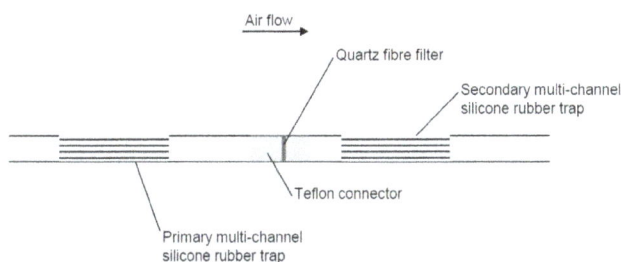

Figure 2: *Schematic of the multi-channel silicone rubber traps and quartz fibre filter employed in the denuder configuration [Forbes et al., 2012].*

Experimental set-up

In all fire lighting methods, sampling started about 2 min after ignition. The sampler location was 1 m from the combustion device, in the stream of effluent gases. Air samples were taken by means of a portable sampling pump at a flow rate of 0.5 L.min^{-1} [Figure 3].

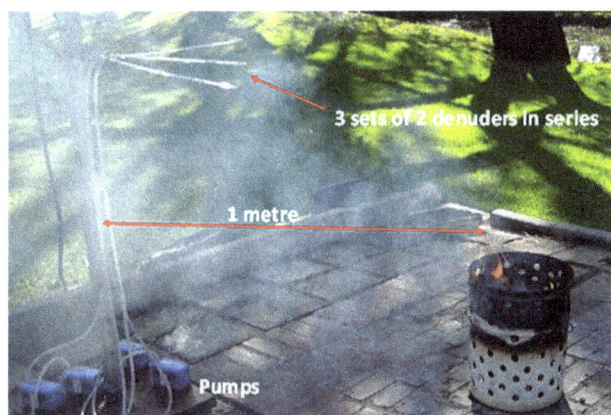

Figure 3: *Experimental set-up for sampling particle emissions from* imbawulas *[Makonese et al., 2014].*

Particle emissions were sampled for 10 min onto Quartz fibre filters, starting 2 min after ignition. These samples were taken in addition to gas phase samples with the denuders. Teflon tubing connections were used in each case to avoid contamination of the sample by leaching organics into the denuder system. After sampling, the traps were end-capped and wrapped in aluminium foil. Samples were then refrigerated prior to analysis.

Analysis of quartz filters

A JSM 5800LV Scanning Electron Microscope with a Thermo Scientific EDS was used to analyse the structure and morphology of different aerosol samples from the quartz fibre filters. Eight samples from the different fire lighting methods were selected for SEM analysis. The punched samples were sputter coated with gold for ~ 15 minutes using a K550 Emitech Sputter Coater.

Results and discussion

Identification of giant carbonaceous soot particles

Figure 4 shows images from an aerosol filter from coal combustion in typical braziers as commonly used in real-world scenarios. Figure 4a shows a conglomerate with a "sponge-like" structure similar to the one observed by Wentzel *et al.* [1999]. The dendritic form gives rise to a material with a high pore volume, and high specific surface. The individual branches of the conglomerate are typically up to 60 nm in width and consist of chains of spherules each smaller than 1 µm in diameter. The low density of the giant aerosols makes them behave aerodynamically like most of the smaller spherical particles of unit density [Wentzel, 2000]. The presence of the giant carbonaceous conglomerates was observed on the filters from both the traditional and the *Basa njengo Magogo* lighting techniques. Major differences include: lower particle loading on the BnM filter, as expected from the lower visible smoke emissions seen at ignition. Typically the BnM conglomerates have a smaller mode diameter compared to conglomerates from the traditional lighting method.

Note on Figure 4b the presence of a heavy loading of fuzzy-textured conglomerates with diameters of 5 µm or more. Like in the Wentzel *et al.* [1999] study, these giant particles are interspersed with a scattering of smaller particles either spherical or irregular in shape. Generally, soot particles are chain-like aggregates of carbon-bearing spheres. One soot aggregate may contain as many as hundreds of carbon spheres with typical diameters that range from 10 to 100 nm, with some up to 150 nm [Li & Shao, 2009]. The spherical primary particles which make up the soot particle are typically 80 nm in diameter, forming chain like agglomerates.

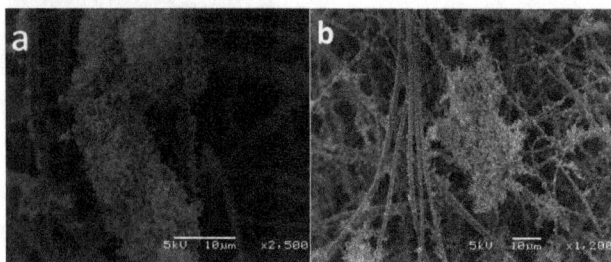

Figure 4: *Micrographs of soot particles (a) and (b) from a high ventilation BnM imbawula fire.*

The optical properties of these giant aerosol particles remain to be investigated. This however, falls outside the scope of this work and we recommend that further work be carried out in this regard. We have shown in this study that the origins of these giant particles are from domestic coal combustion in braziers. This assertion cements arguments put forward in Wentzel *et al.* [1999] that these conglomerates are likely to originate from coal combustion processes in the residential sector.

Identification of spherules and "melted toffee" webs

Figure 5 shows a carbon-dominated material forming an uneven film on the filter material. Such samples are not useful in characterizing single particles. However, they illustrate that the aerosols upon impaction with the filter were in a liquid form. The material then condensed around the quartz fibre material resulting in the formation of web-like conglomerates [Figure 5a]. At low aerosol concentrations and at high temperature (decay phase as shown in Wentzel, 2000), complete or partially solidified spherules can be found on the filters [Figure 5b]. The solidified spheres indicate that they do not arise on impaction with the filter, but that they already exist in the exhaust/flue or in the atmosphere before collection.

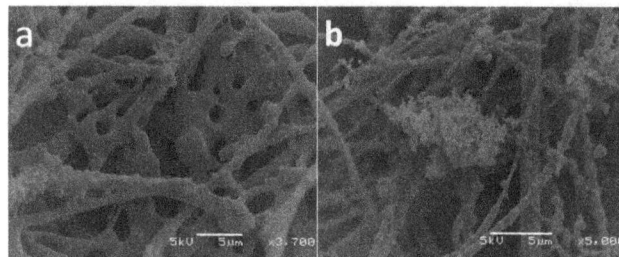

Figure 5: *Showing condensation of particles and formation of spherical particles from (a) traditional fire lighting method at ignition and (b) BnM fire lighting method at ignition.*

The traditional fire lighting method results in lower flame temperatures during the ignition phase compared to the *Basa njengo Magogo* method leading to high levels of smoke particles especially during the ignition phase. The lower temperatures coupled with the lack of oxygen needed for the complete combustion of the fuel results in high smoke emissions and other particles of incomplete combustion. This temperature difference results in differences in particle sizes and morphology emitted to the atmosphere from the two fire lighting methods. During low temperature ignition, the combustible components from the fuel are not completely burned, but condense on particles to form larger droplets which solidify upon cooling. Figure 5b shows that at higher combustion temperatures the particulate emissions appears to be reduced, evidenced by a reduction in the formation of the amorphous (i.e. "melted toffee" like) mesh across overlapping fibres. Figure 5b represents a somewhat pure form of condensation /diffusion growth, with minimal melting, compared to Figure 5a.

Figure 6 show that the liquid aerosol droplets from the coal fire form spherules and bead-like agglomerations along the length of the fibre material. The spherules do not arise on impaction with the filter, but are likely to be present in the exhaust before collection. The general shape of the individual spheroids is quite distinct. Clusters of spherules can be seen and resemble the thread-like basic structure of the giant carbonaceous particles. Such clusters illustrate fine particle growth through nucleation of nanometre size particles and diffusion aggregation toward a more stable mode of 0.3 – 0.5 µmad [Wentzel et al., 1999].

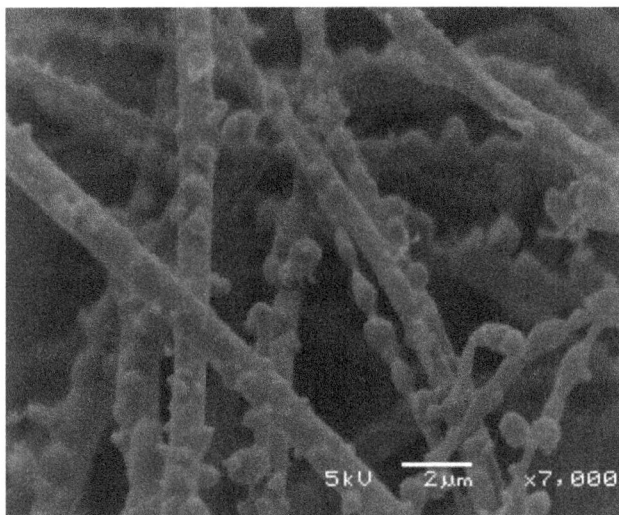

Figure 6: *Details of bead-like particle agglomerates from a traditional fire lighting method. The spherical particles are likely to be present in the exhaust or atmosphere as liquid aerosol droplets.*

Possible modes of formation

In this section we consider some of the circumstantial evidence and postulate possible modes of formation of the large dendritic conglomerates. Unlike gas flames, soot in coal flames is thought to mostly evolve directly from the tar released from the coal during devolatilization. Soot formation is a very complicated process involving hundreds of elemental steps [Fletcher et al., 1997]. Qualitatively, there are three stages that lead to primary soot formation. According to Fletcher et al. [1997:291] the stages include (i) particle inception or nucleation; (ii) surface growth; and (iii) coagulation. After these three stages, however, the primary particles may continue to undergo (iv) agglomeration and (v) aggregation processes. During particle inception, the first condensed-phase material arises from the fuel molecules and their oxidation or pyrolysis products. Surface growth reactions lead to an increase in the mass of soot, but leave the number of particles unchanged. Coagulation also leads to particle growth, where particles collide and coalesce. Although the number of particles is decreased, the soot volume fraction remains constant during the coagulation process. At later stages in the growth process, particles no longer coalesce on collision, but are chemically fused together in chains. Primary particles are discernible in the chains. The growth by non-coalescent collision is known as agglomeration. Usually, agglomerates can sub-sequently become entangled with other agglo-merates through a process known as aggregation [Fletcher et al., 1997:292].

The morphology shown in Figure 6 suggests that the super-

micron particles are primarily generated from the minerals in coal during combustion in the brazier. The formation mechanisms may involve coalescence of inherent minerals, fragmentation of chars, and the melting of excluded minerals [Wu et al., 2011; Linak et al., 2007]. However, the contribution of different mechanisms to the formation of super-micron particles during low temperature domestic combustion events is difficult to distinguish from the present results.

The coal type and fire-lighting methods would have an impact on aerosol formation and can highlight differences in the modes of formation of other soot types. The coals used in this study have relatively high fractions of volatile compounds. When employing the traditional method of lighting a coal fire, an oxygen depleted atmosphere is created in the brazier above the pyrolytic zone and the gradual heating of the brazier allows for the semi-volatile organic compounds to be released from the solid coal and, on cooling, condense. Especially during the initial ignition phase and immediately after refuelling, conditions would favour evaporation and re-condensation, rather than combustion, of volatile and semi-volatile fractions [Wentzel et al., 1999]. The growth into giant conglomerates is likely to occur in the high concentration zones of the stove, as once released into the atmosphere further growth through joining of conglomerates would be inhibited on account of their size and low diffusion velocities [Wentzel, 2000].

The decrease in soot yield with increasing temperature (i.e. high temperatures at ignition when employing the top-lit fire lighting method compared to the bottom up approach) can be explained by the stability of tar molecules at high temperatures and the reactions of tar with gaseous species existing in the post-flame region of the stove. Increases in temperature favour the 'cracking' reaction, which leads to a lower soot yield. Another reason for the decrease of soot with temperature may be due to reactions of the oxygen-containing species, especially OH and O radicals, with tar molecules and polycyclic aromatic hydrocarbons (PAHs). The concentrations of oxygen-containing radicals such as OH and O increase drastically with increases in temperature [Fletcher et al., 1997].

Conclusion

Results presented herein show that coal smoke from domestic burning comprises of aerosols of a range of sizes including those greater than 5 µm in diameter. The experiments performed in this study have shown the potential for generation of large soot agglomerates (5-100 µm in diameter). However, the exact mechanism for formation of such large soot conglomerates remains to be determined. We have shown in this study that the origins of these giant particles are from domestic coal combustion in braziers. This assertion cements arguments put forward in Wentzel et al. [1999] that these conglomerates are likely to originate from coal combustion processes in the residential sector. However, based on the available evidence presented in this paper, it is not possible to establish modes of growth and transformation of the giant conglomerates.

Results have also shown that the fire-lighting methods have an impact on aerosol formation and can highlight differences in the modes of formation of other soot types. A decrease in soot is noticed at higher ignition temperatures (typical of the BnM). This shows that the BnM method of lighting a coal fire holds potential for reducing soot particle emissions to the immediate environment, especially during the ignition phase.

Acknowledgements

We thank the Microscopy Unit of the University of Pretoria, for use of the SEM; the University of Johannesburg for financial support through a URC/Faculty of Science grant to the SeTAR Centre; Thokozile Sithole (SeTAR Centre, University of Johannesburg) and Thapelo Chalatsi (University of Pretoria) for assisting with sampling; Daniel Masekameni (University of Johannesburg) for providing the braziers from his field surveys; Antoinette Buys (UP) for assisting with the SEM analyses. This study was supported in part from a grant from the Global Alliance for Clean Cookstoves (GACC) to the SeTAR Centre as a Regional Stove Testing and Development Centre.

References

Finkelman, B.R. 2007, 'Health impacts of coal: facts and fallacies', *Ambio*, 36 (1): 103 – 106.

Fletcher T.H., Mat J., Rigby J.R., Brown A.L., Webb B.W. 1997, 'Soot in coal combustion systems', Prog. *Energy Combust. Sci.*, 23:283-301.

Forbes P.B.C., Karg E.W., Zimmermann R., Rohwer E.R. 2012, 'The use of multi-channel silicone rubber traps as denuders for polycyclic aromatic hydrocarbons', *Analytica Chimica Acta*, 730:71-9.

Li W., Shao L.Y. 2009, 'Transmission electron microscopy study of aerosol particles from the brown hazes in Northern China', J. *Geophys*. Res. 114:D09302

Li W., Shao L.Y., Shen R., Wang Z., Yang S., Tang U. 2010, 'Size, composition and mixing state of individual aerosol particles in South China Coastal City', J. *Environ*. Sci. 22:561-69.

Linak W.P., Yoo J.-I., Wasson S.J., Zhu W., Wendt, J.O.L., Huggins F.E., Chen Y., Shah N., Huffman G.P., Gilmour M.I. 2007, 'Ultrafine ash aerosols from coal combustion: Characterization and health effects', *Proceedings of the Combustion Institute*, 31:1929-37.

Makonese T., Forbes P., Mudau L., Annegarn H.J. 2014, 'Monitoring of polycyclic aromatic hydrocarbon (PAH) emissions from real world uses of domestic coal braziers', *Proceedings of the Domestic Use of Energy Conference*, 31 March – 2 April 2014, Cape Peninsula University, Cape Town, South Africa.

Pachauri T., Singla V., Satsangi A., Lakhani A, Maharaj Kumari K. 2013, 'SEM-EDX charac-terization of individual coarse particles in Agra, India', *Aerosol and Air Quality Research*, 13:523-36.

Wentzel M. 2000. '*Characterization of aerosols from the South African township of Soweto*', PhD thesis, Technische Hochschule Darmstadt, Germany.

Wentzel M., Annegarn H.J., Helas G., Weinbruch S., Balogh A.G., Sithole J.S. 1999, 'Giant dentritic carbonaceous particles in Soweto aerosols', *South African Journal of Science*, 95:141-5.

Wu H., Pedersen A.G., Glarborg P., Frandsen F.J., Dam-Johansen K., Sander B. 2011, 'Formation of fine particles in co-combustion of coal and solid recovered fuel in a pulverized coal-fired power station', *Proceedings of the Combustion Institute*, 33:2845-52.

24

Sampling and analyses of polychlorinated dibenzo dioxins (PCDDs) and polychlorinated dibenzo furans (PCDFs) emissions in South Africa: A practitioner's guide

Deon L. Posthumus and Gerald B. Woollatt

LEVEGO, PO Box 422, Modderfontein, Gauteng, 1645, South Africa,
Email: info@levego.co.za

Abstract

Dioxins and furans are toxic chemicals. A draft report released for public comment in September 1994 by the US Environmental Protection Agency clearly describes dioxin as a serious public health threat. The public health impact of dioxins may rival the impact that dichlorodiphenyltrichloroethane (DDT) had on public health in the 1960's. According to the United States Environmental Protection Agency(USEPA) report, not only does there appear to be no "safe" level of exposure to dioxin, but levels of dioxin and dioxin-like chemicals have been found in the general US population that are "at or near levels associated with adverse health effects." With this in mind the purpose of this paper is to provide an overview of the current dioxin and furan emissions from industry in South Africa, in terms of compliance with the relevant emission limit values (ELVs) and the current challenges faced with the monitoring and analysis thereof.

Keywords

dioxins, furans

Introduction

"The term Dioxin is commonly used to refer to a family of toxic chemicals that share a similar chemical structure and induce harm through a similar mechanism. Dioxins have been characterized by the USEPA as likely human carcinogens and are anticipated to increase the risk of cancer at background levels of exposure. Examples of dioxin include polychlorinated biphenyls (PCBs), polychlorinated dibenzo dioxins (PCDDs), and polychlorinated dibenzo furans (PCD/Fs)" (Energy Justice Network, 2014; USGS, 2014).

PCDD/F's are by-products of incineration, uncontrolled burning and certain industrial processes. Industrial sources of PCDD/F's to the environment include incinerators, metal smelters, cement kilns, paper and pulp industry, manufacture of chlorinated organics, and coal burning power plants (DOW, 2014). Dioxin is also produced by non-industrial sources (now considered by the U.S. Environmental Protection Agency (USEPA) to be the greatest source in the USA (US EPA, 2014), like residential wood burning, backyard burning of household waste, oil heating, and emissions from diesel vehicles.

South African legislation, Government Notice 893 of 22 November 2013 – Listed Activities and Associated Minimum Emission Standards Identified In Terms Of Section 21 of the National Environmental Management: Air Quality Act, 2004 (Act No. 39 of 2004), sets out PCDD/F emission limits for new and existing plants. The emission limit that is set for all dioxin regulated processes is 0.1 ng I-(Toxic equivalence) TEQ/m^3 (Normalised to a temperature of 0°C, pressure of 101.3 kPa, dry and at specified oxygen (O$_2$) concentration).

The method prescribed in government notice 893 for PCDD/F monitoring is US EPA Method 23 - Determination of Polychlorinated Dibenzo-p-dioxins and Polychlorinated Dibenzofurans from Municipal Waste Combustors (US EPA, no date).

There are no laboratories available in South Africa that can perform the analytical work prescribed in US EPA Method 23. All the collected samples are therefore exported to other countries where these facilities are available. As a consequence significant logistical and analytical costs are incurred.

Samples have to be collected, stored, transported and analysed following the requirements of US EPA Method 23 in order to obtain valid, reliable and accurate data.

The results of PCDD/F data are expressed in terms of toxic equivalent factors (TEQ) which provides an estimate of the toxicity of a sample (Keika Ventures, 2014). The total TEQ value is used in risk assessment studies and regulations in the US and Europe set acceptable TEQ levels for PCDD/F in air emissions. Using the TEQ approach, each individual 2,3,7,8-substituted PCDD/F (there are 17) is assigned a Toxicity Equivalency

Factor(TEF). The TEF factor correlates the toxicity of each 2,3,7,8-substituted PCDD/F to 2,3,7,8-TCDD which is considered to be the most toxic of all PCDD/F's.

There are different sets of TEF's that can be used to calculate TEQ, however, the most commonly used set is the International Toxicity Equivalency Factors (I-TEF). I-TEF is the TEFs referenced in US EPA Method 23 and also the TEF's to be used for South African reporting. The World Health Organization (WHO) TEFs also have wide use in risk assessment study data.

To calculate a sample's TEQ, you multiply the concentration of each specific analyte by its corresponding TEF which gives you the TEQ for each 2,3,7,8-substituted D/F. Sum the TEQ for each 2,3,7,8-substituted analyte to get the Total TEQ for the sample.

$TEQ_{sample} = \sum (concentration \times TEF)_{2,3,7,8\text{-substituted analyte}}$

Table 1 details the two commonly used types of TEF's.

Table 1: TEF Factors

Compound	I-TEF	WHO (Mammals/Humans)
2,3,7,8-TCDD	1	1
1,2,3,7,8-PeCDD	0.5	1
1,2,3,4,7,8-HxCDD+	0.1	0.1
1,2,3,6,7,8-HxCDD	0.1	0.1
1,2,3,7,8,9-HxCDD	0.1	0.1
1,2,3,4,6,7,8-HpCDD	0.01	0.01
OCDD	0.001	0.001
2,3,7,8-TCDF	0.1	0.1
1,2,3,7,8-PeCDF	0.05	0.05
2,3,4,7,8-PeCDF	0.5	0.5
1,2,3,4,7,8-HxCDF	0.1	0.1
1,2,3,6,7,8-HxCDF	0.1	0.1
2,3,4,6,7,8-HxCDF	0.1	0.1
1,2,3,7,8,9-HxCDF	0.1	0.1
1,2,3,4,6,7,8-HpCDF	0.01	0.01
1,2,3,4,7,8,9-HpCDF	0.01	0.01
OCDF	0.001	0.001

Sampling methodology

US EPA Method 23 requires specialised sampling equipment including skilled and trained test personnel. All the glass components/sample exposed components upstream of and including the XAD resin trap shall be cleaned following prescribed cleaning protocols. Filters need to be pre-cleaned following solvent extraction procedure as detailed in the method. XAD resin traps are prepared and spiked prior to usage. The absorbent trap must be used within 4 weeks of cleaning. The XAD traps should also be clearly labelled with expiration date and a unique number.

Careful consideration should therefore be given to timelines when ordering and preparing traps. Traps need to be ordered in advance to allow the laboratory to prepare, spike and courier the components in time for the sampling campaign. New traps should be ordered in the event that the traps surpass the expiration date.

It is important that the XAD-2 adsorbent resin temperature do not exceed 50°C because thermal decomposition will occur. During testing, the XAD-2 temperature must not exceed 20°C for the efficient capture of the PCDD/F's to take place. Consideration should be given to the transportation, storage and handling of reagents on site. High ambient temperatures found in South Africa could easily have an effect on the traps if not stored away from sources of heat or direct sunlight.

Recovery of the sample train should take place immediately after the test is completed. All sample-exposed surfaces (the stack gas is exposed to sampling components (glass liners, nozzles, filter holders, etc) prior to being trapped on the filter and XAD resin) should be sealed and transported to a suitable location for the clean-up/recovery process once the probe is cool enough to handle. The recovery area should be free of dust, smoke and other potential sources of contamination.

A rigorous sample clean-up/recovery procedure is detailed in the method and should be adhered to. The solvents used have to be pesticide grade and only Teflon wash bottles should be utilised for recoveries.

All samples must be extracted within 30 days of collection and analysed within 45 days of extraction. Samples could be detained in customs or redirected and may not be received and extracted within the allowed time window. Samples also need to be stored at temperatures ≤ 4°C.

Figure 1 is a schematic of the EPA method 23 sampling train.

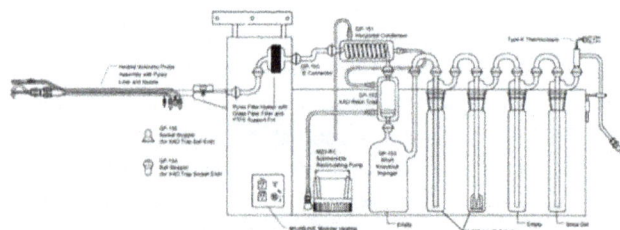

Figure 1: Schematic of sampling system

In Figures 2 and 3, sampling equipment in use is shown.

Figure 2: Typical sample train set up

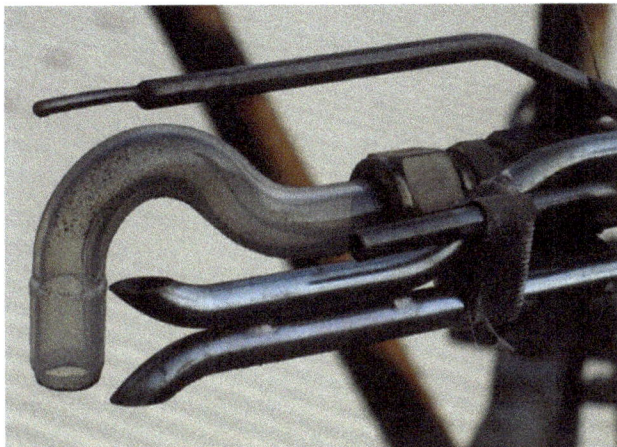

Figure 3: *Glass nozzle, pitot tube and thermocouple*

Reporting criteria

It is important to understand the analytical data and reporting criteria. The report should detail the criteria utilised from the analytical reports.

Symbols are used in the analytical report and also in the presentation of the test results. The symbols indicate results that have special significance and require different procedures in calculations and data interpretation. The data reporting procedures outlined in US EPA Method 23 are used in presenting all analytical results. Any values flagged should be considered and addressed in the final report as the reported values may have statistical significance.

Analytical results that are below detection limits (ND) need special mention as the results could be interpreted differently and therefore reports could vary from one test report/test house to another.

The US EPA has different ways of reporting the data, depending on the intended use of the results. Users may decide to substitute ND with zero, 0.5x ND or use the ND value depending on the intended use of the results.

The following example could be used to demonstrate the above,
- Risk analyses from specific plants – Substitute the ND result with the limit of detection (DL) value. This approach will provide the worst case or highest value. This will generally be indicated on the report as ITEF TEQ (ND=DL; EMPC=EMPC)
- Developing emission factors – Substitute the ND results with half the limit of detection value. This approach may provide an average emission. This will generally be indicated on the report as ITEF TEQ (ND=DL/2; EMPC=EMPC/2)
- Setting emission limits and compliance testing – Substitute ND result with zero. This will generally be indicated on the report as ITEF TEQ (ND=0; EMPC=0). EPA also make mention that applying zero should only be done for tests with a sample time of more than 4 hours

Table 2: *Example of certain laboratory reporting qualifiers/attributes*

Data Qualifiers/Data Attributes	
>	Indicates high recoveries. Shown with the numeric value at the top of the range
B	The analyte is found in the method blank, at a level that is <=10x the sample concentration
C	Two or more congeners co-elute. In EDDs C denotes the lowest IUPAC congener in a coelution group and additional co-eluters for the group are shown with the number of the lowest IUPAC co-eluter
E	The reported concentration exceeds the calibration range (upper point of the calibration curve)
EMPC	Represents an Estimated Maximum Possible Concentration. EMPC's arise in cases where the signal/noise ratio is not sufficient for peak identification (the determined ion-abundance ratio is outside the allowed theoretical range), where there is co-eluting interference, or where a single ion is utilised for quantification due to PFK interference)
J	Indicates that an analyte has a concentration below the reporting limit (lowest point of the calibration curve)
ND	Indicates a non-detect

Table 3 represents an example of a laboratory report showing the difference in concentration levels adopting the aforementioned criteria.

Table 3: *Example of laboratory data*

Compound/Analyte	Method Blank (picogram)	Sample 1 (picogram)	Sample 2 (picogram)
2,3,7,8-TCDD	(1.37)	22.1	(2.19)
1,2,3,7,8-PeCDD	(1.36)	62.1	(2.36)
1,2,3,4,7,8-HxCDD	(1.27)	55.6	(2.19)
1,2,3,6,7,8-HxCDD	(1.26)	93.7	(2.17)
1,2,3,7,8,9-HxCDD	(1.37)	84.8	(2.44)
1,2,3,4,6,7,8-HpCDD	(1.54)	755	[2.6]
OCDD	6.04	1650	[8.15]
2,3,7,8-TCDF	(0.868)	167	(1.89)
1,2,3,7,8-PeCDF	(0.996)	234	(1.87)
2,3,4,7,8-PeCDF	(0.969)	311	(1.7)
1,2,3,4,7,8-HxCDF	(0.758)	347	2.26
1,2,3,6,7,8-HxCDF	(0.725)	366	1.86
2,3,4,6,7,8-HxCDF	(0.739)	428	[1.77]
1,2,3,7,8,9-HxCDF	(0.961)	56.2	(1.6)
1,2,3,4,6,7,8-HpCDF	(0.931)	1480	[3.24]
1,2,3,4,7,8,9-HpCDF	(1.51)	243	(2.29)
OCDF	(2.92)	1340	(5.24)
ITEF TEQ (ND=0; EMPC=0)	0.00604	408	0.412
ITEF TEQ (ND=0; EMPC=EMPC)	0.00604	408	0.656
ITEF TEQ (ND=DL/2; EMPC=0)	1.72	408	3.18
ITEF TEQ (ND=DL/2; EMPC=EMPC)	1.72	408	3.34
ITEF TEQ (ND=DL; EMPC=EMPC)	3.43	408	6.02

Current legislation does not specify the criteria for reporting PCDD/F results other than that I-TEQ should be utilised. In the end, the licensing authority must decide on the most suitable way to consider results reported as ND and detail the requirement in the respective air emissions licence.

In the United Kingdom the worst case scenario is applied to their results and their sampling time stipulation for PCDD/F is six (6) hours per test. Each country will establish its own criteria for sampling and reporting.

Not all laboratories use the same criteria as detailed in Table 2 and Table 3. Certain laboratories will only report ND=0 and ND=DL.

Actual emissions data

Table 4 details actual PCDD/F emission concentrations measured by Levego at various South African processes; Concentration @10% oxygen.

Table 4: *PCDD/F emission concentrations at various South African processes*

Different Industry Types	Concentration; ng/Nm³ (dry) #
Type A - Medical waste without cleaning	4533.70
Type A - Medical waste with poor pollution abatement	3.19
Type A - Medical waste with good pollution abatement	0.04
Type B - Cement	0.0012 or 1.2E-03
Type C - Drum reconditioning	101.67
Type D - Metal industry	1.57

ITEF TEQ (ND=DL;EMPC=EMPC)

The industry types are made up of various processes within each industry. The industry types are limited to the listed activities requiring PCDD/F emission measurements.

From the above it is evident that most emitters are well above the allowable limits. The average concentrations reported above were based on the worst case scenario. Referring to Table 3, sample 2, it becomes clear how many different concentration levels could have been reported.

Conclusion

Considering the toxicity of PCDD/F's, emission concentration levels in South Africa and the various possibilities of reporting PCDD/F concentration it becomes imperative to establish a national format for reporting.

In South Africa, we should as a minimum report both the worst case scenario (ND=DL) as well as the lower bound (ND=0) for reporting PCDD/F emissions. Testing laboratories in South Africa cannot decide on their own reporting criteria. The regulator

need to adopt specific criteria to report to and all the testing laboratories need to apply the same criteria when reporting.

Failure to establish common reporting criteria will create significant inconsistences in terms of legal compliance demonstration, environmental impact assessment studies, and toxicology studies to name a few.

References

Andrew G. Clarke, Industrial Air Pollution Monitoring, Chapman and Hall, ISBN 0 412 63390 6.

DOW (2014) How Dioxins and Furans are Formed. Available at http://www.dow.com/sustainability/debates/dioxin/definitions/how.htm. Accessed on the 19 November 2014.

Energy Justice Network (2014). Dioxins & Furans: The most toxic chemicals known to science. Available at http://www.ejnet.org/dioxin/. Accessed on the 20 November 2014.

Keika Ventures, 2014. Dioxin/Furan (D/F) Solution Page: Method 23 Flow Chart. Available at http://www.keikaventures.com/s_method23.php. Accessed on 19 November 2014.

US EPA (2014) Emission Measurement Centre: Frequent Questions. Available at http://www.epa.gov/ttn/emc/facts.html Accessed on 20 November 2014.

USGS (2014) Environmental Health – Toxic Substances. Available at http://toxics.usgs.gov/definitions/dioxins.html. Accessed on 20 November 2014.

US EPA (No date) Method 23 – Determination of polychlorinated dibenzo-p-dioxins and polychlorinated dibenzofurans from stationary sources. Available at http://www.epa.gov/ttn/emc/promgate/m-23.pdf. Accessed 20 November 2014.

PERMISSIONS

LIST OF CONTRIBUTORS

Petra Maritz, Johan P. Beukes, Pieter G. van Zyl, Elne H. Conradie, Andrew D. Venter and Jakobus J. Pienaar
Unit for Environmental Sciences and Management, North-West University, Potchefstroom Campus, Private Bag X6001, Potchefstroom, 2520, South Africa

Catherine Liousse, Corinne Galy-Lacaux and Pierre Castéra
Laboratoire d'Aérologie, Université Paul Sabatier-CNRS, OMP, 14 Avenue Edouard Belin, 31400 Toulouse, France

Avishkar Ramandh
Sasol Technology R&D, Sasolburg, 1947, South Africa, Avishkar

Gabi Mkhatshwa
Sustainability & Innovation Department, Eskom Corporate Service Division, Rosherville, South Africa

Lynette Herbst
Department of Engineering and Technology Management, University of Pretoria, Cnr Lynnwood Rd and Roper Str, Hatfield, Pretoria, 0001, South Africa.
Laboratory for Atmospheric Sciences, Department of Geography, Geoinformatics and Meteorology, University of Pretoria

Hannes Rautenbach
Laboratory for Atmospheric Sciences, Department of Geography, Geoinformatics and Meteorology, University of Pretoria

M.A. Oosthuizen, M. Matooane and N. Phala
Council for Scientific and Industrial Research, Natural Resources and the Environment, Climate Studies, Modelling and Environmental Health Research Group, Pretoria, 0001, South Africa

C.Y. Wright
Environment and Health Research Unit, South African Medical Research Council, Pretoria, South Africa
Department of Geography, Geoinformatics and Meteorology, University of Pretoria, Pretoria, South Africa

Gregor T. Feig, Beverley Vertue, Seneca Naidoo, Nokulunga Ncgukana and Desmond Mabaso
South African Weather Service 442 Rigel Ave South Erasmusrand Pretoria South Africa

Bianca Wernecke, Brigitte Language, Stuart J. Piketh and Roelof P. Burger
Eskom Holdings SOC Ltd, Megawatt Park, Maxwell Drive, Sunninghill, 2001, Unit for Environmental Sciences and Management, North West University, Potchefstroom, 2520, South Africa

Sarel J. Gates, Gerrit Kornelius, Steven C. Rencken and Neil M. Fagan
University of Pretoria, Dept of Chemical Engineering,Environmental Engineering Group, Private Bag X20 Hatfield, Pretoria, South Africa, 0028

Peter Cowx
Eramet Norway, Sauda, Norway

Luther Els
Resonant Environmental Technologies, Centurion, South Africa, 0046

Brigitte Language, Stuart J. Piketh and Roelof P. Burger
Unit for Environmental Sciences and Management, North West University, Potchefstroom, 2520, South Africa

Lethukuthula Masondo and Kenneth Mohapi
Department of Environmental Health: University of Johannesburg

Daniel Masekameni
Department of Environmental Health: University of Johannesburg
SeTAR Centre, Faculty of Engineering and the Built Environment
Department of Geography, Environmental Management and Energy Studies, Faculty of Science, University of Johannesburg, Auckland Park 2006, Johannesburg

Tafadzwa Makonese
Department of Environmental Health: University of Johannesburg
SeTAR Centre, Faculty of Engineering and the Built Environment

Harold J Annegarn
Energy Institute, Cape Peninsula University of Technology, Cape Town, 8001

Gregor T. Feig
Now at the Council for Scientific and Industrial Research

Seneca Naidoo
Now at the Council for Scientific and Industrial Research

Nokulunga Ncgukana
South African Weather Service, 442 Rigel Ave South, Erasmusrand, Pretoria, South Africa

Belinda L. Garnham and Kristy E. Langerman
Eskom Holdings SOC Limited, Megawatt Park, 1Maxwell Drive, Sunninghill, Sandton

Samantha Keen and Katye Altieri
Energy Research Centre, University of Cape Town, Private Bag X3, Rondebosch, 7700, South Africa

Jared Lodder, Martin A. van Nierop and Elanie van Staden
Gondwana Environmental Solutions, 562 Ontdekkers Road, Florida, Roodepoort, 1716, South Africa

Stuart J. Piketh
Unit for Environmental Sciences and Management, North-West University, Potchefstroom, 2520, South Africa

Aidan J. Henri, Martin Van Nierop, Elanie van Staden and Jared Lodder
Gondwana Environmental Solutions, 562 Ontdekkers Road, Florida, Roodepoort, 1716, South Africa

Luanne B. Stevens
Gondwana Environmental Solutions, 562 Ontdekkers Road, Florida, Roodepoort, 1716, South Africa, Unit for Environmental Sciences and Management, North-West University, Potchefstroom, 2520, South Africa

Stuart Piketh
Unit for Environmental Sciences and Management, North-West University, Potchefstroom, 2520, South Africa

Rebecca M. Garland
Natural Resources and the Environment Unit, Council for Scientific and Industrial Research, Pretoria, South Africa
Climatology Research Group, North West University, Potchefstroom, South Africa

Mogesh Naidoo, Bheki Sibiya and Riëtha Oosthuizen
Natural Resources and the Environment Unit, Council for Scientific and Industrial Research, Pretoria, South Africa

Modupe O. Akinola
Environmental Biology Unit, Department of Cell Biology and Genetics, Faculty of Science,University of Lagos, Nigeria
Department of Environmental Science, University of Botswana, Botswana

M. Lekonpane
Department of Environmental Science, University of Botswana, Botswana
Aqualogic (Pty) Ltd., Plot 182, Unit 1, Commerce Park, Gaborone, Botswana

Ebenezer O. Dada
Environmental Biology Unit, Department of Cell Biology and Genetics, Faculty of Science University of Lagos, Nigeria

Martin A. van Nierop, Elanie van Staden and Jared Lodder
Gondwana Environmental Solutions, 562 Ontdekkers Road, Florida, Roodepoort, 1716, South Africa

Priyanka deSouza
Research Fellow, Senseable City Lab MIT, Cambridge MA 02139

Victor Nthusi
Science Division, United Nations Environment Program(UNEP), UN Avenue, Nairobi, Kenya

Jacqueline M. Klopp
Center for Sustainable Urban Development, Earth Institute, Columbia University, 475 Riverside Dr. Suite 520 New York NY 10115

Bruce E. Shaw
Lamont Doherty Earth Observatory, Columbia University, Palisades, New York, USA

Wah On Ho
Senior Research Scientist, Alphasense Ltd, Sensor Technology House, 300 Avenue West, Skyline 120, Great Notley, Essex CM77 7AA, UK

John Saffell
Technical Director, Alphasense Ltd, Sensor Technology House, 300 Avenue West, Skyline 120, Great Notley, Essex CM77 7AA, UK

Roderic Jones
University of Cambridge, UK

Carlo Ratti
Director, Senseable City Lab MIT, Cambridge MA 02139

Hassan Y. Sulaiman
Yusuf Maitama Sule University Kano, Nigeria

Sedef Çakir
Cyprus International University, Mersin 10, Nicosia, Northern Cyprus

Hanlie Liebenberg-Enslin, Reneé von Gruenewaldt and Lucian Burger
Airshed Planning Professionals, Midrand, South Africa

Hannes Rauntenbach
Climate Change and Variability, South African Weather Service, Pretoria, South Africa
School of Health Systems and Public Health, University of Pretoria, South Africa

Pierru Roberts and Luther Els
Resonant Environmental Technologies, Centurion, 0046

Gerrit Kornelius
Resonant Environmental Technologies, Centurion, 0046
Department of Chemical Engineering, University of Pretoria, Pretoria, 0002

Johann P. Schoeman
28 Maroela Street, Mossel Bay, Western Cape, 6500, South Africa

De Wet Schutte
Cape Peninsula University of Technology, Cape Town 8000, South Africa

Olusegun Oguntoke
Department of Geography, Environmental Management and Energy studies, University of Johannesburg Auckland Park, 2006, Republic of South Africa
Department of Environmental Management and Toxicology, University of Agriculture PMB 2240 Abeokuta, Federal Republic of Nigeria

Harold J. Annegarn
Department of Geography, Environmental Management and Energy studies, University of Johannesburg Auckland Park, 2006, Republic of South Africa

Tafadzwa Makonese
University of Johannesburg, Dept. of Geography, Environmental Management & Energy Studies, Auckland Park 2006, Johannesburg, South Africa

Patricia Forbes and Lorraine Mudau
Laboratory for Separation Science, Department of Chemistry, University of Pretoria, Pretoria, 0002

Harold J. Annegarn
University of Johannesburg, Dept. of Geography, Environmental Management & Energy Studies, Auckland Park 2006, Johannesburg, South Africa
Energy Institute, Cape Peninsula University of Technology, Cape Town

Deon L. Posthumus and Gerald B. Woollatt
LEVEGO, Modderfontein, Gauteng, 1645, South Africa

Index